Neuroprogression in Psychiatric Disorders

Modern Trends in Pharmacopsychiatry

Vol. 31

Series Editor

B.E. Leonard Galway

Neuroprogression in Psychiatric Disorders

Volume Editors

Angelos Halaris Chicago, IL
Brian E. Leonard Galway

31 figures, 8 in color, and 3 tables, 2017

KARGER

Basel · Freiburg · Paris · London · New York · Chennai · New Delhi · Bangkok · Beijing · Shanghai · Tokyo · Kuala Lumpur · Singapore · Sydney

Modern Trends in Pharmacopsychiatry
(Formerly published as 'Modern Problems in Pharmacopsychiatry')

Angelos Halaris
Department of Psychiatry and Behavioral Neuroscience
Loyola University Stritch School of Medicine
and
Loyola University Medical Center
Chicago, IL (USA)

Prof. Brian E. Leonard
Department of Pharmacology
Centre for Pain Research and
Galway Neuroscience Centre
National University of Ireland
Galway (Ireland)

Library of Congress Cataloging-in-Publication Data

Names: Halaris, Angelos, editor. | Leonard, B. E., editor.
Title: Neuroprogression in psychiatric disorders / volume editors, Angelos Halaris, Brian E. Leonard.
Other titles: Modern trends in pharmacopsychiatry ; v. 31. 1662-2685
Description: Basel ; New York : Karger, 2017. | Series: Modern trends in pharmacopsychiatry, ISSN 1662-2685 ; vol. 31 | Includes bibliographical references and index.
Identifiers: LCCN 2017025712| ISBN 9783318060508 (hard cover : alk. paper) | ISBN 9783318060515 (electronic version)
Subjects: | MESH: Mental Disorders--physiopathology | Disease Progression
Classification: LCC RC454.4 | NLM WM 140 | DDC 616.89--dc23 LC record available at https://lccn.loc.gov/2017025712

Bibliographic Indices. This publication is listed in bibliographic services, including Current Contents® and PubMed/MEDLINE.

www.karger.com
Printed on acid-free and non-aging paper (ISO 9706)
ISSN 1662–2685
e-ISSN 1662–4505
ISBN 978–3–318–06050–8
e-ISBN 978–3–318–06051–5

Contents

Preface

Psychiatric and neurological disorders are chronic illnesses characterized by recurrence and relapses and progressively increasing dysfunction. The term "neuroprogression" refers to the temporal clinical progression in most of these disease entities. One of the main factors associated with neuroprogression and, consequently, with disease prognosis, is the frequency of mood episodes, as, for instance, occurs in bipolar and unipolar affective illnesses, but also in psychotic and cognitive disorders including schizophrenic and schizoaffective conditions and the dementias. It is believed the higher the frequency of mood or psychotic episodes, the faster will be the course of neuroprogressive changes rendering the prognosis unfavorable. The biological mechanisms underlying neuroprogression are not fully elucidated and likely involve complex interactions among multiple genes, environmental factors, and epigenetic changes ultimately resulting in impairment in several physiological systems. Neurobiological mechanisms and possibly associated neuroanatomical brain changes in patients with multiple mood episodes appear to include a pro-inflammatory state, increased oxidative stress, and a deficit in neuroprotection and neuroplasticity. Immune system dysfunction and brain-immune interactions have already been identified in numerous studies as crucial contributory factors to neurogression. Acute mood episodes have been associated with significant systemic toxicity, cognitive and functional impairment, and biological changes. These effects are cumulative, being much more prominent after multiple episodes. The neuroprogressive nature of most psychiatric and neuropsychiatric disorders thus induces allostatic states with a steadily increasing allostatic load.

Neuroprogression has important clinical implications, both in terms of early detection and prevention, as well as in the selection of appropriate treatment strategies, given that early and late stages of the specific disorder may present different biological features and substrates. Therefore, it is important to conduct longitudinal studies in evaluating the effects of illness progression on neuronal structures and psychoneuroimmune functions. We believe these findings will soon begin to have significant clinical implications by establishing reliable tests for monitoring the impact of treatments, and by supporting more aggressive and earlier therapeutic and preventive interventions to minimize affective and cognitive symptomatology and clinical deterioration. The possibility of identifying neuroanatomical, psychoneuroimmunological, and neurotransmitter abnormalities along with the detection of genetic and epigenetic vulnerability will ultimately unravel some of the potential pathophysiological mechanisms involved in illness progression. Diagnostic accuracy, precise treatment choices, remission, and relapse prevention, along with early detection of illness vulnerability, are the gold standards we need to be aiming for as clinicians and scientists.

Based on their particular area of expertise and endeavor, the contributors to this volume critically review recent advances in neuroprogression research and its relevance to the impact of chronic psychiatric and neurological disorders on brain structure and function. In addition to the long-term outcome of such disorders, the epigenetic consequences of early childhood abuse on subsequent adverse changes in brain function, which may prelude major psychiatric disorders, have also been considered. The importance of recent postmortem studies and the status of biomarkers provide much of the evidence in support of neuroprogression in psychiatric and neurological disorders.

Angelos Halaris
Brian E. Leonard

Halaris A, Leonard BE (eds): Neuroprogression in Psychiatric Disorders.
Mod Trends Pharmacopsychiatry. Basel, Karger, 2017, vol 31, pp 1–9 (DOI: 10.1159/000470802)

Neuroprogression in Schizophrenia and Psychotic Disorders: The Possible Role of Inflammation

Norbert Müller

Department of Psychiatry and Psychotherapy, Ludwig Maximilian University, Munich, Germany

Abstract

Schizophrenia is a disorder that shows a progressive course in 30–50% of the people concerned. The biology of chronification and progression is unclear. Genetic aspects may play a role, but details are unresolved. The fact that immune-mediated and autoimmune disorders such as rheumatoid arthritis or multiple sclerosis have a very similar course as schizophrenia has focused the interest on the immunopathogenesis of schizophrenia. A clear immune marker for neuroprogression in schizophrenia or psychosis could not be identified up to now, but a proinflammatory immune state (increased markers of cellular immunity) is regularly found in schizophrenia, e.g., increased levels of cytokines such as interleukin-6 (IL-6). Moreover, the tryptophan/kynurenine metabolism is regulated via pro- and anti-inflammatory cytokines and is closely related to the glutamatergic neurotransmission. Certain molecules of this metabolism, such as quinolinic acid or 3OH-kynurenine, have neurotoxic effects and seem to play a role in chronification. Studies with immune/anti-inflammatory-based therapeutic approaches show that acuity or chronicity of the inflammation influence the outcome of therapeutic interventions.

Introduction

Neuroprogression is a not clearly defined concept describing the progression of the disorder. This may include clinical and morphological aspects of progression. There is no doubt that clinicians see a progression of schizophrenia in many patients; from a scientific point of view, it is important to evaluate the mechanisms of progression, especially in patients suffering from psychotic disorders, most importantly schizophrenia, because the pathogenesis of this disorder which affects about 1% of the population still needs to be resolved. Morphological studies of the brain in schizophrenia indicate a progressive loss of brain substance over time [1]. The nature of this loss, however, is unclear. The question of whether it is a loss of neurons, microglia, and/or astrocytes, the type of brain cells preferably disturbed and the mechanism underlying the atrophic process remain to be determined.

Clinical Aspects of Neuroprogression in Schizophrenia and Psychotic Disorders

The modern classification of psychiatric disorders is based on the work of Emil Kraepelin. In 1899, Kraepelin [2] reported in the 6th edition of his textbook of psychiatry the dichotomy of "manic-depressive madness" and "dementia praecox." In contrast to Eugen Bleuler's concept of schizophrenia – both Kraepelin and Bleuler were scholars of Bernhard von Gudden – Kraepelin focused on the course and prognosis of the disorder and emphasized the chronic progressive course, while Bleuler concentrated on the cross-sectional symptomatology. Today, we know that from a clinical point of view schizophrenia can be restricted to one single episode without any later deficit in cognition, affect, or drive, or to more episodes without deficit. However, similar to other studies, the Munich follow-up study, which examined patients 15 years after their hospitalization, showed that only a minority of schizophrenic patients have a single episode (3%), while 40% show a course with relapses and complete remissions, and 57%, i.e., by far the majority, develop a chronic course [3]. In other types of psychoses, the rates of chronic courses are lower, e.g., in delusional disorders still 50%, in acute transient psychosis 20%, and in schizoaffective psychoses 15% [4]. A chronic course was defined as a course with persisting residual symptoms. This example shows that in different types of psychoses there are always risks for chronification, but in schizophrenia this risk is particularly high, albeit modern therapeutic developments such as second-generation antipsychotics, psychoeducation, and systematic behavioral psychotherapy. A deficit syndrome, found in 34% of the patients at their first admission, was relocated in 28% of the patients 15 years later [5].

Neuroimaging Aspects of Neuroprogression in Schizophrenia and Psychotic Disorders

A lot of studies using different techniques describe a progressive loss of brain volume in schizophrenia. A meta-analysis including 27 studies with 928 patients and 867 control subjects involving 32 different brain regions of interest showed convincingly that schizophrenia is associated with progressive structural brain abnormalities, affecting both gray and white matter [1]. Subjects with schizophrenia showed significantly greater volume decreases over time in the whole brain, whole-brain gray matter, frontal gray and white matter, parietal white matter, and temporal white matter, as well as larger increases in lateral ventricular volume than healthy control subjects. This big meta-analysis demonstrates convincingly the progressive loss of brain volume during the course of schizophrenia. However, data also indicate that small abnormalities in the brain volume can already be observed at the first manifestation of schizophrenia [6]. After a first episode, a progressive loss of brain volume was found during the first 2 years, although a volume increase in certain regions could be shown, possibly an effect of antipsychotic treatment [7]. Studies in twin pairs discordant for schizophrenia revealed both morphological changes in the nonaffected twins, possibly reflecting a genetic vulnerability, and smaller brain volumes in the affected twins compared to the nonaffected, probably reflecting a disease-related loss of volume [8].

CNS Volume Loss in Imaging Studies – A Consequence of an Inflammatory Process?

Gross inflammatory changes have not been found in neuroimaging or neuropathological studies of schizophrenia. However, there is no doubt that a decreased CNS volume can be observed already during the first episode, and a progressive loss of CNS volume occurs during the further course of the disease [9–12]. Moreover, a relationship was described between volume loss and an increased genetic risk for higher levels of the immune marker interleukin (IL)-1β [13]; the relationship between maternal IL-8 levels and CNS volume has been mentioned before [14].

In positron emission tomography, PK11195, a ligand for peripheral benzodiazepine receptor binding sites, is used to estimate microglial activation [15]. In schizophrenia, increased PK11195 expression was found to be a marker of inflammatory processes in the CNS [16, 17]. Moreover, a positive correlation of schizophrenic positive symptoms as well as duration of disease with the expression of the microglial activation marker DAA1106 in PET was also noted [18].

Relationship between Cytokines, Infection, and the Risk of Schizophrenia

IL-1β, which can induce the conversion of rat mesencephalic progenitor cells into a dopaminergic phenotype [19–21], and IL-6, which is highly effective in decreasing the survival of fetal brain serotonergic neurons [22], seem to have an important influence on the development of the neurotransmitter systems involved in schizophrenia. Moreover, it has been demonstrated that the administration of IL-1β after birth affects the dopaminergic neurotransmission in adulthood. IL-1 and IL-6, cytokines mainly released from monocytes of the "innate immune system," may only be examples for the immune activation state, and the specificity of these cytokines is a matter of discussion.

Evidence for pre- or perinatal exposure to infections as a risk factor for schizophrenia has not only been obtained from animal models [23, 24]. In humans, studies of infections as risk factors for schizophrenia have been performed on several viral disorders [25–27]. An increased risk for schizophrenia in the offspring was observed after respiratory infections [28, 29], genital infections, and reproductive tract infections [29, 30]. Specifically, women infected with *Toxoplasma gondii* during pregnancy were intensely studied regarding the risk for schizophrenia [31].

In humans, increased maternal levels of the proinflammatory cytokine IL-8 during pregnancy were shown to be associated with an increased risk for schizophrenia in the offspring, whatever the reason for the increase in IL-8 [32]. Moreover, increased maternal IL-8 levels in pregnancy were also significantly related to decreased brain volume, i.e., lower volumes of the right posterior cingulum and left entorhinal cortex and higher volumes of the ventricles in the schizophrenic offspring [14].

Progression of Inflammation Increases the Risk for Schizophrenia

A recent study, the first large-scale epidemiological study in psychiatry, showed, however, that severe infections and autoimmune disorders increase additively the risk of schizophrenia and schizophrenia spectrum disorders [33]. This is an important finding, since mostly maternal infections during pregnancy had been studied before (in animal models). Infections after birth or during childhood and adolescence in patients later diagnosed with schizophrenia, i.e., lifetime infections of schizophrenia patients, have only rarely been studied [33, 34]. The sensitivity of the study in recording infections was not high, because only infections leading to hospital admission were recorded, and, normally, only extraordinarily severe infections lead to a hospital contact. Therefore, despite the large scale of the study, it may have clearly identified only the "tip of the iceberg" of risk factors [34].

Cytokine Alterations and Neuroprogression in Schizophrenia

High levels of proinflammatory substances such as cytokines have been described in the blood and cerebrospinal fluid of schizophrenia patients. Reviews on the imbalance in schizophrenia types 1 and 2 and pro- and anti-inflammatory immune systems as well as innate immunity, including the monocyte/macrophage system [35], have recently been published and indicate that inflammatory processes play an important role in the pathophysiology of at least a subgroup of schizophrenia patients [36–38].

Regarding neuroprogression, a specific focus has to be placed on the cytokine levels in different stages of the disease and treatment. Data clearly reveal that both the stage of the disease and antipsychotic therapy have an influence on the immune function [39, 40].

A meta-analysis which included 40 studies investigating cytokines at different stages of schizophrenia and under different treatment conditions showed significantly higher levels of proinflammatory cytokines. This meta-analysis differentiated between drug-naïve first-episode psychosis, an acute relapse of psychosis, and a third group of schizophrenia patients in the state of stable medicated outpatients and treatment-resistant psychosis. First-episode psychosis and acute-relapse patients had also higher blood cytokine levels than controls for IL-6, tumor necrosis factor (TNF)-α, and interferon (IFN)-γ [38]. The third group under antipsychotic treatment, however, showed significant decreases in IL-6, IL-1β, and IFN-γ, but increases in IL-12 and serum IL-2 receptor [38]. Whether this is an effect of antipsychotic treatment or chronicity, or both, remains to be determined. It is well known, however, that antipsychotic treatment has an important impact on cytokine levels.

Several further interfering clinical variables have to be assessed, including weight gain, body mass index, smoking, aging, age at disease onset, duration and severity of the disorder, psychopathological symptoms, and hospitalization. However, laboratory parameters such as sensitivity of the tests, storage of the samples, and the mode of release and the mechanism of action (IFN-γ acts paracrine by cell-cell contact while TNF-α and IL-6 act in an endocrine way) are also important.

These data show that antipsychotic treatment elicits a different cytokine release pattern which drives most of the cytokine levels of schizophrenic patients into the direction of "normalization." The data are consistent albeit the influence of several interfering variables, e.g. higher levels of proinflammatory cytokines in schizophrenia. Less clear, however, is the impact of neuroprogression or chronification of the schizophrenic state on immune variables.

The Indoleamine 2,3-Dioxygenase-Related Pathway in Schizophrenia: Neuroprotective or Neurotoxic?

Immune alterations influence the dopaminergic, serotonergic, noradrenergic, and glutamatergic neurotransmission. The metabolism of TRP influences the serotonergic and glutamatergic neurotransmission via the activation or inhibition of the enzyme indoleamine 2,3-dioxygenase (IDO). Cytokines of the activated immune system activate IDO of the TRP/kynurenine (KYN) metabolism; other cytokines have an inhibitory effect. The inhibition of IDO by the catalyzing step from serotonin to f50HKYM (formyl-5-hydroxykynuramine) increases serotonin availability, while neuroactive metabolites of the TRP/KYN metabolism, kynurenic acid (KYNA), and quinolinic acid (QUIN) act as NMDA receptor antagonist or agonist, respectively. The only known naturally occurring NMDA receptor antagonist in the human CNS is KYNA, one of the neuroactive intermediate products of the KYN pathway. KYN is the primary major degradation product of tryptophan (TRP). While the excitatory KYN metabolites 3-hydroxy-KYN (3HK) and QUIN are synthesized from KYN en route to NAD, KYNA is formed in a dead-end side arm of the pathway [41]. KYNA acts as both a blocker of the glycine coagonistic site of the NMDA receptor [42] and a noncompetitive inhibitor of the α_7-nicotinic acetylcholine receptor [43]. The production of KYNA is regulated by IDO and TRP Tryptochan 2,3 Dioxygenase (TDO). Both enzymes catalyze the first step in the pathway: the degradation of TRP to KYN. Cytokines such as IFN-γ or TNF-α stimulate the activity of IDO [44]. QUIN, however, is an NMDA receptor agonist and exhibits neurotoxic effects. In accordance with the glutamatergic depletion hypothesis of schizophrenia, studies point to an overweight of KYNA over QUIN in the cerebrospinal fluid [45] and in microglial cells of the hippocampus [46].

The described loss of CNS volume and the activation of microglia, both of which have been clearly demonstrated in neuroimaging studies of schizophrenia patients, match the assumption of a (low-level) inflammatory neurotoxic process.

The Effects of Anti-Inflammatory Treatment Underline the Neuroprogression Theory of Schizophrenia: Therapeutic Benefit in Early Stages

An important argument that inflammation plays a role in schizophrenia comes from the therapeutic benefit of anti-inflammatory medication. The cyclooxygenase-2 (COX-2) inhibitor celecoxib was studied in a prospective, randomized, double-blind study of acute exacerbations of schizophrenia. The patients receiving celecoxib add-on to risperidone showed a statistically significantly better outcome than the patients receiving risperidone alone [47]; the clinical effects of COX-2 inhibition in schizophrenia were especially pronounced in cognition [48]. Pooled data (n = 90) of a further study of risperidone and celecoxib add-on revealed that patients with a disease duration of 2 years or less benefited from the 6-week celecoxib treatment, while patients with a longer duration of disease did not differ from the risperidone and placebo group (Fig. 1).

This result underlines that the efficacy of therapy with a COX-2 inhibitor seems most pronounced in the first years of the schizophrenic disease process and confirms the findings of Rapaport et al. [49], who did not find a benefit of the COX-2 inhibitor in chronic schizophrenia. Therefore, a celecoxib add-on study was designed in patients with a first manifestation of schizophrenia, which also showed a beneficial effect of celecoxib add-on treatment (add-on to amisulpride) in schizophrenia, not only on the PANSS total score, but also on the positive symptom, negative symptom, and the general psychopathology scores [50, 51]. A recent study also demonstrated a beneficial effect of acetylsalicylic acid in schizophrenic spectrum disorders [52]. A meta-analysis of the clinical effects of nonsteroidal anti-inflammatory drugs in schizophrenia revealed significant effects on schizophrenic total, positive, and negative symptoms, which were not restricted to COX-2 inhibition [53], while another meta-analysis found a significant benefit only in schizophrenia patients with a short duration of disease or first manifestation of schizophrenia [54].

These results can be interpreted in the way that chronification (and possibly neuroprogression) are associated with worse effects of anti-inflammatory treatment results in schizophrenia, and the effect of anti-inflammatory treatment is related to the stage of the disease. On the other hand, it is well known from the therapy with first- or second-generation antipsychotics that chronification has a negative impact on the outcome in general. Since the discussed studies of anti-inflammatory treatment are short-term studies over several weeks, it has also to be taken into account that short-term anti-inflammatory treatment has only weak effects on chronic inflammatory diseases. Longer-term anti-inflammatory treatment might show different effects in chronic schizophrenia [55].

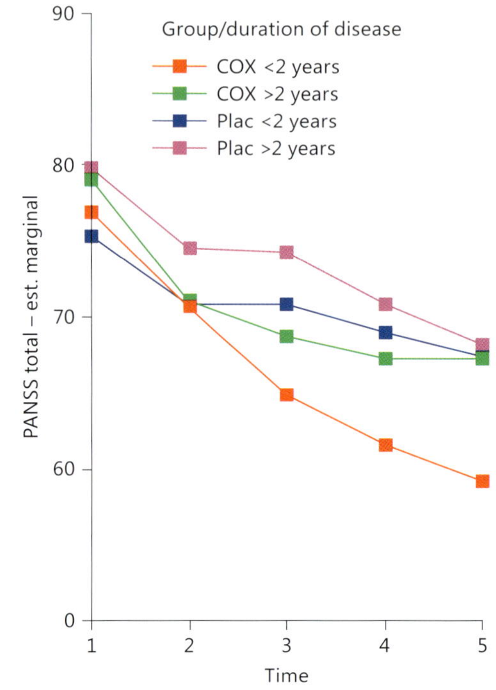

Fig. 1. Comparison of disease duration on the effects of celecoxib (COX) add-on therapy to risperidone. Patients with a disease duration <2 years and celecoxib treatment had a better outcome than patients with 2 years of disease duration and placebo and both groups of patients with a disease duration >2 years (nonsignificant difference).

Further Immune-Related Substances in the Therapy of Schizophrenia

Because of the role of microglial activation in inflammation, minocycline, an antibiotic and inhibitor of microglial activation, is an interesting substance for the treatment of schizophrenia. The improvement in cognition by minocycline has been described in animal models of schizophrenia [56] and in 2 double-blind, placebo-controlled add-on therapy trials in schizophrenia patients [57, 58]. In clinical studies, positive effects on negative schizophrenia symptoms were also noted [58]. Case reports documented positive effects of minocycline on the whole symptom spectrum in schizophrenia [59].

Acetylcysteine and other substances, including omega-3 fatty acids, which have anti-inflammatory and other effects also provide some benefit to schizophrenia patients [for a review, see 60].

First pilot experiences with the cytokine IFN-γ, which stimulates monocytic type 1 immune responses, as a therapeutic approach in schizophrenia are encouraging [61], although side effects, including unwanted immune effects, have to be carefully monitored, and results are only preliminary. On the other hand, such a hypothesis-driven therapeutic approach opens interesting perspectives for the development of therapeutic substances based on etiopathology.

Regarding the use of monoclonal antibodies in schizophrenia, they are an upcoming topic. So far, no convincing data underline the use of monoclonal antibodies in schizophrenia, but looking at cytokine data, there is a rationale for the development of compelling, carefully planned clinical trials with monoclonal antibodies [62].

Acknowledgement

This review was in part supported by the foundation “Immunität und Seele”.

References

1 Olabi B, Ellison-Wright I, McIntosh AM, Wood SJ, Bullmore E, Lawrie SM: Are there progressive brain changes in schizophrenia? A meta-analysis of structural magnetic resonance imaging studies. Biol Psychiatry 2011;70: 88–96.

2 Kraepelin E: Psychiatry: A Textbook for Students and Physicians. Canton, Watson Publishing International, 1899/1990.

3 Möller HJ, Bottlender R, Gross A, Hoff P, Wittmann J, Wegner U, Strauss A: The Kraepelinian dichotomy: preliminary results of a 15-year follow-up study on functional psychoses: focus on negative symptoms. Schizophr Res 2002;56: 87–94.

4 Jäger MDM, Hintermayr M, Bottlender R, Strauss A, Moller HJ: Course and outcome of first-admitted patients with acute and transient psychotic disorders (ICD-10:F23). Focus on relapses and social adjustment. Eur Arch Psychiatry Clin Neurosci 2003;253:209–215.

5 Bottlender R, Wegner U, Wittmann J, Strauss A, Möller HJ: Deficit syndromes in schizophrenic patients 15 years after their first hospitalisation: preliminary results of a follow-up study. Eur Arch Psychiatry Clin Neurosci 1999;249(suppl 4):27–36.

6 Cahn W, Hulshoff Pol HE, Bongers M, Schnack HG, Mandl RC, Van Haren NE, Durston S, Koning H, Van Der Linden JA, Kahn RS: Brain morphology in antipsychotic-naive schizophrenia: a study of multiple brain structures. Br J Psychiatry Suppl 2002;43:s66–s72.

7 Cahn W, Van Haren NE, Hulshoff Pol HE, Schnack HG, Caspers E, Laponder DA, Kahn RS: Brain volume changes in the first year of illness and 5-year outcome of schizophrenia. Br J Psychiatry 2006;189:381–382.

8 Baaré WF, van Oel CJ, Hulshoff Pol HE, Schnack HG, Durston S, Sitskoorn MM, Kahn RS: Volumes of brain structures in twins discordant for schizophrenia. Arch Gen Psychiatry 2001;58:33–40.

9 Gogtay N, Lu A, Leow AD, Klunder AD, Lee AD, Chavez A, Greenstein D, Giedd JN, Toga AW, Rapoport JL, Thompson PM: Three-dimensional brain growth abnormalities in childhood-onset schizophrenia visualized by using tensor-based morphometry. Proc Natl Acad Sci USA 2008;105:15979–15984.

10 Steen RG, Mull C, McClure R, Hamer RM, Lieberman JA: Brain volume in first-episode schizophrenia: systematic review and meta-analysis of magnetic resonance imaging studies. Br J Psychiatry 2006; 188:510–518.

11 Job DE, Whalley HC, McIntosh AM, Owens DG, Johnstone EC, Lawrie SM: Grey matter changes can improve the prediction of schizophrenia in subjects at high risk. BMC Med 2006; 4:29.

12 Chakos MH, Schobel SA, Gu H, Gerig G, Bradford D, Charles C, Lieberman JA: Duration of illness and treatment effects on hippocampal volume in male patients with schizophrenia. Br J Psychiatry 2005;186:26–31.

13 Meisenzahl EM, Rujescu D, Kirner A, Giegling I, Kathmann N, Leinsinger G, Maag K, Hegerl U, Hahn K, Moller HJ: Association of an interleukin-1β genetic polymorphism with altered brain structure in patients with schizophrenia. Am J Psychiatry 2001;158: 1316–1319.

14 Ellman LM, Deicken RF, Vinogradov S, Kremen WS, Poole JH, Kern DM, Tsai WY, Schaefer CA, Brown AS: Structural brain alterations in schizophrenia following fetal exposure to the inflammatory cytokine interleukin-8. Schizophr Res 2010;121:46–54.

15 Versijpt JJ, Dumont F, Van Laere KJ, Decoo D, Santens P, Audenaert K, Achten E, Slegers G, Dierckx RA, Korf J: Assessment of neuroinflammation and microglial activation in Alzheimer's disease with radiolabelled PK11195 and single photon emission computed tomography. A pilot study. Eur Neurol 2003;50:39–47.

16 van Berckel BN, Bossong MG, Boellaard R, Kloet R, Schuitemaker A, Caspers E, Luurtsema G, Windhorst AD, Cahn W, Lammertsma AA, Kahn RS: Microglia activation in recent-onset schizophrenia: a quantitative (*R*)-[^{11}C] PK11195 positron emission tomography study. Biol Psychiatry 2008;64: 820–822.

17 Doorduin J, de Vries EF, Willemsen AT, de Groot JC, Dierckx RA, Klein HC: Neuroinflammation in schizophrenia-related psychosis: a PET study. J Nucl Med 2009;50:1801–1807.

18 Takano A, Arakawa R, Ito H, Tateno A, Takahashi H, Matsumoto R, Okubo Y, Suhara T: Peripheral benzodiazepine receptors in patients with chronic schizophrenia: a PET study with [^{11}C] DAA1106. Int J Neuropsychopharmacol 2010;13:943–950.

19 Ling ZD, Potter ED, Lipton JW, Carvey PM: Differentiation of mesencephalic progenitor cells into dopaminergic neurons by cytokines. Exp Neurol 1998;149: 411–423.

20 Kabiersch A, Furukawa H, del Rey A, Besedovsky HO: Administration of interleukin-1 at birth affects dopaminergic neurons in adult mice. Ann NY Acad Sci 1998;840:123–127.

21 Potter ED, Ling ZD, Carvey PM: Cytokine-induced conversion of mesencephalic-derived progenitor cells into dopamine neurons. Cell Tissue Res 1999; 296:235–246.

22 Jarskog LF, Xiao H, Wilkie MB, Lauder JM, Gilmore JH: Cytokine regulation of embryonic rat dopamine and serotonin neuronal survival in vitro. Int J Dev Neurosci 1997;15:711–716.

23 Buka SL, Tsuang MT, Torrey EF, Klebanoff MA, Bernstein D, Yolken RH: Maternal infections and subsequent psychosis among offspring. Arch Gen Psychiatry 2001;58:1032–1037.

24 Westergaard T, Mortensen PB, Pedersen CB, Wohlfahrt J, Melbye M: Exposure to prenatal and childhood infections and the risk of schizophrenia: suggestions from a study of sibship characteristics and influenza prevalence. Arch Gen Psychiatry 1999;56:993–998.

25 Pearce BD: Schizophrenia and viral infection during neurodevelopment: a focus on mechanisms. Mol Psychiatry 2001;6:634–646.

26 Buka SL, Cannon TD, Torrey EF, Yolken RH: Maternal exposure to herpes simplex virus and risk of psychosis among adult offspring. Biol Psychiatry 2008;63: 809–815.
27 Brown AS, Begg MD, Gravenstein S, Schaefer CA, Wyatt RJ, Bresnahan M, Babulas VP, Susser ES: Serologic evidence of prenatal influenza in the etiology of schizophrenia. Arch Gen Psychiatry 2004;61:774–780.
28 Brown AS, Cohen P, Susser ES, Greenwald MA: Nonaffective psychosis after prenatal exposure to rubella. Am J Psychiatry 2000;157:438–443.
29 Sorensen HJ, Mortensen EL, Reinisch JM, Mednick SA: Association between prenatal exposure to bacterial infection and risk of schizophrenia. Schizophr Bull 2009;35:631–637.
30 Babulas V, Factor-Litvak P, Goetz R, Schaefer CA, Brown AS: Prenatal exposure to maternal genital and reproductive infections and adult schizophrenia. Am J Psychiatry 2006; 163:927–929.
31 Brown AS, Schaefer CA, Quesenberry CP Jr, Liu L, Babulas VP, Susser ES: Maternal exposure to toxoplasmosis and risk of schizophrenia in adult offspring. Am J Psychiatry 2005;162:767–773.
32 Brown AS, Hooton J, Schaefer CA, Zhang H, Petkova E, Babulas V, Perrin M, Gorman JM, Susser ES: Elevated maternal interleukin-8 levels and risk of schizophrenia in adult offspring. Am J Psychiatry 2004;161: 889–895.
33 Benros ME, Nielsen PR, Nordentoft M, Eaton WW, Dalton SO, Mortensen PB: Autoimmune diseases and severe infections as risk factors for schizophrenia: a 30-year population-based register study. Am J Psychiatry 2011;168:1303–1310.
34 Benros ME, Mortensen PB, Eaton WW: Autoimmune diseases and infections as risk factors for schizophrenia. Ann NY Acad Sci 2012;1262:56–66.
35 Sperner-Unterweger B, Barnas C, Fuchs D, Kemmler G, Wachter H, Hinterhuber H, Fleischhacker WW: Neopterin production in acute schizophrenic patients: an indicator of alterations of cell-mediated immunity. Psychiatry Res 1992;42: 121–128.
36 Müller N, Schwarz MJ: Immune system and schizophrenia. Curr Immunol Rev 2010;6:213–220.
37 Potvin S, Stip E, Sepehry AA, Gendron A, Bah R, Kouassi E: Inflammatory cytokine alterations in schizophrenia: a systematic quantitative review. Biol Psychiatry 2008;63:801–808.
38 Miller BJ, Buckley P, Seabolt W, Mellor A, Kirkpatrick B: Meta-analysis of cytokine alterations in schizophrenia: clinical status and antipsychotic effects. Biol Psychiatry 2011;70:663–671.
39 Müller N, Ackenheil M, Hofschuster E, Mempel W, Eckstein R: Cellular immunity in schizophrenic patients before and during neuroleptic treatment. Psychiatry Res 1991;37:147–160.
40 Müller N, Empl M, Riedel M, Schwarz M, Ackenheil M: Neuroleptic treatment increases soluble IL-2 receptors and decreases soluble IL-6 receptors in schizophrenia. Eur Arch Psychiatry Clin Neurosci 1997;247:308–313.
41 Schwarcz R, Pellicciari R: Manipulation of brain kynurenines: glial targets, neuronal effects, and clinical opportunities. J Pharmacol Exp Ther 2002;303:1–10.
42 Kessler M, Terramani T, Lynch G, Baudry M: A glycine site associated with N-methyl-D-aspartic acid receptors: characterization and identification of a new class of antagonists. J Neurochem 1989;52:1319–1328.
43 Hilmas C, Pereira EF, Alkondon M, Rassoulpour A, Schwarcz R, Albuquerque EX: The brain metabolite kynurenic acid inhibits alpha7 nicotinic receptor activity and increases non-alpha7 nicotinic receptor expression: physiopathological implications. J Neurosci 2001;21:7463–7473.
44 Grohmann U, Fallarino F, Puccetti P: Tolerance, DCs and tryptophan: much ado about IDO. Trends Immunol 2003; 24:242–248.
45 Kegel ME, Bhat M, Skogh E, Samuelsson M, Lundberg K, Dahl ML, Sellgren C, Schwieler L, Engberg G, Schuppe-Koistinen I, Erhardt S: Imbalanced kynurenine pathway in schizophrenia. Int J Tryptophan Res 2014;7:15–22.
46 Gos T, Myint AM, Schiltz K, Meyer-Lotz G, Dobrowolny H, Busse S, Muller UJ, Mawrin C, Bernstein HG, Bogerts B, Steiner J: Reduced microglial immunoreactivity for endogenous NMDA receptor agonist quinolinic acid in the hippocampus of schizophrenia patients. Brain Behav Immun 2014;41:59–64.
47 Müller N, Riedel M, Scheppach C, Brandstätter B, Sokullu S, Krampe K, Ulmschneider M, Engel RR, Möller HJ, Schwarz MJ: Beneficial antipsychotic effects of celecoxib add-on therapy compared to risperidone alone in schizophrenia. Am J Psychiatry 2002;159: 1029–1034.
48 Müller N, Riedel M, Schwarz MJ, Engel RR: Clinical effects of COX-2 inhibitors on cognition in schizophrenia. Eur Arch Psychiatry Clin Neurosci 2005;255:149–151.
49 Rapaport MH, Delrahim KK, Bresee CJ, Maddux RE, Ahmadpour O, Dolnak D: Celecoxib augmentation of continuously ill patients with schizophrenia. Biol Psychiatry 2005;57:1594–1596.
50 Müller N: COX-2 inhibitors as antidepressants and antipsychotics: clinical evidence. Curr Opin Investig Drugs 2010;11:31–42.
51 Müller N, Krause D, Dehning S, Musil R, Schennach-Wolff R, Obermeier M, Möller HJ, Klauss V, Schwarz MJ, Riedel M: Celecoxib treatment in an early stage of schizophrenia: results of a randomized, double-blind, placebo-controlled trial of celecoxib augmentation of amisulpride treatment. Schizophr Res 2010; 121:119–124.
52 Laan W, Grobbee DE, Selten JP, Heijnen CJ, Kahn RS, Burger H: Adjuvant aspirin therapy reduces symptoms of schizophrenia spectrum disorders: results from a randomized, double-blind, placebo-controlled trial. J Clin Psychiatry 2010;71:520–527.
53 Sommer IE, de Witte L, Begemann M, Kahn RS: Nonsteroidal anti-inflammatory drugs in schizophrenia: ready for practice or a good start? A meta-analysis. J Clin Psychiatry 2012;73:414–419.
54 Nitta M, Kishimoto T, Müller N, Weiser M, Davidson M, Kane JM, Correll CU: Adjunctive use of nonsteroidal anti-inflammatory drugs for schizophrenia: a meta-analytic investigation of randomized controlled trials. Schizophr Bull 2013;39:1230–1241.
55 Akhondzadeh S, Tabatabaee M, Amini H, Ahmadi Abhari SA, Abbasi SH, Behnam B: Celecoxib as adjunctive therapy in schizophrenia: a double-blind, randomized and placebo-controlled trial. Schizophr Res 2007;90:179–185.

56 Mizoguchi H, Takuma K, Fukakusa A, Ito Y, Nakatani A, Ibi D, Kim HC, Yamada K: Improvement by minocycline of methamphetamine-induced impairment of recognition memory in mice. Psychopharmacology (Berl) 2008;196: 233–241.
57 Levkovitz Y, Mendlovich S, Riwkes S, Braw Y, Levkovitch-Verbin H, Gal G, Fennig S, Treves I, Kron S: A double-blind, randomized study of minocycline for the treatment of negative and cognitive symptoms in early-phase schizophrenia. J Clin Psychiatry 2010;71:138–149.
58 Chaudhry IB, Hallak J, Husain N, Minhas F, Stirling J, Richardson P, Dursun S, Dunn G, Deakin B: Minocycline benefits negative symptoms in early schizophrenia: a randomised double-blind placebo-controlled clinical trial in patients on standard treatment. J Psychopharmacol 2012;26:1185–1193.
59 Ahuja N, Carroll BT: Possible anti-catatonic effects of minocycline in patients with schizophrenia. Prog Neuropsychopharmacol Biol Psychiatry 2007;31:968–969.
60 Sommer IE, van Westrhenen R, Begemann MJ, de Witte LD, Leucht S, Kahn RS: Efficacy of anti-inflammatory agents to improve symptoms in patients with schizophrenia: an update. Schizophr Bull 2014;40:181–191.
61 Grüber L, Bunse T, Weidinger E, Reichard H, Müller N: Adjunctive recombinant human interferon gamma-1b for treatment-resistant schizophrenia in 2 patients. J Clin Psychiatry 2014;75: 1266–1267.
62 Miller BJ, Buckley PF: The case for adjunctive monoclonal antibody immunotherapy in schizophrenia. Psychiatr Clin North Am 2016;39:187–198.

Prof. Dr. med. Dipl.-Psych. Norbert Müller
Department of Psychiatry and Psychotherapy
Ludwig Maximilian University
Nussbaumstrasse 7
DE–80336 Munich (Germany)
E-Mail Norbert.Mueller@med.uni-muenchen.de

Halaris A, Leonard BE (eds): Neuroprogression in Psychiatric Disorders.
Mod Trends Pharmacopsychiatry. Basel, Karger, 2017, vol 31, pp 10–26 (DOI: 10.1159/000470803)

The Link between Refractoriness and Neuroprogression in Treatment-Resistant Bipolar Disorder

Isabelle E. Bauer · Jair C. Soares · Salih Selek · Thomas D. Meyer
Department of Psychiatry and Behavioral Sciences, University of Texas Health Science Center at Houston, Houston, TX, USA

Abstract

Treatment refractoriness remains one of the biggest challenges in the field of bipolar disorder (BD) as treatments are often suboptimal or unsatisfactory. Recent evidence points towards a potential link between the progressively evolving nature of BD, increased inflammation, and reduced treatment response. There are several medications and other somatic treatments available, but remission rates are low, and medication compliance is still problematic. Psychotherapeutic techniques appear to be promising in several disease states and in relapse prevention, but additional research is needed to determine who will benefit from what strategy the most. Current knowledge on the link between neuroprogression in BD and poor treatment response promotes the use of anti-inflammatory and neuroprotective strategies in the early phases of BD. In the later stages of BD, mood stabilization and medication adherence would be essential in preventing additional brain changes and loss of cognitive reserve. Additional large-scale, longitudinal, and methodologically robust studies are urgently needed to develop effective therapeutic interventions for treatment-resistant BD.

Introduction

Considered to be one of the top 10 leading causes of disability-adjusted years [1], bipolar disorder (BD) is a mood disorder with profound effects on the individual's cognitive and social functioning. The treatment of BD poses great challenges due to the unpredictable clinical course (e.g., emotional ups and downs), duration of the mood episodes, and the heterogeneous nature of clinical symptoms (e.g., manic, depressive, hypomanic, and mixed). Despite the large range of treatments for BD, approximately 37% of patients relapse within 1 year, and 60% relapse within 2 years after recovering from an episode [2]. Further, in BD, the suicide rate – the 10th leading cause of death in the United States [3] – amounts to 7.8% in men and 4.9% in women [4]. Treatment options for individuals with treatment-resistant BD (TRBD), i.e., those who do not respond well to first-line treatments, are still limited and their long-term efficacy is still unexplored.

This chapter will review current knowledge of pharmacological and psychological interventions for TRBD. Given recent suggestions that elements of inflammation may contribute to the reoccurrence of mood episodes, we will also discuss potential biological mechanisms underlying neuroprogression in TRBD.

Definition of Refractoriness

Current research in BD prevention and its psychopharmacological therapy has primarily targeted the reduction in the burden associated with BD, and the improvement in long-term clinical and social outcomes [5, 6]. Despite major advances, refractoriness and recurrence represent the major treatment challenges in BD research. Treatment refractoriness is traditionally defined as the failure of recovering from an abnormal mood state despite adequate trials of at least 2 treatments [7, 8]. Surprisingly, there is no uniform definition of refractoriness, and no definition of refractoriness specific to BD. In unipolar depression, refractoriness refers to the unsuccessful response to two antidepressants after a treatment period of at least 6 months [9]. Some authors define it as the failure to respond to first-line agents for BD (e.g., lithium) and specify that during treatment serum levels must be between 0.6 and 0.8 mmol/L to verify adherence to medication [10, 11]. Other researchers provide a broader definition of refractoriness by describing it as being characterized by an increasing number of mood episodes along with reduced responsiveness to otherwise efficacious treatments, such as lithium [12] and psychoeducational interventions [13, 14].

Some guidelines are more specific and include a range of criteria:

(1) The patient must have experienced 5 or more mood episodes, and 2 of these episodes should have occurred in the past 3 years despite adequate medication.

(2) The patient failed to respond to 2 different treatments administered either in the form of individual or combined treatments.

(3) The criteria for "adequate medication" must be fulfilled, which was defined as a treatment period of ≥12 month (≥6 months for RBD patients), and blood tests must show adequate serum levels of the psychotropic medication (lithium ≥0.5 mmol/l, carbamazepine ≥5 mg, and valproate ≥50 mg) [15].

Brief Note on Rapid Cycling Bipolar Disorder

Rapid cycling BD can be viewed as a special form of TRBD. This variant of the illness is defined by 4 or more episodes of illness within a 12-month period [16]. TRBD is characterized by recurrent mood episodes over a period of 12 months. These episodes may occur in a random or sequential order, and be manic, depressive, hypomanic, or mixed in nature and interspersed with periods of remission lasting for a maximum of 2 months [17]. Approximately 10–20% of patients with BD present with rapid cycling [18], 70–90% of these patients are women [19], and they often suffer from BD of subtype II [19]. Despite the large number of pharmacological studies focusing on rapid cycling BD, the management of this group of patients remains challenging [20–24]. It is not uncommon for these patients to report frequent changes in medication or polypharmacy [25]. As we will discuss in the following section, evidence for the efficacy of treatments targeting TRBD is limited. Some studies have, however, shown that some treatments may be successful when administered in combination, while other treatment regimens rely heavily on the clinicians' expertise and their knowledge of the patients' needs.

Assessment of Treatment Response in Treatment-Resistant Bipolar Disorder

There are several treatment guidelines in BD but most of them are not specific to "treatment refractoriness." Most of them have a stepwise approach in which dose, number of medications, and frequency of administration are increased or decreased as necessary, to achieve and ideally maintain treatment response. Thus, the mention of "third-line suggestions for BD" is often viewed as an intervention in case of treatment resistance. Poor compliance with drug treatment is, however, a common problem among patients with TRBD. Given that evidence-based data and guidelines on criteria and timing for medication changes remain limited, some patients are at risk of failing to respond to medication because they discontinue treatment or take medication on an irregular basis [26]. As a result, some clinicians may recommend a premature change in medication by interpreting the lack of response to medication as a sign of "refractoriness" rather than an indication of noncompliance to the treatment [27]. As such, during the initial phases of a new treatment, the occurrence of a mild or moderate mood episode should not trigger a change in medication, as medication may take some time to take effect. Clinical judgment is therefore extremely important to assess the cause of refractoriness and/or poor compliance, to manage medication, and to avoid unnecessary side effects.

Current guidelines recommend to review the diagnosis on a regular basis and rule out possible organic and substance-induced or -triggered causes and discontinue antidepressants, if any were given before [28]. For instance, endocrinological (e.g., hypothyroidism), neurological and autoimmune diseases, and neurodegenerative illnesses should always be considered, as they may compromise the effectiveness of treatments [10]. The treatment of comorbidities such as substance use and the evaluation of psychosocial factors are also of relevance. Furthermore, the duration of the medication trial is of relevance: each medication trial should last for a minimum of 3 weeks and be provided at an adequate dose excluding the titration period in mania [29]. Notably, some treatments that have been found to trigger rapid cycling, such as tricyclic antidepressants (TCAs) and stimulants, should be discontinued [15]. Medication compliance should also be assessed based upon the clinical interview and the serum levels of the medications, notably mood stabilizers.

Current Treatment Options for Treatment-Resistant Bipolar Disorder and Their Efficacy

Pharmacotherapy

The American Psychiatric Association (APA) [30] recommends lithium as one of the first-line treatments for BD. Lithium nonresponders are often administered lamotrigine, a second mood stabilizer, or olanzapine if they have a history of severe or refractory mania. In particular, the combination of lithium plus valproate appears to be an appropriate approach for patients with BD subtype I with a history of severe or refractory manic episodes. In nonresponders to valproate or carbamazepine, the addition of lithium or lamotrigine can also be implemented. Another option is bupropion or a selective serotonin reuptake inhibitor (SSRI) in BD patients with a history of mild-moderate mania or hypomania. Overall, clinicians often first consider optimizing antidepressant dosage or lengthening therapy. In patients who do not respond to medication at all, antidepressant drug substitution is often efficacious. Combining 2 or more medications can be considered but needs constant monitoring to avoid side effects [31]. In the next section, we will review the literature on treatment optimization, combination, and substitution, and summarize their efficacy in TRBD patients.

Optimization

An open study reported that the addition of lithium to carbamazepine led to improvement in depression in 6 of 13 BD patients who did not respond to carbamazepine [32]. A double-blind study compared the efficacy of adding a second mood stabilizer versus the addition of paroxetine in the treatment of BD [33]. Twenty-seven patients with BD on either lithium or valproate were randomly assigned to either a second mood stabilizer (lithium for those on valproate and valproate for those on lithium) or paroxetine for 6 weeks [10]. The severity of mood symptoms improved to the same extent in both groups. Notably, there were significantly more noncompleters in the mood stabilizer combination group compared with the paroxetine and mood stabilizer group. This suggests that medication adherence to the lithium-valproate combination may be reduced compared to that to the mood stabilizer-paroxetine combination. These findings should, however, be interpreted with caution due to the small sample size in each group.

Combination

The Canadian Network for Mood and Anxiety Treatments (CANMAT) and the International Society of Bipolar Disorders (ISBD) recommended the following combinations in treatment-resistant bipolar depression: (1) lithium with carbamazepine, pramipexole, monoamine oxidase inhibitors, TCAs, venlafaxine, SSRI (except paroxetine), or lamotrigine, (2) divalproex with venlafaxine, TCA, SSRI (except paroxetine), or lamotrigine (with cautious titration), and (3) quetiapine with lamotrigine [28]. To date, 4 double-blind trials have examined the efficacy of antidepressant add-ons to mood stabilizers [34–37]. The double-blind trial by Sachs et al. [37] compared the addition of bupropion or the TCA desipramine to either lithium, an anticonvulsant, or a mood stabilizers in 15 patients with acute BD. In this 12-month trial, the response rate was 50% for the desipramine group and 55% for the bupropion group. The switch to mania or hypomania occurred in 50% of the patients on desipramine and in 11% on bupropion. There was, however, no difference in the antidepressant efficacy between both treatments. In a study of acute BD, the addition of paroxetine was found to be as effective as the addition of a second mood stabilizer [35]. The manic switch occurred in 8% of the lithium plus imipramine group and in 2% of the lithium plus placebo group. Notably, none of the individuals on paroxetine and lithium switched. Further, in individuals with high serum lithium levels, the switch rate was 6% for the imipramine plus lithium group but 0% in the other two groups. A 10-week double-blind study examined the efficacy of bupropion, sertraline, or venlafaxine add-on to mood stabilizers in the treatment of BD patients that had a breakthrough depressive episode while on adequate doses of mood stabilizers. Overall, improved mood state was observed in 35 (37%) of the 95 participants, and only 13 individuals (14%) showed manic/hypomanic switches. Among the 48 patients that took part in the continuation part of the trial (12 months), 16 (33%) had manic/hypomanic switches [36]. Overall, the addition of an antidepressant is an effective strategy for treating BD. However, the manic/hypomanic switch may occur in spite of the mood stabilizer cotherapy, particularly with TCAs.

A single-blind trial compared the efficacy of the anticonvulsant, topiramate, and bupropion used as add-ons in 36 outpatients with BD [38]. Response rates were 56% for the topiramate group and 59% for the bupropion group. These results suggest that topiramate is as effective as bupropion. However, given the small sample size and lack of placebo control in this study, additional work focusing on these treatments is needed. Notably, in 6 patients on topiramate and 4 on bupropion, treatment was discontinued because of adverse events. The response rate to gabapentin monotherapy was found to be higher than that to placebo [39]. Thus, gabapentin may not be an ideal add-on in treating acute BD.

Another single-blind trial compared the addition of the SSRI paroxetine (n = 30) versus the serotonin-norepinephrine reuptake inhibitor venlafaxine (n = 30) to mood stabilizers for acute bipolar depression over a 6-week period [40]. There were no differences in efficacy between both groups, with 43% of patients in the paroxetine group and 48% in the venlafaxine group responding to treatment. Notably, the manic/hypomanic switch rate was slightly higher, albeit not statistically significant, in the venlafaxine group (13%) than in the paroxetine group (3%).

However, given the evidence of the efficacy of olanzapine monotherapy in acute BD [41] and reports of the beneficial effects of atypical antipsychotics in treating depressive symptoms [42], one could speculate that the addition of an atypical antipsychotic, in particular olanzapine, may be of help for some refractory bipolar depressed patients. A study compared the efficacy of the TCA imipramine, the SSRI paroxetine, and a placebo treatment used as adjunctive treatments to lithium. The sample included patients with TRBD with high serum lithium levels (≥0.8 mmol/L). The placebo add-on was found to be as effective as both antidepressant add-ons in treating depressive symptoms in BD [34]. Although no clinical trial tested the efficacy of lamotrigine along with lithium, this type of combination may be promising as lamotrigine does not lead to a manic switch and is efficacious in BD [43].

Substitution

The "drug substitution" strategy involves replacing a failed medication with another medication. Examples of this strategy include: substituting lithium with lamotrigine, carbamazepine or valproate, or replacing an antidepressant with another antidepressant (i.e., replacing paroxetine with bupropion or vice versa), or with a novel anticonvulsant such as topiramate. The double-blind trial of Thase et al. [44] found that 75% of the patients who were refractory to imipramine (n = 12) responded to double-blind treatment with tranylcypromine. If not tried before, quetiapine or lurasidone may also be given, since they are approved first-line treatments for BD. Other studies using either randomized, double-blind, placebo-controlled, or crossover designs found beneficial effects of ketamine in TRBD patients both as a monotherapy and as an add-on treatment [45, 46]. For treatment-resistant mania, haloperidol, clozapine, oxcarbazepine, tamoxifen, and cariprazine may also be included as options. Recent evidence also supports the use of novel agents, including zotepine, levetiracetam, phenytoin, mexiletine, ω-3-fatty acids, calcitonin, rapid tryptophan depletion, allopurinol, amisulpride, folic acid, and memantine [28].

Other Somatic Therapies

The field of nonpharmacological therapies for TRBD is rapidly evolving, and new somatic therapies are now considered to be valuable options for patients who have failed numerous other treatments. A major challenge for clinicians (and patients alike) is how to integrate the results from published clinical trials in the clinical decision-making process. In the following section, we will briefly discuss electroconvulsive therapy (ECT), transcranial magnetic stimulation (TMS), magnetic seizure treatment, and deep brain stimulation (DBS).

Electroconvulsive Therapy and Other Neurostimulation Treatments

ECT has been used for more than 80 years for treating both depressive and manic symptoms. Most of the reviews published to date highlight its beneficial effects in TRBD [47–49]. A randomized, controlled trial compared the efficacy of ECT to an algorithm-based pharmacological treatment. They found a significantly higher response rate in ECT (73.9 vs. 35.0%), but remission rates were comparable between groups (34.8 vs. 30.0%) [50, 51]. Two retrospective studies conducted in TRBD patients and patients with unipolar and bipolar depression showed reduced rates of hospitalization with maintenance ECT [52, 53]. Despite these promising findings, ECT is often not offered as a first-step treatment because of the risk of permanent memory or cognitive impairment often associated with the waveform of the pulse (e.g., sine wave ECT vs. brief pulse), the bilateral and dominant hemisphere electrode placement, and the use of high-energy ECT [54, 55]. Another source of concern is that ECT is performed under general anesthesia. Although this approach ensures a safe and pain-free experience, this may be seen as risky. The efficacy of ECT requires deep understanding of the interaction between anesthetic drugs and seizure activity, and an awareness of the physiologic effects of ECT as well as the treatment of those effects.

In sum, ECT could be a promising method to help individuals with TRBD. Unlike other treatments that alleviate bipolar depression but can trigger a manic/hypomanic switch, ECT is more likely to stabilize mood for a longer period of time. Despite concerns related to permanent cognitive deficits, better knowledge of pulse waveform and anesthetic procedures has improved the safety of ECT and reduced its side effects. There is, however, a lack of studies comparing ECT to pharmacological treatments. These studies would be essential to assess the efficacy and safety of ECT in preventing recurrence and relapse associated with BD. Although repetitive medication trial failures are likely to be less cost-effective than ECT [56], no published study has assessed the cost-effectiveness of ECT in depression or schizophrenia. Additional research in this field may be helpful to include ECT in the standard treatment plan for BD, possibly increasing recovery rates and reducing treatment costs for the patients.

Transcranial Magnetic Stimulation

Despite the small number of studies focusing on deep TMS [57], this intervention is considered to be a promising nonpharmacological intervention to treat various psychiatric disorders such as BD [58]. To date, only 1 published clinical trial focused on the efficacy of TMS in BD [59]. Nineteen BD patients received daily prefrontal deep TMS every weekday for 4 consecutive weeks using 20-Hz pulses. The authors reported that the response was a 50% decrease in the Hamilton Depression Rating Scale score 1 week after the last treatment session; remission was considered a score less than 10; 63.2% (12/19) of patients were classified as responders and 52.6% (10/19) as "in remission." Despite the small sample size, concomitant medications, and lack of placebo stimulations, this study provides preliminary evidence of the efficacy of deep TMS compared to standard TMS for BD. An add-on study using low-frequency standard TMS targeting the right dorsolateral prefrontal cortex in 11 TRBD type I or II patients was found to be well tolerated and effective (54% response rate), and remission rates approached 36.3% [60].

Despite these promising findings, large-scale studies are needed to determine the effectiveness of this intervention, specifically for TRBD. It is noteworthy that the magnetic seizure treatment – an intervention that integrates the therapeutic aspects of ECT and TMS – is gaining popularity for its potential to achieve the efficacy of ECT with safety rates similar to those of TMS. This treatment is still viewed as experimental, but preliminary evidence suggests that the magnetic seizure treatment is less likely to impair memory and is associated with faster recovery times compared with ECT [61].

Deep Brain Stimulation

Neuromodulation techniques are established as effective treatments for neurological conditions, such as Parkinson disease and chronic pain [62]. DBS is a novel neuromodulation technique that has recently been found to be helpful for treatment-resistant neurologic and psychiatric disorders [63–65]. It involves the stereotactic implantation of electrodes in neuroanatomical targets where stimulation is applied via a stimulator device implanted subcutaneously [66]. DBS is currently being tested for the treatment of treatment-resistant depression [63]. DBS seems to produce a significant reduction in symptoms and high rates of remission [67]. Recently, the medial forebrain bundle has emerged as an additional plausible target [68, 69].

The pilot study by Schlaepfer et al. [70] demonstrated the safety and efficacy of DBS for treatment-resistant depression. In particular, this study showed how short- and long-term local stimulation of neural network activity within the ventral-tegmental area and the prefrontal cortex (including the prefrontal medial bundle) and specific stimulation parameters in selected targets can produce response and remission rates over previously resistant and refractory mood symptoms. In line with the results of Schlaepfer et al. [70], Fenoy et al. [71] showed that the stimulation of the medial forebrain bundle was associated with rapid antidepressant effects within the 1st week of stimulation. Another study used subcallosal cingulate DBS in 10 unipolar and 7 bipolar treatment-resistant patients. The authors reported a decrease in depression and an increase in functionality [72]. Most important, no participant included in this study relapsed. A recent review concluded that the superolateral medial forebrain bundle may be an ideal site for DBS because it has a large number of projections to the frontal cortex. This region is involved in cognitive control and emotional regulation [73]. In sum, despite the promising results, there is an incomplete understanding of the mechanisms of action and optimal neuroanatomical targets. Additional studies are needed to gather clinical data on target selection, eligibility criteria, and long-term outcomes, and ensure the safety of the patients.

Psychosocial Adjunctive Therapies

There is well-established evidence of the efficacy of adjunctive psychological therapies, such as family-focused treatment (FFT), cognitive therapy (CT), group psychoeducation, and interpersonal and social rhythm therapy (IPSRT) in BD [74–77]. Psychological support and medication appear to be more effective than medication alone in terms of long-term outcomes, increased compliance, better social functioning, and reduced symptom severity [78–80]. FFT is the most studied intervention in BD. It involves psychoeducation and training on how to deal with family stress and promotes the development of appropriate communication and problem-solving skills. When compared to a control group receiving 2 educational sessions and emergency counseling sessions as needed, BD patients receiving FFT showed fewer relapses and longer time to relapse and greater improvement in depressive symptoms (but not manic symptoms) over 1 year [81]. The ability of families to show emotions and provide constructive criticism increased the effectiveness of FFT. Critically, a 2-year follow-up showed continuation of these beneficial treatment effects with FFT in terms of relapses, suicidal risk, medication adherence, and mood symptoms. Another study compared FFT to an individual therapy [82], and the authors found that, over 2 years, FFT did not decrease time to first relapse, but led to less relapses in general and fewer hospitalizations than individual therapy.

Other psychological treatment options for BD include CT. When combined with medication, a 12-month CT resulted in reduced numbers of mood episodes, hospitalizations, a decrease in depressive symptoms, and increased social functioning in a group of BD patients [83]. The effectiveness of this intervention was found to be stronger in the first 12 months of treatment when compared to the 24-month follow-up. Similarly, Meyer and Hautzinger [84] found a trend that CT was preventing relapses during treatment but was not superior to a supportive treatment condition during follow-up. However, the supportive therapy condition cannot be considered treatment as usual (TAU) since it includes a number of psychological treatment elements such as psychoeducation and completing daily mood charts. Scott et al. [13] randomized a group of patients with severe, recurrent BD to CT or TAU. While CT was not superior to TAU in preventing recurrences, those with fewer lifetime mood episodes benefited from CT compared to those with more former episodes. Peters et al. [85] also found some evidence that the clinical course of BD may affect responsiveness to psychological treatment. However, in the STEP-BD (Systematic Treatment Enhancement Program for Bipolar Disorder) study, the association was not linear, i.e., those with very few or a lot of prior depressive episodes benefited less from intensive psychotherapy. Perhaps motivation for psychological input is different when patients have experienced that the problems do not go away by just taking the medication. Group psychoeducation was found to be more effective in terms of number of relapses, number of mood episodes, and hospitalizations compared to an unstructured therapeutic treatment [86].

IPSRT, an individual therapy derived from interpersonal therapy, focuses on resolution of interpersonal conflicts and regularity in daily routines, and helps the patient gain insight into the link between mood and social functioning [87]. IPSRT was compared to intensive clinical management, a therapy focusing on education, symptoms, medication review, and nonspecific support. Frank et al. [87] had a sophisticated design and were able to show that if IPSRT is introduced during the acute phase of BD and was associated with a longer time to relapse and better social functioning. However, if IPSRT was just introduced during the maintenance phase of BD, the efficacy of IPSRT was comparable to that of intensive clinical management.

In summary, there is evidence that psychosocial therapies help decrease the risk of relapse and improve clinical outcomes in BD. Current findings support psychotherapeutic interventions that target mood symptoms, improve coping mechanisms, and enhance protective factors such as social support. This field is still in its infancy, and, therefore, studies focusing on special groups, such as TRBD, have not been published yet. Group therapy may be more cost-effective, while family therapy may be of assistance in family environments with hostile or intrusive families [29]. Overall, a flexible and individualized approach may be beneficial for TRBD (Table 1).

Table 1. Potential treatment options for treatment-resistant bipolar disorder

Mood stabilizers
Lithium
Carbamazepine
Valproate
Other agents
Combination of mood stabilizers
Anticonvulsants (first/second generation)
Antidepressants
Other somatic therapies
Electroconvulsive therapy
Transcranial magnetic stimulation
Deep brain stimulation
Psychosocial adjunctive therapies
Family-focused treatment
Cognitive therapy
Group psychoeducation
Interpersonal and social rhythm therapy

Neuroprogression and Bipolar Disorder

Recently, BD has been reconceptualized by some authors as a progressive disorder, presenting high levels of morbidity and mortality [88]. This is based on evidence showing an increase in clinical, immune-inflammatory, and neuroanatomical abnormalities when comparing patients with first mood episodes to those who experienced multiple mood episodes [78, 79, 89]. Moreover, multiple mood episodes [80, 90] and episode recurrence [80] have been associated with a poor clinical and functional outcome. Hence, current models hypothesize that BD and treatment refractoriness may be linked to neuroprogression.

Viewing BD as both a psychiatric and a progressive disorder may contribute to a better understanding and management of BD emphasizing both the importance of prevention and early intervention, and the selection of medication targeting inflammation to improve mood symptoms and maintain functionality [91–93]. Empirical evidence supporting the concept of neuroprogression in bipolar disorder is mixed. In the next few sections, we will briefly review the biological, neuroimaging, and cognitive findings in relation to this concept.

Inflammatory Markers and Bipolar Disorder

Dysregulation of the immune system is postulated to be involved in the pathophysiology of BD.

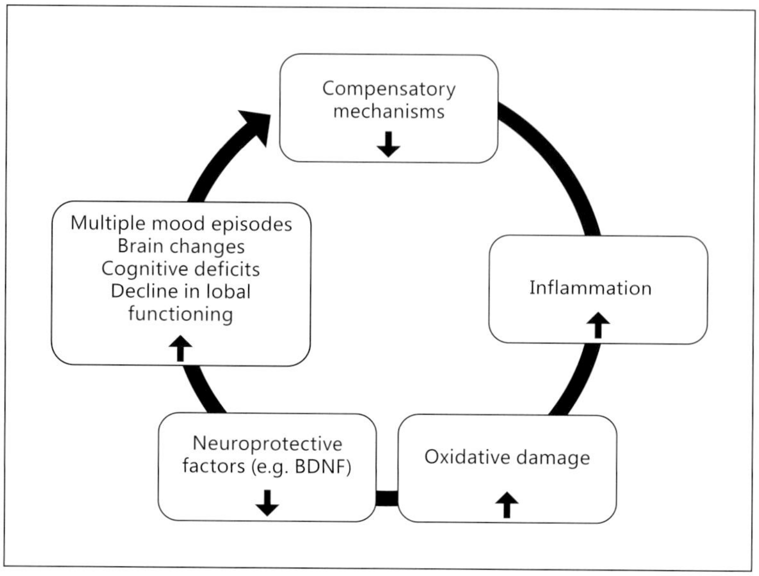

Fig. 1. Factors playing a role in neuroprogression and refractoriness in bipolar disorder. ↑, increasing; ↓, decreasing.

There is evidence of increased immunological response including microglial alterations [94, 95] and increased tumor necrosis factor (TNF)-α, soluble TNF-α receptor-1, interleukin (IL)-4, IL-6, soluble IL-6 receptor, IL-10, IL-1 receptor antagonist, and soluble IL-2 receptor in the blood of patients with BD [96–98] compared to healthy control individuals. Levels of cytokines appear to fluctuate across mood phases. For instance, a study found that depressed BD patients displayed lower levels of soluble TNF-α receptor-1 compared to patients in manic/hypomanic and euthymic states [99]. Notably, mood stabilizers have been hypothesized to increase the activity of antioxidants and decrease inflammatory processes associated with reduced levels of IL-1β, IL-6, and TNF-α [100]. Similarly, lithium and divalproex – which have proven antioxidant and anti-inflammatory properties – may help prevent and reduce the decline observed during the course of the disease [101].

Changes in immunological responses and inflammation appear to lead to an increase in reactive oxidative species and potentially damage to lipids and proteins [102, 103]. These changes may lead to early apoptosis, reduced neuronal and glial density, and eventually structural changes in the brain as well as cognitive decline (Fig. 1) [89, 92, 104, 105]. Additional evidence of the progressive nature of BD comes from a study showing stage-dependent changes in the levels of brain-derived neurotrophic factor (BDNF) and cytokines. The anti-inflammatory IL-10 was found to be elevated in the early but not in the late stages of BD [106]. Both proinflammatory IL-6 and TNF-α were elevated throughout the course of illness. There are numerous biomarker studies in BD, but they have been mostly conducted in patients who are not refractory to treatment. In summary, 4 areas have emerged for BD biomarkers: mitochondrial dysfunction and oxidative stress, trophic factors (i.e., BDNF), inflammatory cytokines, and urinary metabolites. Additional research on the predictive power of inflammatory markers and the development of medication targeting the immune-inflammatory system may, therefore, help health professionals establish more appropriate therapeutic options for patients with BD, and possibly those showing treatment resistance.

Neuroimaging Findings

Postmortem studies have revealed changes in the number, density, and size of neurons and glial cells in BD patients, especially in frontal and subcortical areas [105]. A recent postmortem study showed an increase in oxidative stress in the anterior cingulate cortex of BD patients, which may suggest that oxidative stress is a leading factor in brain abnormalities [107]. Structural brain studies have also shown that patients with BD display decreased gray matter volumes of the orbital and medial prefrontal cortex, ventral striatum, and mesotemporal cortex [108, 109] in addition to abnormal amygdala volumes [110, 111].

Functional brain imaging studies provide evidence that BD may be linked to abnormalities in brain regions responsible for emotional regulation and cognitive control [112]. Amygdala reactivity to emotional stimuli outweighs that of prefrontal cortical regions and leads to reduced emotion regulation [113, 114]. It has been suggested that reduced frontolimbic activation may lead to mania [112]. A robust body of evidence supports the concept of amygdala dysfunction as a stable feature of BD. Increased bilateral amygdala activation was reported in euthymic BD [115, 116]. Findings are, however, mixed, as a review of functional neuroimaging findings across mood states concluded that the amygdala was overall hyperactive during manic phases, fluctuating during depression, and comparable to that of healthy controls during the euthymic period. Other regions involved in monitoring and emotion processing, such as the anterior cingulate cortex and the striatum, showed a similar state-dependent profile. By comparison, the ventral prefrontal cortex appears to be hypoactivated across mood phases [116–118].

A number of studies have found white matter (WM) abnormalities in brain networks responsible for emotional control. Microstructural abnormalities in WM fiber tracts connecting to the limbic-striatal, cingulate, thalamus, corpus callosum, and prefrontal regions have been observed in children, adolescents, and adults with BD [119–126]. Notably, corpus callosum and thalamic radiation of middle-aged and drug-naïve BD populations showed reduced WM integrity. The reduction in WM integrity observed in BD has been linked to processes of demyelination, cerebral hypoperfusion, neuroinflammation, and reduced mitochondrial metabolism [127, 128], but there is little empirical evidence supporting either of these biological hypotheses. A study showed increased WM hyperintensities and decreased callosal myelination in at-risk individuals even prior to any affective symptoms or behavioral abnormalities [129]. According to different data, these alterations may exist well before an affective episode, underlying and contributing to additional developmental abnormalities during adolescence and ultimately leading to emotional dysregulation and illness onset [130]. Some investigators observed that structural abnormalities were correlated to age and temporal illness progression, such as decreases in subgenual cingulate cortex volumes, ventral prefrontal cortex reductions, and increases in amygdala [131–133].

In sum, structural and functional abnormalities in the corticolimbic network appear to be markers of BD that may be involved in emotion processing and regulation. Findings indicate a decline in functionality during the course of BD. Little is known about the longitudinal structural brain changes in BD, and most of the studies were conducted in patients who were not treatment resistant. Considering the neuroprogression hypothesis, additional research is needed to investigate brain structural changes along with immunological changes over time in both BD patients and vulnerable individuals.

Cognitive Impairment

Cognitive deficits are discussed as being prevalent in BD [134–137]. For example, overall manic patients display poorer functional cognitive abilities in verbal memory, verbal fluency, and cognitive estimation when compared with depressed and remitted patients [138]. By comparison, bipolar depressive patients present primarily with verbal memory deficits [136, 137, 139, 140]. However, it is still unclear whether cognitive performance fluctuates over the course of the disease, or whether changes are mostly related to the course of illness along with the aging process. While the majority of studies in BD have focused on heterogeneous BD populations, a small number of cross-sectional studies compared cognitive functioning across the acute and euthymic phases of BD [138, 141, 142]. The evaluation of a group of BD participants at discharge and 6–8 weeks after discharge revealed that manic individuals maintained the same pattern of reduced psychomotor speed than that presented at discharge [143]. Notably, a study comparing patients with BD in manic/mixed episode, depressive episode, and patients with unipolar depression showed that manic patients had deficits in episodic and working memory, spatial attention, and problem solving. In contrast, depressed bipolar und unipolar patients demonstrated impairment only in episodic memory [144]. Thus, the individual's current mood phase appears to have a significant impact on cognitive performance. Further, little is still known about the trajectory of cognitive decline during the lifetime course and whether deficits observed in individuals with BD who have experienced multiple mood episodes are age or illness dependent [145]. This is probably due to the lack of longitudinal studies exploring the link between neuroprogression and neurocognitive changes over time [146, 147]. The concept of "cognitive decline" in BD has, therefore, yet to be verified [146].

Medication and Neuroprogression

Consistent with the neuroprogression hypothesis, increasing evidence points to the neuroprotective effects of medications such as lithium [149, 150]. Mood stabilizers modulate the expression of potent neurotrophic factors and show the potential to counteract brain atrophy and cell death [151, 152]. A neuroimaging study conducted before and after a 4-week lithium treatment showed an increase in gray matter volume in the prefrontal cortex in BD subjects [150]. Moreover, a spectrophotometry postmortem study concluded that lithium stimulates mitochondrial respiratory chain enzyme activities [149]. The authors hypothesized that reduced oxidative damage may underlie the beneficial effects of lithium. Indeed, lithium therapy was associated with an increase in whole brain volume and cortical gray matter but no striatal changes [151]. Another intriguing finding is that while haloperidol-induced changes faded after 8 weeks of drug withdrawal, lithium-treated rodents maintained the positive effects of lithium [151]. Although informed inferences from animal studies should be drawn with caution, these findings may indicate that the neuroprotective properties of lithium may underlie its beneficial effects on treatment response and mood recurrence.

Clinical Implications of Neuroprogression and Refractoriness

The concept of "neuroprogression" has been linked to "refractoriness" for a number of reasons. A number of Danish studies showed that a higher number of episodes is associated with recurrence and rapid cycling in BD [153, 154]. Multiple episodes have been found to be highly associated with treatment resistance [155]. Another study reported that response to lamotrigine was negatively correlated with the number of previous episodes [156]. Similarly, a late nonresponse to lithium prophylaxis was shown in patients with higher number of previous episodes and hospitalizations [157]. These findings may indirectly support the hypothesis that neuroprogression increases neurotoxicity and may reduce the individual's ability to respond to treatment [158]. Studies using psychosocial interventions showed a similar relationship between treatment outcome and the number of mood episodes as reported above [159]. Results are still controversial as other studies did not detect an association between the number of previous episodes and pharmacological or psychosocial treatments [160, 161].

Taken together, neuroimaging and cognitive and clinical findings suggest a potential link between the progressively evolving neurotoxic nature of BD and the severity of the clinical course of BD. Overall, a high number of mood episodes and hospitalizations and rapid cycling BD appear to be predictors of a poor outcome of BD. It could be argued that the increasing neurotoxic effects of multiple mood episodes underlie treatment resistance or mood recurrence [157]. However, given that the majority of current studies in neuroprogression and BD are correlational, a causality link between neuroprogression and refractoriness cannot be established as yet. The review of current knowledge does, however, suggest that integrating more biologic information, e.g., inflammation, into clinical data may assist in refining diagnosis, prevention, and intervention approaches for BD.

Conclusions

Treatment refractoriness remains one of the biggest challenges in the field of BD, and treatments are often suboptimal or unsatisfactory. An increasing body of research has investigated the link between neuroprogression and treatment resistance. Potential biological mechanisms underlying this link are oxidative stress, proinflammatory mediators, and alterations in neurotrophins. Evidence points towards the relationship between multiple mood episodes, increased inflammation, oxidative damage, and decreased levels of neuroprotective factors such as BDNF. These biological factors are hypothesized to lead to permanent damage (e.g., brain atrophy and cognitive deficits) and reduced treatment response. TRBD may, therefore, be characterized by faster neuroprogression than non-TRBD. Our ability to effectively treat refractory BD remains, however, problematic. There are several medications and other somatic treatments available, but remission rates are low and medication adherence remains a big problem. Psychotherapeutic treatments appear to be promising, but these might need to be more specifically targeted to the stage of BD as well. Promising interventions may include anti-inflammatory and neuroprotective medications in the early phases of BD. In the later stages of BD, strategies targeting mood stabilization and medication adherence would be essential to prevent additional brain atrophy, and loss of cognitive reserve could be implemented.

Acknowledgment

This work was partly supported by the Dunn Foundation and Pat Rutherford, Jr. Chair in Psychiatry at UTHealth.

References

1 Mathers CD, Loncar D: Projections of global mortality and burden of disease from 2002 to 2030. PLoS Med 2006; 3:e442.
2 Gitlin MJ, Swendsen J, Heller TL, Hammen C: Relapse and impairment in bipolar disorder. Am J Psychiatry 1995; 152:1635.
3 Xu J, Murphy SL, Kochanek KD, Bastian BA: Deaths: final data for 2013. Natl Vital Stat Rep 2016;64:1–118.
4 Nordentoft M, Mortensen PB, Pedersen CB: Absolute risk of suicide after first hospital contact in mental disorder. Arch Gen Psychiatry 2011;68:1058–1064.
5 Malhi GS, Berk M: Diagnosing bipolar disorder. Defining thresholds and setting boundaries. Aust NZ J Psychiatry 2014;48:500–504.
6 Malhi GS, Tanious M, Das P, Coulston CM, Berk M: Potential mechanisms of action of lithium in bipolar disorder. Current understanding. CNS Drugs 2013;27:135–153.

7 Gitlin MJ: Treatment-resistant bipolar disorder (review). Bull Menninger Clin 2001;65:26–40.
8 Poon SH, Sim K, Sum MY, Kuswanto CN, Baldessarini RJ: Evidence-based options for treatment-resistant adult bipolar disorder patients. Bipolar Disord 2012;14:573–584.
9 Fava M: Diagnosis and definition of treatment-resistant depression. Biol Psychiatry 2003;53:649–659.
10 Yatham LN, Calabrese JR, Kusumakar V: Bipolar depression: criteria for treatment selection, definition of refractoriness, and treatment options. Bipolar Disord 2003;5:85–97.
11 Goodwin GM: Evidence-based guidelines for treating bipolar disorder: revised second edition – recommendations from the British Association for Psychopharmacology. J Psychopharmacol 2009;23:346–388.
12 Kessing LV, Hellmund G, Geddes JR, Goodwin GM, Andersen PK: Valproate v. lithium in the treatment of bipolar disorder in clinical practice: observational nationwide register-based cohort study. Br J Psychiatry 2011;199:57–63.
13 Scott J, Paykel E, Morriss R, et al: Cognitive-behavioural therapy for severe and recurrent bipolar disorders. Br J Psychiatry 2006;188:313–320.
14 Torrent C, Bonnin Cdel M, Martínez-Arán A, et al: Efficacy of functional remediation in bipolar disorder: a multicenter randomized controlled study. Am J Psychiatry 2013;170:852–859.
15 Bauer M, Ströhle A: Behandlungsstrategien bei prophylaxeresistenten bipolaren Störungen. Nervenarzt 1999;70: 587–599.
16 APA, Diagnostic and Statistical Manual of Mental Disorders, ed 5. Arlington, American Psychiatric Association, 2013.
17 Sachs G, Graves M: Investigational strategies: treatment of rapid cycling, mixed episodes, and atypical bipolar mood disorder; in Marneros A, Goodwin F (eds): Bipolar Disorders: Mixed States, Rapid Cycling and Atypical Forms. Cambridge, Cambridge University Press, 2005, p 369.
18 Kupka RW, Luckenbaugh DA, Post RM, Leverich GS, Nolen WA: Rapid and non-rapid cycling bipolar disorder: a meta-analysis of clinical studies. J Clin Psychiatry 2003;64:1483–1494.
19 Papadimitriou GN, Calabrese JR, Dikeos DG, Christodoulou GN, et al: Rapid cycling bipolar disorder: biology and pathogenesis. Int J Neuropsychopharmacol 2005;8:281–292.
20 Fountoulakis KN, Kontis D, Gonda X, Yatham LN: A systematic review of the evidence on the treatment of rapid cycling bipolar disorder. Bipolar Disord 2013;15:115–137.
21 Sachs GS, Thase ME: Bipolar disorder therapeutics: maintenance treatment. Biol Psychiatry 2000;48:573–581.
22 Bowden CL, Calabrese JR, McElroy SL, Gyulai L, Wassef A, Petty F, et al: A randomized, placebo-controlled 12-month trial of divalproex and lithium in treatment of outpatients with bipolar I disorder. Divalproex Maintenance Study Group. Arch Gen Psychiatry 2000;57: 481–489.
23 Calabrese JR, Shelton MD, Rapport DJ, Kujawa M, Kimmel SE, Caban S: Current research on rapid cycling bipolar disorder and its treatment. J Affect Disord 2001;67:241–255.
24 Koukopoulos A, Sani G, Koukopoulos AE, Minnai GP, Girardi P, Pani L, et al: Duration and stability of the rapid-cycling course: a long-term personal follow-up of 109 patients. J Affect Disord 2003;73:75–85.
25 Frye MA, Ketter TA, Leverich GS, Huggins T, Lantz C, Denicoff KD, et al: The increasing use of polypharmacotherapy for refractory mood disorders: 22 years of study. J Clin Psychiatry 2000;61:478–415.
26 Kane JM, Kishimoto T, Correll CU: Non-adherence to medication in patients with psychotic disorders: epidemiology, contributing factors and management strategies. World Psychiatry 2013;12: 216–226.
27 Sabaté E: Adherence to Long-Term Therapies: Evidence for Action. Geneva, World Health Organization, 2003.
28 Yatham LN, et al: Canadian Network for Mood and Anxiety Treatments (CANMAT) and International Society for Bipolar Disorders (ISBD) collaborative update of CANMAT guidelines for the management of patients with bipolar disorder: update 2013. Bipolar Disord 2013;15:1–44.
29 Gitlin M: Treatment-resistant bipolar disorder. Mol Psychiatry 2006;11:227–240.
30 American Psychiatric Association: Publication Manual of the American Psychological Association, ed 6. Washington, American Psychological Association, 2010.
31 Gotto J, Rapaport MH: Treatment options in treatment-resistant depression. Prim Psychiatry 2005;12:42.
32 Kramlinger KG, Post RM: The addition of lithium to carbamazepine: antidepressant efficacy in treatment-resistant depression. Arch Gen Psychiatry 1989; 46:794.
33 Goldberg JF, Burdick KE, Endick CJ: Preliminary randomized, double-blind, placebo-controlled trial of pramipexole added to mood stabilizers for treatment-resistant bipolar depression. Am J Psychiatry 2004;161:564–566.
34 Nemeroff CB, Evans DL, Gyulai L, Sachs GS, Bowden CL, Gergel IP, et al: Double-blind, placebo-controlled comparison of imipramine and paroxetine in the treatment of bipolar depression. Am J Psychiatry 2001;158:906–912.
35 Young LT, Joffe RT, Robb JC, MacQueen GM, Marriott M, Patelis-Siotis I: Double-blind comparison of addition of a second mood stabilizer versus an antidepressant to an initial mood stabilizer for treatment of patients with bipolar depression. Am J Psychiatry 2000;157: 124–126.
36 Post RM, Altshuler LL, Frye MA, Suppes T, Rush AJ, Keck PE Jr, et al: Rate of switch in bipolar patients prospectively treated with second-generation antidepressants as augmentation to mood stabilizers. Bipolar Disord 2001;3:259–265.
37 Sachs GS, Lafer B, Stoll AL, Banov M, Thibault AB, Tohen M, et al: A double-blind trial of bupropion versus desipramine for bipolar depression. J Clin Psychiatry 1994;55:391–393.
38 McIntyre RS, Mancini DA, McCann S, Srinivasan J, Sagman D, Kennedy SH: Topiramate versus bupropion SR when added to mood stabilizer therapy for the depressive phase of bipolar disorder: a preliminary single-blind study. Bipolar Disord 2002;4:207–213.
39 Frye MA, Ketter TA, Kimbrell TA, Dunn RT, Speer AM, Osuch EA, et al: A placebo-controlled study of lamotrigine and gabapentin monotherapy in refractory mood disorders. J Clin Psychopharmacol 2000;20:607–614.

40 Vieta E, Martinez-Arán A, Goikolea JM, Torrent C, Colom F, Benabarre A, et al: A randomized trial comparing paroxetine and venlafaxine in the treatment of bipolar depressed patients taking mood stabilizers. J Clin Psychiatry 2002;63:478–512.
41 Tohen M, Vieta E, Calabrese J, Ketter TA, Sachs G, Bowden C, et al: Efficacy of olanzapine and olanzapine-fluoxetine combination in the treatment of bipolar I depression. Arch Gen Psychiatry 2003; 60:1079–1088.
42 Yatham LN, Goldstein JM, Vieta E, Bowden CL, Grunze H, Post RM, et al: Atypical antipsychotics in bipolar depression: potential mechanisms of action. J Clin Psychiatry 2005;66(suppl 5):40–48.
43 Kusumakar V, Yatham LN: An open study of lamotrigine in refractory bipolar depression. Psychiatry Res 1997;72: 145–148.
44 Thase ME, Sachs GS: Bipolar depression: pharmacotherapy and related therapeutic strategies. Biol Psychiatry 2000;48: 558–572.
45 Lally N, Nugent AC, Luckenbaugh DA, Ameli R, Roiser JP, Zarate CA: Anti-anhedonic effect of ketamine and its neural correlates in treatment-resistant bipolar depression. Transl Psychiatry 2014;4:e469.
46 Hui Poon S, Sim K, Baldessarini RJ: Pharmacological approaches for treatment-resistant bipolar disorder. Curr Neuropharmacol 2015;13:592–604.
47 Pacchiarotti I, Bond DJ, Baldessarini RJ, Nolen WA, Grunze H, Licht RW, et al: The International Society for Bipolar Disorders (ISBD) task force report on antidepressant use in bipolar disorders. Am J Psychiatry 2013;170:1249–1262.
48 Nivoli AMA, Colom F, Murru A, Pacchiarotti I, Castro-Loli P, González-Pinto A, et al: New treatment guidelines for acute bipolar depression: a systematic review. J Affect Disord 2011;129:14–26.
49 Medda P, Toni C, Perugi G: The mood-stabilizing effects of electroconvulsive therapy. J ECT 2014;30:275–282.
50 Goodwin FK, Jamison KR: Manic-Depressive Illness: Bipolar Disorders and Recurrent Depression. Oxford, Oxford University Press, 2007.
51 Schoeyen HK, Kessler U, Andreassen OA, Auestad BH, Bergsholm P, Malt UF, et al: Treatment-resistant bipolar depression: a randomized controlled trial of electroconvulsive therapy versus algorithm-based pharmacological treatment. Am J Psychiatry 2015;172:41–51.
52 Vaidya NA, Mahableshwarkar AR, Shahid R: Continuation and maintenance ECT in treatment-resistant bipolar disorder. J ECT 2003;19:10–16.
53 Post T, Kemmler G, Krassnig T, Brugger A, Hausmann A: Wirksamkeit einer EKT-Erhaltungstherapie (EEKT) bei Patienten mit therapieresistenten affektiven Störungen – Ergebnisse einer retrospektiven Datenanalyse. Neuropsychiatrie 2015;29:133.
54 Robertson H, Pryor R: Memory and cognitive effects of ECT: informing and assessing patients. Adv Psychiatr Treat 2006;12:228–237.
55 Food and Drug Administration: Meeting to Discuss the Classification of Electroconvulsive Therapy Devices (ECT). Executive Summary. Silver Spring, FDA, 2011.
56 Markowitz J, Brown R, Sweeney J, Mann JJ: Reduced length and cost of hospital stay for major depression in patients treated with ECT. Am J Psychiatry 1987; 144:1025–1029.
57 Bersani FS, Minichino A, Enticott PG, Mazzarini L, Khan N, Antonacci G, et al: Deep transcranial magnetic stimulation as a treatment for psychiatric disorders: a comprehensive review. Eur Psychiatry 2013;28:30–39.
58 Kedzior KK, Gierke L, Gellersen HM, Berlim MT: Cognitive functioning and deep transcranial magnetic stimulation (DTMS) in major psychiatric disorders: a systematic review. J Psychiatr Res 2016;75:107–115.
59 Harel EV, Zangen A, Roth Y, Reti I, Braw Y, Levkovitz Y: H-coil repetitive transcranial magnetic stimulation for the treatment of bipolar depression: an add-on, safety and feasibility study. World J Biol Psychiatry 2011;12:119–126.
60 Dell'Osso B, Ketter TA: Assessing efficacy/effectiveness and safety/tolerability profiles of adjunctive pramipexole in bipolar depression: acute versus long-term data. Int Clin Psychopharmacol 2013;28:297–304.
61 Cretaz E, Brunoni AR, Lafer B: Magnetic seizure therapy for unipolar and bipolar depression: a systematic review. Neural Plast 2015;2015:521398.
62 Gildenberg PL: Evolution of neuromodulation. Stereotact Funct Neurosurg 2005;83:71–79.
63 Williams NR, Okun MS: Deep brain stimulation (DBS) at the interface of neurology and psychiatry. J Clin Invest 2013;123:4546–4556.
64 Mayberg HS, Lozano AM, Voon V, McNeely HE, Seminowicz D, Hamani C, et al: Deep brain stimulation for treatment-resistant depression. Neuron 2005;45:651–660.
65 Kennedy SH, Giacobbe P, Rizvi SJ, Placenza FM, Nishikawa Y, Mayberg HS, et al: Deep brain stimulation for treatment-resistant depression: follow-up after 3 to 6 years. Am J Psychiatry 2011; 168:502–510.
66 Tye SJ, Frye MA, Lee KH: Disrupting disordered neurocircuitry: treating refractory psychiatric illness with neuromodulation. Mayo Clin Proc 2009;84: 522–532.
67 Anderson RJ, Frye MA, Abulseoud OA, Lee KH, McGillivray JA, Berk M, et al: Deep brain stimulation for treatment-resistant depression: efficacy, safety and mechanisms of action. Neurosci Biobehav Rev 2012;36:1920–1933.
68 Coenen VA, Mädler B, Schiffbauer H, Urbach H, Allert N: Individual fiber anatomy of the subthalamic region revealed with diffusion tensor imaging: a concept to identify the deep brain stimulation target for tremor suppression. Neurosurgery 2011;68:1069–1076.
69 Coenen VA, Schlaepfer TE, Allert N, Mädler B: Diffusion tensor imaging and neuromodulation: DTI as key technology for deep brain stimulation. Int Rev Neurobiol 2012;107:207–234.
70 Schlaepfer TE, Bewernick BH, Kayser S, Mädler B, Coenen VA: Rapid effects of deep brain stimulation for treatment-resistant major depression. Biol Psychiatry 2013;73:1204–1212.
71 Fenoy AJ, Schulz P, Selvaraj S, Burrows C, Spiker D, Cao B, et al: Deep brain stimulation of the medial forebrain bundle: distinctive responses in resistant depression. J Affect Disord 2016;203: 143–151.
72 Holtzheimer PE, Kelley ME, Gross RE, Filkowski MM, Garlow SJ, Barrocas A, et al: Subcallosal cingulate deep brain stimulation for treatment-resistant unipolar and bipolar depression. Arch Gen Psychiatry 2012;69:150–158.
73 Galvez JF, Keser Z, Mwangi B, Ghouse AA, Fenoy AJ, Schulz PE, et al: The medial forebrain bundle as a deep brain stimulation target for treatment resistant depression: a review of published data. Prog Neuropsychopharmacol Biol Psychiatry 2015;58:59–70.

74 Colom F, Vieta E: A perspective on the use of psychoeducation, cognitive-behavioral therapy and interpersonal therapy for bipolar patients. Bipolar Disord 2004;6:480–486.
75 Scott J, Gutierrez MJ: The current status of psychological treatments in bipolar disorders: a systematic review of relapse prevention. Bipolar Disord 2004;6:498–503.
76 Gutierrez MJ, Scott J: Psychological treatment for bipolar disorders. Eur Arch Psychiatry Clin Neurosci 2004;254:92–98.
77 Miklowitz D, Goldstein M: Bipolar Disorder: A Family-Focused Treatment Approach. New York, Guilford, 1997.
78 Strakowski SM, DelBello MP, Sax KW, Zimmerman ME, Shear PK, Hawkins JM, Larson ER: Brain magnetic resonance imaging of structural abnormalities in bipolar disorder. Arch Gen Psychiatry 1999;56:254–260.
79 Kapczinski F, Dias VV, Kauer-Sant'Anna M, Brietzke E, Vázquez GH, Vieta E, Berk M: The potential use of biomarkers as an adjunctive tool for staging bipolar disorder. Prog Neuropsychopharmacol Biol Psychiatry 2009;33:1366–1371.
80 Berk M, Hallam KT, McGorry PD: The potential utility of a staging model as a course specifier: a bipolar disorder perspective. J Affect Disord 2007;100:279–281.
81 Miklowitz DJ, Simoneau TL, George EL, Richards JA, Kalbag A, Sachs-Ericsson N, et al: Family-focused treatment of bipolar disorder: 1-year effects of a psychoeducational program in conjunction with pharmacotherapy. Biol Psychiatry 2000;48:582–592.
82 Rea MM, Tompson MC, Miklowitz DJ, Goldstein MJ, Hwang S, Mintz J: Family-focused treatment versus individual treatment for bipolar disorder: results of a randomized clinical trial. J Consult Clin Psychol 2003;71:482–492.
83 Lam DH, Hayward P, Watkins ER, Wright K, Sham P: Relapse prevention in patients with bipolar disorder: cognitive therapy outcome after 2 years. Am J Psychiatry 2005;162:324–329.
84 Meyer TD, Hautzinger M: Cognitive behaviour therapy and supportive therapy for bipolar disorders: relapse rates for treatment period and 2-year follow-up. Psychol Med 2012;42:1429–1439.
85 Peters A, Sylvia LG, Magalhães PV, Miklowitz DJ, Frank E, Otto MW, et al: Age of onset, course of illness and response to psychotherapy in bipolar disorder: results from the Systematic Treatment Enhancement Program for Bipolar Disorder (STEP-BD). Psychol Med 2014;44:3455–3467.
86 Colom F, Vieta E, Martinez-Aran A, Reinares M, Goikolea JM, Benabarre A, Torrent C, Comes M, Corbella B, et al: A randomized trial on the efficacy of group psychoeducation in the prophylaxis of recurrences in bipolar patients whose disease is in remission. Arch Gen Psychiatry 2003;60:402–407.
87 Frank E, Kupfer DJ, Thase ME, Mallinger AG, Swartz HA, Fagiolini AM, et al: Two-year outcomes for interpersonal and social rhythm therapy in individuals with bipolar I disorder. Arch Gen Psychiatry 2005;62:996–1004.
88 Crump C, Sundquist K, Winkleby M: Comorbidities and mortality in bipolar disorder: a Swedish national cohort study. JAMA Psychiatry 2013;70:931–939.
89 Gildengers AG, Chung KH, Huang SH, Begley A, Aizenstein HJ, Tsai SY: Neuroprogressive effects of lifetime illness duration in older adults with bipolar disorder. Bipolar Disord 2014;16:617–623.
90 Berk M, Berk L, Dodd S, Cotton S, Macneil C, Daglas R, et al: Stage managing bipolar disorder. Bipolar Disord 2014;16:471–477.
91 Berk M, Hallam K, Malhi GS, Henry L, Hasty M, et al: Evidence and implications for early intervention in bipolar disorder. J Ment Health 2010;19:113–126.
92 Rizzo LB, Hallam K, Malhi GS, Henry L, Hasty M, Macneil C, et al: The theory of bipolar disorder as an illness of accelerated aging: implications for clinical care and research. Neurosci Biobehav Rev 2014;42:157–169.
93 McNamara RK, Nandagopal JJ, Strakowski SM, DelBello MP: Preventative strategies for early-onset bipolar disorder: towards a clinical staging model. CNS Drugs 2010;24:983–996.
94 Schroeter ML, Steiner J, Mueller K: Glial pathology is modified by age in mood disorders – a systematic meta-analysis of serum S100B in vivo studies. J Affect Disord 2011;134:32–38.
95 Breunis MN, Kupka RW, Nolen WA, Suppes T, Denicoff KD, Leverich GS, et al: High numbers of circulating activated T cells and raised levels of serum IL-2 receptor in bipolar disorder. Biol Psychiatry 2003;53:157–165.
96 Modabbernia A, Taslimi S, Brietzke E, Ashrafi M: Cytokine alterations in bipolar disorder: a meta-analysis of 30 studies. Biol Psychiatry 2013;74:15–25.
97 Munkholm K, Braüner JV, Kessing LV, Vinberg M: Cytokines in bipolar disorder vs. healthy control subjects: a systematic review and meta-analysis. J Psychiatr Res 2013;47:1119–1133.
98 Munkholm K, Vinberg M, Kessing LV: Cytokines in bipolar disorder: a systematic review and meta-analysis. J Affect Disord 2013;144:16–27.
99 Bai Y-M, Su TP, Tsai SJ, Wen-Fei C, Li CT, Pei-Chi T, et al: Comparison of inflammatory cytokine levels among type I/type II and manic/hypomanic/euthymic/depressive states of bipolar disorder. J Affect Disord 2014;166:187–192.
100 de Sousa RT, Zarate CA Jr, Zanetti MV, Costa AC, Talib LL, Gattaz WF, et al: Oxidative stress in early stage bipolar disorder and the association with response to lithium. J Psychiatr Res 2014;50:36–41.
101 Shao L, Sun X, Xu L, Young LT, Wang JF: Mood stabilizing drug lithium increases expression of endoplasmic reticulum stress proteins in primary cultured rat cerebral cortical cells. Life Sci 2006;78:1317–1323.
102 Pandya CD, Howell KR, Pillai A: Antioxidants as potential therapeutics for neuropsychiatric disorders. Prog Neuropsychopharmacol Biol Psychiatry 2013;46:214–223.
103 Soeiro-de-Souza MG, Andreazza AC, Carvalho AF, Machado-Vieira R, Young LT, Moreno RA: Number of manic episodes is associated with elevated DNA oxidation in bipolar I disorder. Int J Neuropsychopharmacol 2013;16:1505–1512.
104 Fries GR, Vasconcelos-Moreno MP, Gubert C, Santos BT, da Rosa AL, Eisele B, et al: Early apoptosis in peripheral blood mononuclear cells from patients with bipolar disorder. J Affect Disord 2014;152:474–477.

105 Gigante AD, Young LT, Yatham LN, Andreazza AC, Nery FG, Grinberg LT, et al: Morphometric post-mortem studies in bipolar disorder: possible association with oxidative stress and apoptosis. Int J Neuropsychopharmacol 2011;14:1075–1089.
106 Kauer-Sant'Anna M, Kapczinski F, Andreazza AC, Bond DJ, Lam RW, Young LT, et al: Brain-derived neurotrophic factor and inflammatory markers in patients with early- vs late-stage bipolar disorder. Int J Neuropsychopharmacol 2009;12:447–458.
107 Wang JF, Shao L, Sun X, Young LT: Increased oxidative stress in the anterior cingulate cortex of subjects with bipolar disorder and schizophrenia. Bipolar Disord 2009;11:523–529.
108 Frey BN, Zunta-Soares GB, Caetano SC, Nicoletti MA, Hatch JP, Brambilla P, et al: Illness duration and total brain gray matter in bipolar disorder: evidence for neurodegeneration? Eur Neuropsychopharmacol 2008;18:717–722.
109 Moorhead TWJ, McKirdy J, Sussmann JE, Hall J, Lawrie SM, Johnstone EC, et al: Progressive gray matter loss in patients with bipolar disorder. Biol Psychiatry 2007;62:894–900.
110 Lisy ME, Jarvis KB, DelBello MP, Mills NP, Weber WA, Fleck D, et al: Progressive neurostructural changes in adolescent and adult patients with bipolar disorder. Bipolar Disord 2011;13:396–405.
111 Konarski JZ, McIntyre RS, Kennedy SH, Rafi-Tari S, Soczynska JK, Ketter TA: Volumetric neuroimaging investigations in mood disorders: bipolar disorder versus major depressive disorder. Bipolar Disord 2008;10:1–37.
112 Strakowski SM, Adler CM, Almeida J, Altshuler LL, Blumberg HP, Chang KD, et al: The functional neuroanatomy of bipolar disorder: a consensus model. Bipolar Disord 2012;14:313–325.
113 Phillips ML, Ladouceur CD, Drevets WC: A neural model of voluntary and automatic emotion regulation: implications for understanding the pathophysiology and neurodevelopment of bipolar disorder. Mol Psychiatry 2008;13:833–857.
114 Phillips ML, Drevets WC, Rauch SL, Lane R: Neurobiology of emotion perception. II. Implications for major psychiatric disorders. Biol Psychiatry 2003;54:515–528.
115 Strakowski SM, Adler CM, Holland SK, Mills N, DelBello MP: A preliminary FMRI study of sustained attention in euthymic, unmedicated bipolar disorder. Neuropsychopharmacology 2004;29:1734–1740.
116 Chen CH, Suckling J, Lennox BR, Ooi C, Bullmore ET: A quantitative meta-analysis of fMRI studies in bipolar disorder. Bipolar Disord 2011;13:1–15.
117 Townsend J, Altshuler LL: Emotion processing and regulation in bipolar disorder: a review. Bipolar Disord 2012;14:326–339.
118 Liu J, Blond BN, van Dyck LI, Spencer L, Wang F, Blumberg HP: Trait and state corticostriatal dysfunction in bipolar disorder during emotional face processing. Bipolar Disord 2012;14:432–441.
119 Adler CM, Adams J, DelBello MP, Holland SK, Schmithorst V, Levine A, et al: Evidence of white matter pathology in bipolar disorder adolescents experiencing their first episode of mania: a diffusion tensor imaging study. Am J Psychiatry 2006;163:322–324.
120 Adler CM, Holland SK, Schmithorst V, Wilke M, Weiss KL, Pan H, et al: Abnormal frontal white matter tracts in bipolar disorder: a diffusion tensor imaging study. Bipolar Disord 2004;6:197–203.
121 Benedetti F, Absinta M, Rocca MA, Radaelli D, Poletti S, Bernasconi A, et al: Tract-specific white matter structural disruption in patients with bipolar disorder. Bipolar Disord 2011;13:414–424.
122 Frazier JA, Breeze JL, Papadimitriou G, Kennedy DN, Hodge SM, Moore CM, et al: White matter abnormalities in children with and at risk for bipolar disorder. Bipolar Disord 2007;9:799–809.
123 Beyer JL, Taylor WD, MacFall JR, Kuchibhatla M, Payne ME, Provenzale JM, et al: Cortical white matter microstructural abnormalities in bipolar disorder. Neuropsychopharmacology 2005;30:2225–2229.
124 Liu JX, Chen YS, Hsieh JC, Su TP, Yeh TC, Chen LF: Differences in white matter abnormalities between bipolar I and II disorders. J Affect Disord 2010;127:309–315.
125 Nortje G, Stein DJ, Radua J, Mataix-Cols D, Horn N: Systematic review and voxel-based meta-analysis of diffusion tensor imaging studies in bipolar disorder. J Affect Disord 2013;150:192–200.
126 Mahon K, Wu J, Malhotra AK, Burdick KE, DeRosse P, Ardekani BA, et al: A voxel-based diffusion tensor imaging study of white matter in bipolar disorder. Neuropsychopharmacology 2009;34:1590–1600.
127 Mahon K, Burdick KE, Szeszko PR: A role for white matter abnormalities in the pathophysiology of bipolar disorder. Neurosci Biobehav Rev 2010;34:533–554.
128 Canales-Rodríguez EJ, Pomarol-Clotet E, Radua J, Sarró S, Alonso-Lana S, Del Mar Bonnín C, et al: Structural abnormalities in bipolar euthymia: a multi-contrast molecular diffusion imaging study. Biol Psychiatry 2013;76:239–248.
129 Sprooten E, Fleming KM, Thomson PA, Bastin ME, Whalley HC, Hall J, et al: White matter integrity as an intermediate phenotype: exploratory genome-wide association analysis in individuals at high risk of bipolar disorder. Psychiatry Res 2013;206:223–231.
130 Lyoo IK, Kim MJ, Stoll AL, Demopulos CM, Parow AM, Dager SR, et al: Frontal lobe gray matter density decreases in bipolar I disorder. Biol Psychiatry 2004;55:648–651.
131 Blumberg HP, Krystal JH, Bansal R, Martin A, Dziura J, Durkin K, et al: Age, rapid-cycling, and pharmacotherapy effects on ventral prefrontal cortex in bipolar disorder: a cross-sectional study. Biol Psychiatry 2006;59:611–618.
132 Kalmar JH, Wang F, Chepenik LG, Womer FY, Jones MM, Pittman B, et al: Relation between amygdala structure and function in adolescents with bipolar disorder. J Am Acad Child Adolesc Psychiatry 2009;48:636–642.
133 Sanches M, Sassi RB, Axelson D, Nicoletti M, Brambilla P, Hatch JP, et al: Subgenual prefrontal cortex of child and adolescent bipolar patients: a morphometric magnetic resonance imaging study. Psychiatry Res 2005;138:43–49.
134 Bora E, Yucel M, Pantelis C: Cognitive endophenotypes of bipolar disorder: a meta-analysis of neuropsychological deficits in euthymic patients and their first-degree relatives. J Affect Disord 2009;113:1–20.

135 Glahn DC, Bearden CE, Barguil M, Barrett J, Reichenberg A, Bowden CL, et al: The neurocognitive signature of psychotic bipolar disorder. Biol Psychiatry 2007;62:910–916.
136 Quraishi S, Frangou S: Neuropsychology of bipolar disorder: a review. J Affect Disord 2002;72:209–226.
137 Martínez-Arán A, Vieta E, Colom F, Torrent C, Sánchez-Moreno J, Reinares M, et al: Cognitive impairment in euthymic bipolar patients: implications for clinical and functional outcome. Bipolar Disord 2004;6:224–232.
138 Dixon T, Kravariti E, Frith C, Murray RM, McGuire PK: Effect of symptoms on executive function in bipolar illness. Psychol Med 2004;34:811–821.
139 Albus M, Hubmann W, Wahlheim C, Sobizack N, Franz U, Mohr F: Contrasts in neuropsychological test profile between patients with first-episode schizophrenia and first-episode affective disorders. Acta Psychiatr Scand 1996;94:87–93.
140 Martinez-Aran A, Vieta E, Colom F, Torrent C, Reinares M, Goikolea JM, et al: Do cognitive complaints in euthymic bipolar patients reflect objective cognitive impairment? Psychother Psychosom 2005;74:295–302.
141 Basso MR, Lowery N, Neel J, Purdie R, Bornstein RA: Neuropsychological impairment among manic, depressed, and mixed-episode inpatients with bipolar disorder. Neuropsychology 2002;16:84–91.
142 Martínez-Arán A, Vieta E, Reinares M, Colom F, Torrent C, Sánchez-Moreno J, et al: Cognitive function across manic or hypomanic, depressed, and euthymic states in bipolar disorder. Am J Psychiatry 2004;161:262–270.
143 Gruber S, Rathgeber K, Bräunig P, Gauggel S: Stability and course of neuropsychological deficits in manic and depressed bipolar patients compared to patients with major depression. J Affect Disord 2007;104:61–71.
144 Sweeney JA, Kmiec JA, Kupfer DJ: Neuropsychologic impairments in bipolar and unipolar mood disorders on the CANTAB neurocognitive battery. Biol Psychiatry 2000;48:674–684.
145 Bauer IE, Meyer TD, Sanches M, Zunta-Soares G, Soares JC: Does a history of substance abuse and illness chronicity predict increased impulsivity in bipolar disorder? J Affect Disord 2015; 179:142–147.
146 Martino DJ, Strejilevich SA, Manes F: Neurocognitive functioning in early-onset and late-onset older patients with euthymic bipolar disorder. Int J Geriatr Psychiatry 2013;28:142–148.
147 Cao B, Passos IC, Mwangi B, Bauer IE, Zunta-Soares GB, Kapczinski R, et al: Hippocampal volume and verbal memory performance in late-stage bipolar disorder. J Psychiatr Res 2016;73:102–107.
148 Bo C, Bauer IE, Sharma AN, Mwangi B, Frazier T, Lavagnino L, et al: Reduced hippocampus volume and memory performance in bipolar disorder patients carrying the *BDNF* val66met met allele. J Affect Disord 2016;198: 198–205.
149 Maurer IC, Schippel P, Volz HP: Lithium-induced enhancement of mitochondrial oxidative phosphorylation in human brain tissue. Bipolar Disord 2009;11:515–522.
150 Moore GJ, Cortese BM, Glitz DA, Zajac-Benitez C, Quiroz JA, Uhde TW, et al: A longitudinal study of the effects of lithium treatment on prefrontal and subgenual prefrontal gray matter volume in treatment-responsive bipolar disorder patients. J Clin Psychiatry 2009;70:699–705.
151 Vernon AC, Natesan S, Crum WR, Cooper JD, Modo M, Williams SC, et al: Contrasting effects of haloperidol and lithium on rodent brain structure: a magnetic resonance imaging study with postmortem confirmation. Biol Psychiatry 2012;71:855–863.
152 Kim S, Zavitsanou K, Gurguis G, Webster MJ: Neuropathology markers and pathways associated with molecular targets for antipsychotic drugs in postmortem brain tissues: exploration of drug targets through the Stanley Neuropathology Integrative Database. Eur Neuropsychopharmacol 2012;22:683–694.
153 Kessing LV, Hellmund G, Andersen PK: Predictors of excellent response to lithium: results from a nationwide register-based study. Int Clin Psychopharmacol 2011;26:323–328.
154 Kessing LV, Hansen MG, Andersen PK: Course of illness in depressive and bipolar disorders. Br J Psychiatry 2004; 185:372–377.
155 Swann AC, Bowden CL, Calabrese JR, Dilsaver SC, Morris DD: Differential effect of number of previous episodes of affective disorder on response to lithium or divalproex in acute mania. Am J Psychiatry 1999;156:1264–1266.
156 Obrocea GV, Dunn RM, Frye MA, Ketter TA, Luckenbaugh DA, Leverich GS, et al: Clinical predictors of response to lamotrigine and gabapentin monotherapy in refractory affective disorders. Biol Psychiatry 2002;51:253–260.
157 Maj M: The impact of lithium prophylaxis on the course of bipolar disorder: a review of the research evidence. Bipolar Disord 2000;2:93–101.
158 Berk M, Kapczinski F, Andreazza AC, Dean OM, Giorlando F, Maes M, et al: Pathways underlying neuroprogression in bipolar disorder: focus on inflammation, oxidative stress and neurotrophic factors. Neurosci Biobehav Rev 2011;35:804–817.
159 Scott J, Paykel E, Morriss R, Bentall R, Kinderman P, Johnson T, et al: Cognitive-behavioural therapy for severe and recurrent bipolar disorders: randomised controlled trial. Br J Psychiatry 2006;188:313–320.
160 Baldessarini RJ, Tondo L, Hennen J: Lithium treatment and suicide risk in major affective disorders: update and new findings. J Clin Psychiatry 2003; 64(suppl 5):44–52.
161 Magalhaes PV, Kapczinski F, Nierenberg AA, Deckersbach T, Weisinger D, Dodd S, et al: Illness burden and medical comorbidity in the Systematic Treatment Enhancement Program for Bipolar Disorder. Acta Psychiatr Scand 2012;125:303–308.

Jair C. Soares, MD, PhD
Department of Psychiatry and Behavioral Science
University of Texas Health Science Center at Houston
1941 East Road, Houston, TX 77054 (USA)
E-Mail Jair.C.Soares@uth.tmc.edu

Halaris A, Leonard BE (eds): Neuroprogression in Psychiatric Disorders.
Mod Trends Pharmacopsychiatry. Basel, Karger, 2017, vol 31, pp 27–36 (DOI: 10.1159/000470804)

Neuroprogression and Immune Activation in Major Depressive Disorder

Jeffrey H. Meyer

Department of Psychiatry, University of Toronto, and Research Imaging Centre, Centre for Addiction and Mental Health, Campbell Family Mental Health Research Institute at CAMH, Toronto, ON, Canada

Abstract

Traditionally, the neurobiology of major depressive disorder (MDD) has been largely considered from the perspective of the state of major depressive episodes (MDE) versus being in remission, but the current accumulation of disease markers, largely acquired cross-sectionally, is strongly suggestive of neuroprogressive aspects of MDD. This chapter focuses on the changes in disease markers involved in the reorganization of the nervous system in MDD, including the translocator protein (TSPO; an index of microglial activation), glial fibrillary acidic protein (GFAP; an index of astroglial activation), [^{11}C]harmine (a marker of monoamine oxidase A; MAO-A), and several other indices (metabotropic glutamate receptor 5 [mGluR5], excitatory amino acid transporters, and magnetic resonance imaging spectroscopy measurements) of glutamate dysregulation. These are markers of processes involved in immune activation, oxidative stress, and chronic glucocorticoid exposure. Positron emission tomography studies of the TSPO distribution volume, a marker of microglial activation, provide strong evidence for microglial activation throughout the gray matter of the brain during MDE of MDD. In postmortem studies, GFAP reductions in the orbitofrontal cortex, anterior cingulate cortex, and hippocampus indicate a deficit in reactive astroglia. Elevated MAO-A levels are present throughout the gray matter of the brain, including affect-modulating brain regions, starting in high-risk states for MDE such as the early postpartum period, perimenopause, heavy cigarette smoking, heavy alcohol intake, and prior to MDE recurrence. Evidence is accumulating for glutamate dysregulation, with some findings of reduced glutamate transporter density in the orbitofrontal cortex, and decreased mGluR5 density. Collectively, these changes suggest an imbalance in the immune system with increased microglial activation and decreased astroglial activation, continued elevations of the MAO-A level, and, likely, the development of extracellular glutamate dysregulation. Many of these imbalances involve processes implicated in increased oxidative stress, apoptosis, and neurodegeneration. Future studies are required to assess potential therapeutics targeting these processes to ameliorate progression of MDD.

Introductory Comments

Neuroprogression can be defined as the pathological reorganization of the nervous system along the course of a major depressive disorder (MDD). Given the extensive interactions between the supporting structures of glia and neurons, such reorganization is interpreted to include glial alterations in MDD. For MDD, an illness spanning decades, neuroprogression is largely restricted to cross-sectional investigations because postmortem studies investigate a single timepoint, and the newer neuroimaging technologies have not been available for sufficient periods of time to be able to evaluate disease progression over long periods. This review focuses on several important markers that are dysregulated in MDD as well as their implications for neuroprogression, including those for microglial activation, astroglial activation, monoamine oxidase A (MAO-A) level, and glutamate dysregulation.

Microglial Activation

Microglia are present throughout the brain and are important in the immune surveillance system in the central nervous system (CNS) where they detect inflammatory signals, such as damage- and pathogen-associated molecular patterns, and cytokines [1]. In response to such stimuli, microglia transform from this detection state into a response state [1, 2], changing morphology from longer slender dendrites to thickened shorter fewer dendrites, having a larger cell body volume, and sometimes becoming ameboid, while migrating to and engulfing the site of insult possibly in response to cytokine release [3, 4]. Activated microglia thus represent an important component of neuroinflammation [5].

However, activated microglia are also implicated in neurodegeneration. The initial response system of microglial activation has the potential to contribute towards chronic pathological processes of neurodegeneration. Microglia responding to stimuli may produce reactive oxygen and nitrogen species, which may damage neuronal and glial tissue and lead to damage- and pathogen-associated molecular patterns and continued microglial activation [6]. Similarly, cytokine secretion of activated microglia can have paracrine effects to perpetuate microglial activation [6]. Hence, activated microglia may, in some circumstances, become a neurodegenerative process.

Initial studies of microglial activation in MDD tended to have small sample sizes reflecting an exploratory nature or were oriented towards suicide rather than MDD per se. Van Otterloo et al. [7] reported no difference in the density of activated microglia in the white matter of the orbitofrontal region in 10 MDD subjects. Steiner et al. [8] reported increased density of quinolinic acid-positive cells, a marker influenced by microglial activation, in the anterior cingulate cortex (ACC) of 7 patients with major depressive episode (MDE). Amongst investigations in suicide victims, one study reported greater HLA-DR staining, a marker of microglial activation, in the dorsolateral prefrontal cortex (PFC) and ACC [9] but no relationship to MDD; however, the sample size was not oriented toward MDD having less than 9 patients with MDD. In a sample of 10 patients with MDD, some of whom were in a current MDE, an [^{11}C] PBR28 positron emission tomography (PET) study found no difference between patients and controls across a range of gray matter regions [10]. Torres-Platas et al. [11] investigated the white matter in the dorsal anterior cingulate in a sample of 24 MDE and 17 control subjects and reported an increased ratio of primed/ramified microglia, where the primed state is intermediary to the fully activated state, although other ratios of activated to ramified microglia were not significant. Results across studies were not definitive for ruling out microglial activation in MDD since none of the investigations of microglial activation in gray matter of MDD included more than 10 subjects, and the only study which investigated a

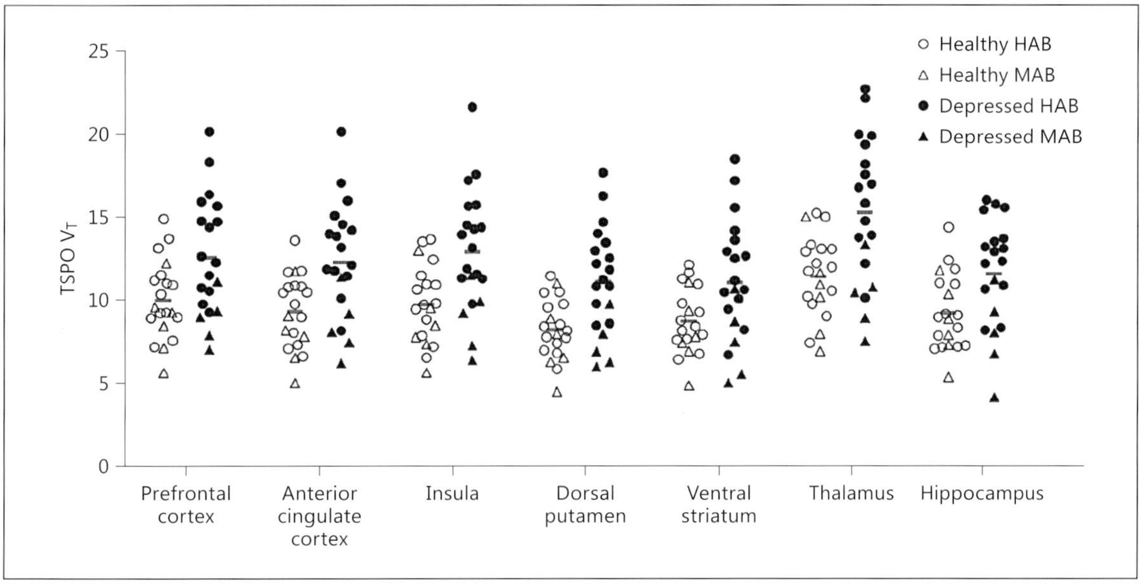

Fig. 1. Elevated translocator protein (TSPO) density (V_T) during a major depressive episode (MDE) secondary to major depressive disorder (MDD) (adapted from Setiawan et al. [12]). TSPO V_T was significantly greater in MDE of 20 MDD subjects (15 high-affinity binders [HAB] and 5 mixed affinity binders [MAB]) compared to 20 controls (healthy: 14 HAB and 6 MAB); all second-generation TSPO radioligands, such as [^{18}F]FEPPA, show differential binding according to the SNP rs6971 of the TSPO gene resulting in HAB and MAB. Bars indicate means in each group. Analysis of variance results are as follows: prefrontal cortex, $F_{1,37} = 8.07$, $p = 0.007$; anterior cingulate cortex, $F_{1,37} = 12.24$, $p = 0.001$; insula, $F_{1,37} = 12.34$, $p = 0.001$; dorsal putamen, $F_{1,37} = 14.1$, $p = 0.001$; ventral striatum, $F_{1,37} = 6.9$, $p = 0.013$; thalamus, $F_{1,37} = 13.6$, $p = 0.001$; hippocampus, $F_{1,37} = 7.5$, $p = 0.009$.

larger number of subjects was restricted to white matter sampled in the anterior cingulate [11].

In a definitive [^{18}F]FEPPA PET study of 20 unmedicated MDE subjects with no active comorbid psychiatric illnesses and 20 controls, a marker of microglial activation, the translocator protein (TSPO)-specific distribution volume (TSPO V_T), was significantly elevated in the primary regions of the PFC, ACC, and insular cortex by a substantial magnitude of 30% (Fig. 1) [12]. While the finding was prominent in the a priori selected regions, TSPO V_T was also greater throughout all the other gray matter regions examined. Consistent with this finding, in a [^{11}C]PBR28 PET imaging study, Innis et al. [13] presented elevated TSPO V_T within the gray matter regions assessed, including subregions of the PFC and ACC, in 11 unmedicated MDE subjects.

Since microglial activation may include components harmful to neurons and glia (termed M1 responses) but may also include helpful components (termed M2 responses), such as clearing of cellular debris, inducing angiogenesis, and promoting tissue repair [14], these studies would suggest that medications under investigation for shifting activated microglia from an M1 to an M2 state, such as minocycline and bexarotene [15], should be investigated as strategies to intervene in the neuroprogressive course of MDD.

Loss of Activated Astroglia

Protoplasmic astrocytes, characterized by a dense network of branched dendritic processes, are found throughout the gray matter. The processes

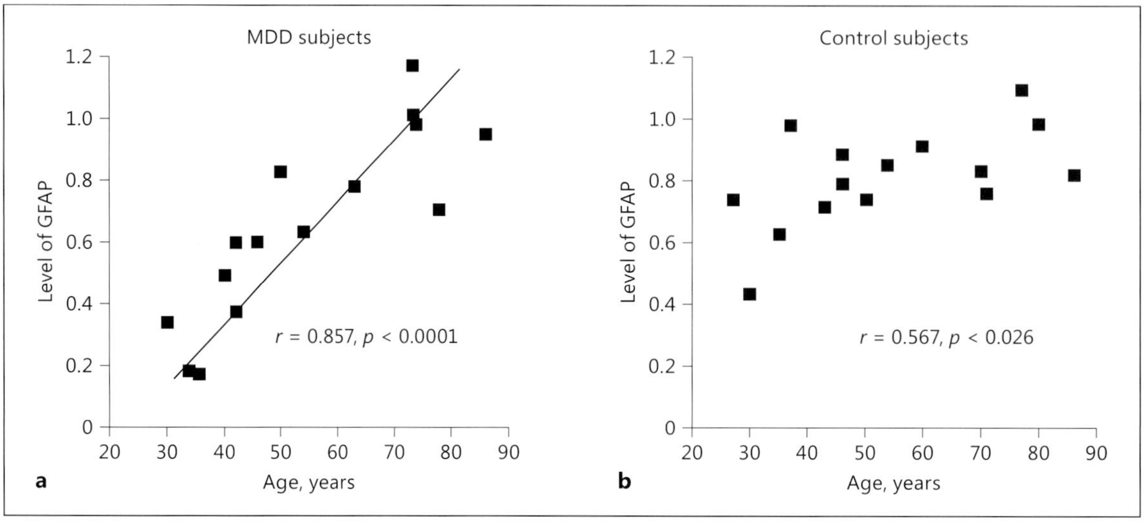

Fig. 2. Scatter plots showing positive significant correlations between the glial fibrillary acidic protein (GFAP) level and age at the time of death in subjects with major depressive disorder (MDD) (**a**) and controls (**b**). The mean GFAP level was significantly lower in the MDD group ($t_{14} = 2.8$, $p = 0.014$); however, the difference in GFAP levels in subjects ≤60 years was highly significant ($t_8 = 3.855$, $p = 0.005$) (adapted from Si et al. [26]).

of protoplasmic astrocytes envelope synapses, interdigitate with each other to form gap junctions, and may extend along blood vessels [16]. The processes extending from a single astrocyte may extend to envelope a tremendous number of synapses, influencing the regulation of neurotransmitters, such as monoamines and glutamate, at these sites [17]. As will be discussed in the next subsection, MAO-A levels are elevated throughout gray matter in MDE, and astrocytes represent a significant cellular compartment of MAO-A in MDE [18]. Activated astrocytes also synthesize and release neurotrophic factors important for neuronal survival, growth, and differentiation, maintenance of synaptic plasticity, and synaptic efficiency [17]. The foot processes of astrocytes also extend to the walls of capillaries, have access to glucose uptake, may influence blood flow, and are implicated in the regulation of the blood-brain barrier.

Contrasting increased astrogliosis, which is often associated with some types of disease pathologies, MDD is associated with a loss of activated astrocytes [19]. Levels of glial fibrillary acidic protein (GFAP), a commonly applied marker of activated astrocytes, are decreased at least in several key regions in MDD, including the dorsolateral PFC, subgenual and supracallosal ACC, orbitofrontal cortex, and in CA1 and CA2 regions of the hippocampus [20–24]. Interestingly, astroglial loss in the PFC resulting from toxins is associated with depressive behavior in rodents [25]. In an intriguing study of younger and older MDD subjects, GFAP-immunoreactive astrocyte density in the dorsolateral PFC was lower in subjects younger than age 60 years as compared to age-matched controls without psychiatric illness [21, 26] (Fig. 2). Older subjects, some of whom had late-onset MDD, had elevated GFAP in the dorsolateral PFC as compared to age-matched controls without psychiatric illness [21]. Since the age-related findings of GFAP loss in humans occur predominantly before 60 years of age and since chronic stress can precipitate loss of GFAP levels and mRNA in the PFC in rodents [27, 28], GFAP loss is best interpreted as a

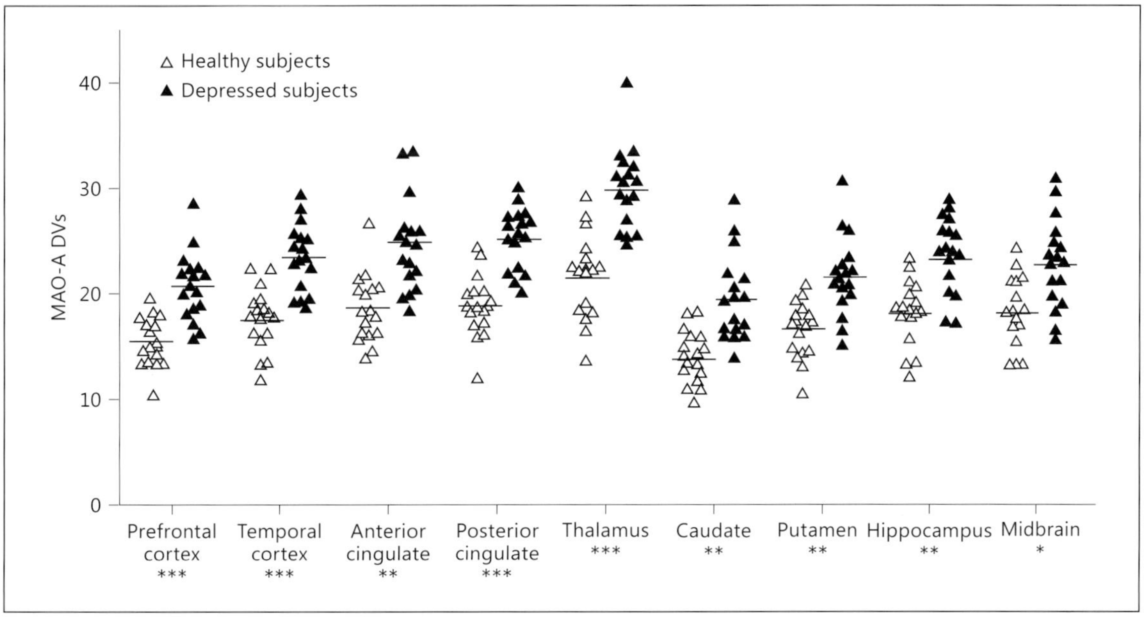

Fig. 3. Monoamine oxidase A (MAO-A)-specific distribution volume (V_S) was significantly increased (on average, 34% or 2 SD) in subjects with major depressive disorder (MDD; $n = 20$) compared to healthy controls ($n = 17$) in each region (adapted from Meyer et al. [39]). Bars indicate means in each group. * $p = 0.001$, ** $p = 0.0001$, *** $p = 0.00001$.

marker of early-stage disease in early-onset MDD.

An interesting hypothesis is that loss of astroglial cells might contribute towards reduced EAAT1 and EAAT2 glutamate transporters, leading towards excessively high levels of extracellular glutamate, which could contribute towards toxicity to GABA-releasing neurons and glutamatergic pyramidal cell neurons in depressed elderly subjects [19]. The model of reduced glutamate uptake and cycling due to loss and abnormal functioning of astroglia has also been proposed for neurodevelopmental/neurodegenerative disorders and addiction [29], but it is an intriguing match to MDD, given the ongoing clinical developments of N-methyl-D-aspartate receptor (NMDAR) antagonists for reducing MDE symptoms. Glutamate dysfunction and MDD will be discussed further in the last subsection.

Elevated Monoamine Oxidase A

MAO-A is an important enzyme found on the outer mitochondrial membranes in neurons and glia [30]. In human brain, MAO-A density is highest in the locus coeruleus but also high in the cortex, hippocampus, and striatum; lower in the cerebellar cortex; and minimal in white matter [30, 31]. MAO-A has an important role related to mood disorders since it metabolizes serotonin, norepinephrine, and dopamine [32, 33]. However, MAO-A is also important in neurodegeneration due to its role in generating oxidative stress through the production of hydrogen peroxide and in predisposing towards intrinsic apoptosis [32, 33].

There has been some misunderstanding of MAO-A as solely being a therapeutic target for MAO inhibitors in mood disorders, because the first several postmortem studies of MAO-A levels and activity had not selectively sampled early-onset MDD. The first study examined suicide victims irrespective of diagnosis and the following two sampled older MDD subjects [34–36]. Since only 50% of suicide victims have MDD [37], and late-onset MDD is often associated with Parkinson and Alzheimer disease [38], the issue of whether MAO-A levels or activity were altered in the most common early-onset MDD was unclear until more recently.

In 2006, applying [^{11}C]harmine PET, MAO-A-specific distribution volume (V_S), an index of MAO-A density, was measured in medication-free MDE secondary to early-onset MDD [39] and MAO-A V_S was highly significantly elevated ($p < 0.001$) in each region, average magnitude 34% (or 2 SD) during MDE (Fig. 3). This was a definitive study of MAO-A binding in MDE in early-onset MDD. The samples excluded psychiatric and medical comorbidity to focus on the differences attributable to diagnosis. Subjects with MDE were drug free for at least 5 months and most were antidepressant naive. The effect size was large, and the radiotracer has outstanding quality in terms of its specificity, high affinity, reversibility, validity, and reliability [reviewed in 40]. In 2008, Barton et al. [41] reported a consistent finding of elevated brain serotonin turnover in unmedicated depressed patients. In 2009, the finding of greater MAO-A binding in MDE was replicated with [^{11}C]harmine PET. In 2011, the finding was replicated in antidepressant-free MDE subjects in a postmortem study of the orbitofrontal cortex applying Western blot [42, 43]; a similar elevation in MAO-A activity was found.

Several aspects of MAO-A phenomenology in MDD exhibit a relationship to neuroprogression. R1 is a nuclear transcription factor that inhibits MAO-A synthesis and, when reduced, is implicated in apoptosis. R1 is reduced in the PFC in MDD and its reduction correlates with the elevation in MAO-A density [43]. TIEG2 is a nuclear transcription factor that regulates genes participating in apoptosis and promotes MAO-A synthesis. TIEG2 is elevated in the PFC in MDD, and its elevation correlates with the elevation in MAO-A density in MDD [18, 44]. Also, at a clinical level, in an extended sample of 42 MDE subjects, MAO-A binding in the PFC and ACC was more prominently elevated in those with greater severity and reversed neurovegetative symptoms (hypersomnia, hyperphagia, or weight gain), which are problematic aspects of MDD progression, given that severity is associated with greater burden of illness and weight gain is associated with greater medical morbidity [45].

While MAO-A elevations in gray matter binding occur during MDE, they are also prominent during prodromal phases of MDE. Mechanistically, MAO-A transcription, level, and/or activity are increased in cell lines and animal models through several processes, including chronic glucocorticoid exposure, since glucocorticoids bind to the MAO-A promotor and promote nuclear transcription factor production [46] and toxin exposure [47], and declines in estrogens [reviewed in 48]. It is intriguing that the same processes that influence MAO-A in cell lines and animal models occur in conditions associated with a high risk for MDE, such as early postpartum period, perimenopause, and during heavy cigarette smoking and alcohol dependence [reviewed in 40, 48–50]. What is even more interesting is that in humans, particularly in the PFC and ACC, MAO-A V_T, an index of the MAO-A level, is elevated in each of these states implicating a high risk for onset of MDE symptoms particularly in the PFC and ACC [40, 48–50]. After selective serotonin reuptake inhibitor treatment for MDE, elevated MAO-A V_T may persist, and when it is elevated in the PFC and ACC during recovery, it is associated with recurrence [42]; thus, collectively, elevated MAO-A levels play an important role in many prodromal/high-risk states for MDE [40, 48–50].

Since MAO-A metabolizes several monoamines such as serotonin, norepinephrine, and dopamine, elevations in the MAO-A level assessed in PET studies using [^{11}C]harmine, which binds to the center of the functional pocket of MAO-A [51], would be expected to drive down ongoing levels of these monoamines. This is consistent with a central theme of the monoamine theory in MDD that depletion of monoamines leads to depressed mood, as originally demonstrated in the 1950s during treatment with reserpine-based antihypertensives: the chronic monoamine lowering after reserpine was associated with subsequent MDE onset, which typically occurred 2 weeks to 4 months later [52]. In addition, elevated MAO-A levels have further implications given the role of MAO-A in generating oxidative stress and facilitating apoptosis [33]. Hence, elevated MAO-A levels, particularly in the PFC and ACC, can be viewed as a common pathological phenotype that occurs during high-risk prodromes, persists through common selective serotonin reuptake inhibitor treatments, and is implicated in recurrent MDE.

Glutamate Dysregulation

An appealing model of neuroprogression in MDD is that loss of activated astrocytes may reflect a loss of EAAT2 sites available for the reuptake of glutamate [19, 29]. EAAT2 receptors, found on astrocytes, are a major route of glutamate uptake in the gray matter of the brain [19, 29]. Reduced EAAT2 immunoreactivity was reported in the orbitofrontal cortex in MDD [20], and some animal paradigms of stress and/or increased glucocorticoid exposure are associated with reduced expression of EAAT2 in the hippocampus and cortical regions [53]. An investigation of the metabotropic glutamate receptor 5 (mGluR5) in primarily early-onset MDD applying [^{11}C]ABP688 PET in vivo reported relatively decreased mGluR5 binding in the PFC, cingulate cortex, insula, thalamus, and hippocampus. This study was combined with a postmortem investigation of the PFC applying Western blot, which also found reduced protein levels in that region [54]. Both results are consistent with a model of extracellular glutamate dysregulation leading to a downregulation of this receptor type [54]. A subsequent [^{11}C]ABP688 PET investigation in older MDD subjects did not find a difference between groups [55]. Since the methods applied should provide comparable results, the most likely explanation for the discrepant findings would be the difference in samples, namely that the difference at younger age is either not prominent at very late disease stages or in late-onset MDD (with the latter seeming more plausible).

Evidence of dysregulation of Glx, a composite measure of intracellular glutamate, extracellular glutamate, and glycine, is supported by applications of proton magnetic resonance spectroscopy in reasonably large samples of MDD, but the specific relationship to progression of disease is not yet clear as this has not been a study focus, and major investigations across groups tend to focus on different regions [56, 57].

The concept of MDD progressing towards elevated glutamate has significant clinical implications given that treatment-resistant MDD subjects, who are typically in a later phase of illness, often experience a rapid, short-term mood elevation after a single dose of ketamine, an NMDAR antagonist [53, 58]. A caution about interpreting the ketamine effect as being solely mediated by ketamine antagonism of NMDAR is that ketamine is nonselective having both 5-HT_{2A} and D_2 agonist properties [59], and metabolites of ketamine demonstrate therapeutic effects in animal models through activation of AMPA receptors [60]. Additional trials of selective NMDAR antagonists continue to hold intriguing promise with some positive results [61], although it may be that clinical trials should be oriented towards a phase-specific level of disease rather than a frequently applied window of nonresponse to a specified number of previous failed clinical trials.

Concluding Comments

Neuroprogression in MDD has a particular combination of increased microglial activation and reduced astroglial activation demonstrated within important affect-modulating regions such as the orbitofrontal cortex, dorsolateral PFC, ACC, and hippocampus [20–24]. Stress-induced changes have been implicated in loss of activated astroglia [28] although, given that MAO-A can facilitate oxidative stress and apoptosis and that elevated MAO-A has been identified in astroglia in MDD, it is possible that there are interconnections with elevated MAO-A levels [32, 33]. Elevated MAO-A levels occur throughout gray matter in high-risk states for MDD and are implicated as a common early change in MDD [62]. Loss of astrocytes is implicated in loss of EAAT2 transporters in MDD [20], which dysregulates glutamate levels. Further investigation is needed in longitudinal studies to ascertain the interrelationships of phase-specific changes across MDD to develop phase-specific treatments and alter progression of MDD.

References

1 Kreutzberg GW: Microglia: a sensor for pathological events in the CNS. Trends Neurosci 1996;19:312–318.
2 Gehrmann J, Matsumoto Y, Kreutzberg GW: Microglia: intrinsic immuneffector cell of the brain. Brain Res Brain Res Rev 1995;20:269–287.
3 Carbonell WS, Murase S, Horwitz AF, Mandell JW: Migration of perilesional microglia after focal brain injury and modulation by CC chemokine receptor 5: an in situ time-lapse confocal imaging study. J Neurosci 2005;25: 7040–7047.
4 Cross AK, Woodroofe MN: Chemokines induce migration and changes in actin polymerization in adult rat brain microglia and a human fetal microglial cell line in vitro. J Neurosci Res 1999;55: 17–23.
5 Venneti S, Wiley CA, Kofler J: Imaging microglial activation during neuroinflammation and Alzheimer's disease. J Neuroimmune Pharmacol 2009;4:227–243.
6 Block ML, Zecca L, Hong JS: Microglia-mediated neurotoxicity: uncovering the molecular mechanisms. Nat Rev Neurosci 2007;8:57–69.
7 Van Otterloo ES, Miguel-Hidalgo JJ, Stockmeier C, Rajkowska G: Microglia immunoreactivity is unchanged in the white matter of orbitofrontal cortex in elderly depressed patients. Annu Meet Soc Neurosci, Washington, 2005.
8 Steiner J, Walter M, Gos T, Guillemin GJ, Bernstein HG, Sarnyai Z, Mawrin C, Brisch R, Bielau H, Meyer zu Schwabedissen L, Bogerts B, Myint AM: Severe depression is associated with increased microglial quinolinic acid in subregions of the anterior cingulate gyrus: evidence for an immune-modulated glutamatergic neurotransmission? J Neuroinflammation 2011;8:94.
9 Steiner J, Bielau H, Brisch R, Danos P, Ullrich O, Mawrin C, Bernstein HG, Bogerts B: Immunological aspects in the neurobiology of suicide: elevated microglial density in schizophrenia and depression is associated with suicide. J Psychiatr Res 2008;42:151–157.
10 Hannestad J, DellaGioia N, Gallezot JD, Lim K, Nabulsi N, Esterlis I, Pittman B, Lee JY, O'Connor KC, Pelletier D, Carson RE: The neuroinflammation marker translocator protein is not elevated in individuals with mild-to-moderate depression: a [^{11}C]PBR28 PET study. Brain Behav Immun 2013;33:131–138.
11 Torres-Platas SG, Cruceanu C, Chen GG, Turecki G, Mechawar N: Evidence for increased microglial priming and macrophage recruitment in the dorsal anterior cingulate white matter of depressed suicides. Brain Behav Immun 2014;42:50–59.
12 Setiawan E, Wilson AA, Mizrahi R, Rusjan PM, Miler L, Rajkowska G, Suridjan I, Kennedy JL, Rekkas PV, Houle S, Meyer JH: Role of translocator protein density, a marker of neuroinflammation, in the brain during major depressive episodes. JAMA Psychiatry 2015;72:268–275.
13 Innis RB, Richards E, PIke V, Zarate C: Neuroinflammation in Depression: PET Imaging of Translocator Protein (TSPO) and Opportunities for COX-1 and COX-2. Society of Biological Psychiatry. Atlanta, Elsevier, 2016, vol 79, pp 2S–3S.
14 Orihuela R, McPherson CA, Harry GJ: Microglial M1/M2 polarization and metabolic states. Br J Pharmacol 2016;173: 649–665.
15 Amantea D, Bagetta G: Drug repurposing for immune modulation in acute ischemic stroke. Curr Opin Pharmacol 2016;26:124–130.
16 Sofroniew MV, Vinters HV: Astrocytes: biology and pathology. Acta Neuropathol 2010;119:7–35.
17 Rajkowska G, Miguel-Hidalgo JJ: Gliogenesis and glial pathology in depression. CNS Neurol Disord Drug Targets 2007;6:219–233.
18 Harris S, Johnson S, Duncan JW, Udemgba C, Meyer JH, Albert PR, Lomberk G, Urrutia R, Ou XM, Stockmeier CA, Wang JM: Evidence revealing deregulation of the KLF11-MAO A pathway in association with chronic stress and depressive disorders. Neuropsychopharmacology 2015;40:1373–1382.
19 Rajkowska G, Stockmeier CA: Astrocyte pathology in major depressive disorder: insights from human postmortem brain tissue. Curr Drug Targets 2013;14:1225–1236.

20 Miguel-Hidalgo JJ, Waltzer R, Whittom AA, Austin MC, Rajkowska G, Stockmeier CA: Glial and glutamatergic markers in depression, alcoholism, and their comorbidity. J Affect Disord 2010; 127:230–240.
21 Miguel-Hidalgo JJ, Baucom C, Dilley G, Overholser JC, Meltzer HY, Stockmeier CA, Rajkowska G: Glial fibrillary acidic protein immunoreactivity in the prefrontal cortex distinguishes younger from older adults in major depressive disorder. Biol Psychiatry 2000;48:861–873.
22 Muller MB, Lucassen PJ, Yassouridis A, Hoogendijk WJ, Holsboer F, Swaab DF: Neither major depression nor glucocorticoid treatment affects the cellular integrity of the human hippocampus. Eur J Neurosci 2001;14:1603–1612.
23 Cotter D, Mackay D, Landau S, Kerwin R, Everall I: Reduced glial cell density and neuronal size in the anterior cingulate cortex in major depressive disorder. Arch Gen Psychiatry 2001;58: 545–553.
24 Ongur D, Drevets WC, Price JL: Glial reduction in the subgenual prefrontal cortex in mood disorders. Proc Natl Acad Sci USA 1998;95:13290–13295.
25 Banasr M, Duman RS: Glial loss in the prefrontal cortex is sufficient to induce depressive-like behaviors. Biol Psychiatry 2008;64:863–870.
26 Si X, Miguel-Hidalgo JJ, O'Dwyer G, Stockmeier CA, Rajkowska G: Age-dependent reductions in the level of glial fibrillary acidic protein in the prefrontal cortex in major depression. Neuropsychopharmacology 2004;29: 2088-2096.
27 Gosselin RD, Gibney S, O'Malley D, Dinan TG, Cryan JF: Region specific decrease in glial fibrillary acidic protein immunoreactivity in the brain of a rat model of depression. Neuroscience 2009;159:915–925.
28 Banasr M, Chowdhury GM, Terwilliger R, Newton SS, Duman RS, Behar KL, Sanacora G: Glial pathology in an animal model of depression: reversal of stress-induced cellular, metabolic and behavioral deficits by the glutamate-modulating drug riluzole. Mol Psychiatry 2010;15:501–511.
29 Blanco-Suarez E, Caldwell AL, Allen NJ: Role of astrocyte-synapse interactions in CNS disorders. J Physiol 2017;595: 1903–1916.
30 Saura J, Bleuel Z, Ulrich J, Mendelowitsch A, Chen K, Shih JC, Malherbe P, Da Prada M, Richards JG: Molecular neuroanatomy of human monoamine oxidases A and B revealed by quantitative enzyme radioautography and in situ hybridization histochemistry. Neuroscience 1996;70:755–774.
31 Saura J, Kettler R, Da Prada M, Richards JG: Quantitative enzyme radioautography with ^{3}H-Ro 41-1049 and ^{3}H-Ro 19-6327 in vitro: localization and abundance of MAO-A and MAO-B in rat CNS, peripheral organs, and human brain. J Neurosci 1992;12:1977–1999.
32 Ou XM, Chen K, Shih JC: Monoamine oxidase A and repressor R1 are involved in apoptotic signaling pathway. Proc Natl Acad Sci USA 2006;103:10923–10928.
33 Youdim MB, Edmondson D, Tipton KF: The therapeutic potential of monoamine oxidase inhibitors. Nat Rev Neurosci 2006;7:295–309.
34 Mann JJ, Stanley M: Postmortem monoamine oxidase enzyme kinetics in the frontal cortex of suicide victims and controls. Acta Psychiatr Scand 1984;69: 135–139.
35 Ordway GA, Farley JT, Dilley GE, Overholser JC, Meltzer HY, Balraj EK, Stockmeier CA, Klimek V: Quantitative distribution of monoamine oxidase A in brainstem monoamine nuclei is normal in major depression. Brain Res 1999; 847:71–79.
36 Sherif F, Marcusson J, Oreland L: Brain gamma-aminobutyrate transaminase and monoamine oxidase activities in suicide victims. Eur Arch Psychiatry Clin Neurosci 1991;241:139–144.
37 Barraclough B, Bunch J, Nelson B, Sainsbury P: A hundred cases of suicide: clinical aspects. Br J Psychiatry 1974; 125:355–373.
38 Krishnan KR: Biological risk factors in late life depression. Biol Psychiatry 2002;52:185–192.
39 Meyer JH, Ginovart N, Boovariwala A, Sagrati S, Hussey D, Garcia A, Young T, Praschak-Rieder N, Wilson AA, Houle S: Elevated monoamine oxidase A levels in the brain: an explanation for the monoamine imbalance of major depression. Arch Gen Psychiatry 2006;63:1209–1216.
40 Rekkas PV, Wilson AA, Lee VW, Yogalingam P, Sacher J, Rusjan P, Houle S, Stewart DE, Kolla NJ, Kish S, Chiuccariello L, Meyer JH: Greater monoamine oxidase A binding in perimenopausal age as measured with carbon 11-labeled harmine positron emission tomography. JAMA Psychiatry 2014;71:873–879.
41 Barton DA, Esler MD, Dawood T, Lambert EA, Haikerwal D, Brenchley C, Socratous F, Hastings J, Guo L, Wiesner G, Kaye DM, Bayles R, Schlaich MP, Lambert GW: Elevated brain serotonin turnover in patients with depression: effect of genotype and therapy. Arch Gen Psychiatry 2008;65:38–46.
42 Meyer JH, Wilson AA, Sagrati S, Miler L, Rusjan P, Bloomfield PM, Clark M, Sacher J, Voineskos AN, Houle S: Brain monoamine oxidase A binding in major depressive disorder: relationship to selective serotonin reuptake inhibitor treatment, recovery, and recurrence. Arch Gen Psychiatry 2009;66:1304–1312.
43 Johnson S, Stockmeier CA, Meyer JH, Austin MC, Albert PR, Wang J, May WL, Rajkowska G, Overholser JC, Jurjus G, Dieter L, Johnson C, Sittman DB, Ou XM: The reduction of R1, a novel repressor protein for monoamine oxidase A, in major depressive disorder. Neuropsychopharmacology 2011;36:2139–2148.
44 Grunewald M, Johnson S, Lu D, Wang Z, Lomberk G, Albert PR, Stockmeier CA, Meyer JH, Urrutia R, Miczek KA, Austin MC, Wang J, Paul IA, Woolverton WL, Seo S, Sittman DB, Ou XM: Mechanistic role for a novel glucocorticoid-KLF11 (TIEG2) protein pathway in stress-induced monoamine oxidase A expression. J Biol Chem 2012;287: 24195–24206.
45 Chiuccariello L, Houle S, Miler L, Cooke RG, Rusjan PM, Rajkowska G, Levitan RD, Kish SJ, Kolla NJ, Ou X, Wilson AA, Meyer JH: Elevated monoamine oxidase A binding during major depressive episodes is associated with greater severity and reversed neurovegetative symptoms. Neuropsychopharmacology 2014; 39:973–980.
46 Ou XM, Chen K, Shih JC: Glucocorticoid and androgen activation of monoamine oxidase A is regulated differently by R1 and Sp1. J Biol Chem 2006;281:21512–21525.

47 Fitzgerald JC, Ugun-Klusek A, Allen G, De Girolamo LA, Hargreaves I, Ufer C, Abramov AY, Billett EE: Monoamine oxidase-A knockdown in human neuroblastoma cells reveals protection against mitochondrial toxins. FASEB J 2014;28:218–229.
48 Sacher J, Wilson A, Houle S, Hassan S, Rusjan P, Bloomfield P, Stewart D, Meyer J: Elevated brain monoamine oxidase A binding in the early postpartum period. Arch Gen Psychiatry 2010;67:468–474.
49 Bacher I, Houle S, Xu X, Zawertailo L, Soliman A, Wilson AA, Selby P, George TP, Sacher J, Miler L, Kish SJ, Rusjan P, Meyer JH: Monoamine oxidase A binding in the prefrontal and anterior cingulate cortices during acute withdrawal from heavy cigarette smoking. Arch Gen Psychiatry 2011;68:817–826.
50 Matthews B, Kish S, Rusjan P, Boileau I, Houle S, Meyer J: Monoamine oxidase A binding during alcohol dependence (abstract). Biol Psychiatry 2012;71:8S.
51 Son SY, Ma J, Kondou Y, Yoshimura M, Yamashita E, Tsukihara T: Structure of human monoamine oxidase A at 2.2-A resolution: the control of opening the entry for substrates/inhibitors. Proc Natl Acad Sci USA 2008;105:5739–5744.
52 Freis ED: Mental depression in hypertensive patients treated for long periods with large doses of reserpine. N Engl J Med 1954;251:1006–1008.
53 Sanacora G, Banasr M: From pathophysiology to novel antidepressant drugs: glial contributions to the pathology and treatment of mood disorders. Biol Psychiatry 2013;73:1172–1179.
54 Deschwanden A, Karolewicz B, Feyissa AM, Treyer V, Ametamey SM, Johayem A, Burger C, Auberson YP, Sovago J, Stockmeier CA, Buck A, Hasler G: Reduced metabotropic glutamate receptor 5 density in major depression determined by [^{11}C]ABP688 PET and postmortem study. Am J Psychiatry 2011;168:727–734.
55 DeLorenzo C, Sovago J, Gardus J, Xu J, Yang J, Behrje R, Kumar JS, Devanand DP, Pelton GH, Mathis CA, Mason NS, Gomez-Mancilla B, Aizenstein H, Mann JJ, Parsey RV: Characterization of brain mGluR5 binding in a pilot study of late-life major depressive disorder using positron emission tomography and [^{11}C]ABP688. Transl Psychiatry 2015;5:e693.
56 Hasler G, van der Veen JW, Tumonis T, Meyers N, Shen J, Drevets WC: Reduced prefrontal glutamate/glutamine and γ-aminobutyric acid levels in major depression determined using proton magnetic resonance spectroscopy. Arch Gen Psychiatry 2007;64:193–200.
57 Sanacora G, Gueorguieva R, Epperson CN, Wu YT, Appel M, Rothman DL, Krystal JH, Mason GF: Subtype-specific alterations of γ-aminobutyric acid and glutamate in patients with major depression. Arch Gen Psychiatry 2004;61:705–713.
58 Zarate CA Jr, Singh JB, Carlson PJ, Brutsche NE, Ameli R, Luckenbaugh DA, Charney DS, Manji HK: A randomized trial of an N-methyl-D-aspartate antagonist in treatment-resistant major depression. Arch Gen Psychiatry 2006;63:856–864.
59 Kapur S, Seeman P: NMDA receptor antagonists ketamine and PCP have direct effects on the dopamine D_2 and serotonin 5-HT_2 receptors – implications for models of schizophrenia. Mol Psychiatry 2002;7:837–844.
60 Zanos P, Moaddel R, Morris PJ, Georgiou P, Fischell J, Elmer GI, Alkondon M, Yuan P, Pribut HJ, Singh NS, Dossou KS, Fang Y, Huang XP, Mayo CL, Wainer IW, Albuquerque EX, Thompson SM, Thomas CJ, Zarate CA Jr, Gould TD: NMDAR inhibition-independent antidepressant actions of ketamine metabolites. Nature 2016;533:481–486.
61 Garay RP, Zarate CA Jr, Charpeaud T, Citrome L, Correll CU, Hameg A, Llorca PM: Investigational drugs in recent clinical trials for treatment-resistant depression. Expert Rev Neurother 2017, pp 1–17.
62 Meyer JH: Neuroimaging markers of cellular function in major depressive disorder: implications for therapeutics, personalized medicine, and prevention. Clin Pharmacol Ther 2012;91:201–214.

Jeffrey H. Meyer, MD, PhD, FRCPC
CAMH Research Imaging Centre
College St. Site, CAMH
250 College St., M5T 1R8 Toronto (Canada)
E-Mail jeff.meyer@camhpet.ca

Halaris A, Leonard BE (eds): Neuroprogression in Psychiatric Disorders.
Mod Trends Pharmacopsychiatry. Basel, Karger, 2017, vol 31, pp 37–55 (DOI: 10.1159/000470805)

Inflammation Effects on Glutamate as a Pathway to Neuroprogression in Mood Disorders

Ebrahim Haroon · Andrew H. Miller

Emory Behavioral Immunology Program, Department of Psychiatry and Behavioral Sciences, Winship Cancer Institute, Emory University School of Medicine, Atlanta, GA, USA

Abstract

Neuroprogression is a term used to describe worsening psychopathology, poor treatment response, and declining cognitive and functional outcomes among patients with chronic mental disorders. Chronic inflammatory activation and glutamate-mediated excitotoxicity are two key etiological factors implicated in the development of neuroprogression. In this chapter, we hypothesize that the association between chronic inflammatory activation, neuroprogression, and glutamate dysregulation might be mediated by glial dysfunction. The role played by other mechanisms that increase glutamatergic activity, including oxidative stress and kynurenine pathway activation, will also be discussed, albeit briefly. We will conclude by providing a mechanistic model that draws upon our previous research, to link neuroimmune/neurotransmitter dysregulation to disrupted local neural activity and brain network connectivity eventually leading to a neuroprogressive course and associated clinical outcomes.

Introduction

The term "neuroprogression" refers to a combination of treatment nonresponsiveness, relapsing and declining course of illness, and progressive neuropathological changes commonly seen in several psychiatric disorders [1]. Increased inflammatory activity and brain glutamate dysregulation are two major pathophysiological pathways that contribute to the evolution of these neuroprogressive changes [2]. In this chapter, we advance a model that links increased inflammatory activation, glial pathology, and glutamate dysregulation with progressive neuropathology, symptomatic decline, and treatment resistance. Accordingly, we will begin this chapter by summarizing the evidence supporting immune dysregulation in mood disorders followed by sections devoted to how this dysregulation impacts glutamate metabolism. Along the way, we will examine the role played by glial cell dysfunction in mediating the association between inflammation and glutamate regulation. We will conclude by proposing a working model to synthesize biological data with trajectories of adverse clinical outcomes and integrate novel treatment approaches.

Depression – Scope and Consequences

Mood disorders are chronic, progressive, and difficult-to-treat disorders that are believed to afflict approximately 30 million adults in the US and up to 12% of all individuals worldwide [3]. Suicide represents the 10th leading cause of death in the US with ~40,000 adults dying by suicide each year [4, 5]. A larger number of depressed individuals endure a life filled with severe limitations, and depression is believed to be the single major cause of disease burden and disease-related disability over the life span [3]. Underscoring this tragic fact are data that about a third of depressed patients do not respond to any of the available treatments, including electroconvulsive therapy (ECT) [6]. Thus, there is a pressing need for innovative ideas and novel approaches to target mechanisms underlying nonresponse to treatment and thereby limit disability and persistent suffering. In this context, dysregulation of inflammatory activation and glutamate signaling are two key neurobiological pathways implicated in precipitating treatment nonresponse and neuroprogression. Both these neurobiological pathways will likely yield many targets for new drug development, but research in these areas has progressed largely along independent lines without significant integration or cohesion.

Basis for the Role of Inflammation in the Etiology of Mood Disorders and Depression

Research over the past three decades has convincingly demonstrated that a subgroup of depressed patients (whether bipolar of unipolar) manifest all of the cardinal signs of systemic inflammation [7]. Depressed subjects consistently exhibit elevated plasma and cerebrospinal fluid (CSF) concentrations of inflammatory molecules [8]. Four independent meta-analytic studies have confirmed that plasma concentrations of inflammatory molecules such as cytokines (especially tumor necrosis factor-α and interleukin [IL]-6) and acute phase proteins (such as C-reactive protein [CRP]) are increased among patients with depression [9–12]. In addition, experimental induction of inflammation by experimental administration of endotoxin and typhoid vaccine or therapeutic induction of inflammation by administration of interferon (IFN)-α for treatment of medical disorders has been shown to reliably result in depressive symptoms in humans and in preclinical models [8]. Blockade of inflammatory activity using cytokine antagonists such as infliximab and nonsteroidal anti-inflammatory medications has also been shown to reverse and treat symptoms of depression and fatigue [13, 14]. Taken together, these data strongly support the hypothesis that inflammatory activation might impact negatively on mood, behavior, and cognitive processing.

Mounting evidence supports the etiological role played by inflammation in the development of mood disorders in some individuals. Inflammatory activation is known to impact every known pathophysiological pathway associated with causation of depression and mood disorders [15]. Moreover, patients who demonstrate chronic inflammatory activation are often resistant to the conventional antidepressant or mood stabilizer treatments [16]. Chronic inflammatory activation has also been shown to induce several neuroprogressive pathologies including oxidative/nitrosative stress, mitochondrial dysfunction, hypothalamic-pituitary-adrenal axis dysregulation, epigenetic changes, and neurotrophic disturbances [7, 17, 18]. Interestingly, virtually all of the above pathologies have been interpedently associated with glutamatergic dysfunction [2, 19, 20].

Glutamate Physiology and Pathology in Major Depressive Disorder

Glutamate is a neurotransmitter, which is present in abundant quantities in the brain [21]. Glutamate mediates more than 90% of fast excitatory neurotransmission in the human brain. Fast excitatory neurotransmission is responsible for many essential neural functions involved in the regulation of behavior and leads to synaptic plasticity and experience-based learning [21]. Recovery from fast excitatory neural activity consumes more than 80% of the glucose and oxygen supply to the brain and thus forms the basis for brain activity mapping techniques using ^{18}F-fluorodeoxyglucose (FDG) uptake positron emission tomography (PET) and blood oxygen level-dependent (BOLD) functional magnetic resonance imaging (fMRI) [22]. Glutamate receptors are also expressed on the axon terminals and dendritic spines of neurons associated with other neurotransmitters such as dopamine and γ-amino butyrate, contributing to their neuroadaptation and plasticity [23]. In addition to its role in neurotransmission, glutamate is also a participant in the metabolism of glucose, lipids, and proteins, and is an integral product of the tricarboxylic acid energy cycle [24]. Thus, glutamate changes are seen in multiple disorders involving glucose and lipid metabolism, such as diabetes and cardio/cerebrovascular disease that are also associated with high incidence rates of neuroprogressive mood disorder pathologies [25, 26].

Glutamate-Glutamine Cycling and Glutamate Transporters

Glutamate is synthesized from one of its two precursors, i.e. glutamine released by astroglial cells or from α-ketoglutarate produced by the neuronal tricarboxylic acid cycle [27]. Following its synthesis, glutamate is kept tightly packaged in specialized vesicles inside presynaptic neurons and possibly glial cells from where it is released into the synaptic cleft during neuronal excitation [27]. After its release, glutamate binds to its target receptor sites, and the unbound glutamate is rapidly cleared (within milliseconds) by specialized excitatory amino acid reuptake transporters (EAATs) located on the synaptic surface of astrocytes [28]. The rapid clearance of glutamate from synapses is critical in maintaining synaptic integrity, because overstimulation by uncleared glutamate can result in excitotoxic death of postsynaptic neurons [28]. There are several subtypes of EAATs , and the reader is referred to more extensive reviews in this regard [27–29]. Following uptake into the astrocytic cytoplasm, glutamate is then converted into its inert metabolite glutamine by the enzyme glutamine synthase [27]. Glutamine is then released into the extrasynaptic space from where it is reabsorbed into presynaptic neurons, reconverted back into glutamate by the enzyme glutaminase, repackaged into vesicles, and recycled for neurotransmission [27]. Under physiological conditions, the astrocytic processes encircling the synapse – known as "cradles" – are believed to constitute a functional barrier that prevents glutamate from "spilling over" into the adjoining extrasynaptic space. Pathologies associated with defective "cradling" include inflammatory activation and are known to increase "spillover" of glutamate into the extrasynaptic regions with deleterious consequences [2, 30].

Glutamate Receptors

Glutamate released into the synaptic or extrasynaptic regions binds to two distinct types of postsynaptic receptors – ionotropic and metabotropic receptors. The inotropic receptors are voltage-gated ion channels and include N-methyl D-aspartate (NMDA), α-amino-3-hydroxy-5-methyl-4-isoxazolepropionic acid (AMPA), and kainate (KA) receptors. The ionotropic receptors are directly activated by neuronal depolarization currents and do not require an intermediary signal transducing mechanism. Upon binding to glutamate, AMPA receptors open Na^+ ion channels that further propagate neuronal currents along postsynaptic neurons. AMPA receptors are highly sensitive to neuronal current fluctuations. Consequently, it is considered a high-value target in the treatment of epilepsy. In contrast, NMDA receptors are less voltage sensitive than the AMPA receptors and open both Na^+ and Ca^+ channels. The opening of Ca^+ channels can yield different results in different contexts with activation of neuroplasticity-promoting synaptic protein cascades under homeostatic conditions and oxidative stress-promoting cascades in pathological contexts. The role of the KA receptor is less well defined but has been hypothesized to play a role in neuronal plasticity, inhibitory neurotransmission by γ-aminobutyric acid and genesis of seizure activity. In contrast to ionotropic receptors, metabotropic receptors (mGluR; type 1–5) require an intermediary G-protein-based second-messenger system to transduce information. While ionotropic receptors are activated on an all or none basis by a combination of voltage changes and glutamate release, the activation of mGluRs is more nuanced and controls the intensity of excitatory activity [31]. mGluRs are divided into 3 groups namely group I (mGluRs 1 and 5), group II (mGluRs 2 and 3), and group III (mGluRs 4 and 6).

Intra- versus Extrasynaptic Glutamate Receptors

The ionotropic and metabotropic glutamate receptors and glutamate transporters are located in both intra- and extrasynaptic neuronal sites and on the glial cells [2, 32]. A balance between the intra- and extrasynaptic receptor activity is essential to maintain appropriate neuronal vitality and functioning [33]. Intrasynaptic signaling refers to a classical model of neurotransmission where pre- and postsynaptic neurons communicate with high synaptic fidelity [34]. In contrast, extrasynaptic signaling involves tonic, transsynaptic spread of activation resulting from binding of glutamate spillover onto extrasynaptic receptors [34–37]. At physiological levels, this synaptic cross talk facilitates "tonic" regional activation. The astroglial cells stimulate this tonic regional activity in physiological contexts by releasing small amounts of glutamate into the extrasynaptic space – a process referred to as "gliotransmission" [38]. Astroglial cells also transmit this activity at a transregional level by connecting with each other physically and biochemically to form large-scale astroglial syncytia [39]. However, at extremes, astroglial glutamate release can lead to a chaotic neurotransmission [34–37]. Our working model proposes that inflammation-induced increase in glutamate spillover leads to uncontrolled extrasynaptic glutamate signaling, chaotic neurotransmission, signaling disarray, and ultimately neurocognitive dysfunction and neuroprogression [2, 32].

Synaptogenic Effects of Glia and Glutamate

Through their effects on synaptogenesis and synaptic plasticity, immune/inflammatory activity, glial pathology, and increased glutamate might underlie neuroprogressive changes. Synaptic adaptation and plasticity are vital brain functions that are susceptible to disruption by multiple influences such as variable neurotrophic factor support, effects of sex hormones, and inflammatory activity [19]. The effects of synaptic glutamate on downstream protein cascades leading to the release of synapse-building proteins such as brain-derived neurotrophic factor, mTOR (mammalian target of rapamycin), AMPA A1 receptor subunits, and postsynaptic density protein (PSD)-95 in the postsynaptic neurons forms the cornerstone of brain health, synaptic adaptation, and neural plasticity [19, 40, 41]. Unregulated inflammatory activity and glutamate dysregulation might have a disruptive effect on synaptic health and adaptation in chronic depression, and medications that reverse this dysregulation can exert antidepressant effects [19]. PSD-95 is a key synaptogenic protein which anchors ionotropic glutamate receptors such as NMDA and AMPA to their postsynaptic sites. Transcriptomic and proteomic analyses have demonstrated inflammation-induced deficiencies in PSD-95 in postmortem brain specimens of depressed suicide victims [42]. It is well known that dysfunction or inhibition of postsynaptic ionotropic receptors can lead to reflexive increase in glutamate release by presynaptic neurons thereby increasing the downstream toxic effects on postsynaptic neurons [2].

Buffering and Handling of Extrasynaptic Glutamate Spillover

In physiological contexts, the toxic effects of glutamate spillover are controlled by a combination of "buffering" and "binding" by the EAATs [43, 44]. The binding of glutamate to EAATs occurs at a much faster rate than its ultimate transport into the astrocytic cytoplasm. Hence, EAATs cope with acute surges in extracellular glutamate by repeatedly binding, releasing, and rebinding to glutamate – a process referred to as "buffering" [44]. Thus, under physiological conditions, EAATs use buffering mechanisms to control freely diffusing glutamate just enough to fine-tune extrasynaptic signaling and shape tonic activation [34–37, 43, 44]. Increases in extracellular glutamate and lack of buffering can lead to a wider diffusion radius of glutamate spillover and result in excessive extrasynaptic signaling.

Dysregulation of Glutamate Homeostasis by Inflammatory Activity

Immune activation is one of the key disruptors of glutamate homeostasis, its intrasynaptic containment, and regulation of its spillover. Several mechanisms have been proposed to explain the effects of inflammation upon glutamate neurotransmission. Some of these mechanisms will be reviewed in the following paragraphs.

Decreases in the Functioning of Excitatory Amino Acid Reuptake Transporters
Approximately 85% of glutamate not bound to target receptor sites is cleared by EAATs located on the synaptic surface of astrocytes. EAATs are complex transporter proteins synthesized by genes, whose functioning is under the control of a variety of environmental factors, including immune molecules. Thus, inflammatory molecules can lead to decreases in the functioning, expression, and trafficking of astrocytic EAATs [45–48]. Defective functioning and activity of EAATs can lead to decreased clearance of glutamate leading to its intrasynaptic buildup and eventual spillover. Immune dysfunction appears to impact all known subtypes of EAATs. A detailed discussion of the impact of immune molecules on EAAT functioning and expression is provided elsewhere [2].

Reverse Efflux of Glutamate from Astrocytes
Under physiological conditions, the transport of glutamate is unidirectional – from the synapse into the astrocytes [46]. However, severe astrocytic pathologies such as acute ischemia and inflammation can lead to reverse operation of EAATs resulting in retrograde movement of glutamate from astrocytes into the synaptic space leading to profound neurotoxicity [46].

Exaggerated Release of Vesicular Glutamate
Intracellular glutamate is stored in vesicles by vesicular glutamate transporters [24]. These storage vesicles are seen in presynaptic neurons and released during neural activity. However, these vesicles have also been demonstrated in astrocytes and are known to participate in gliotransmission

[39, 49]. It has been demonstrated that during inflammatory activation this vesicular glutamate release can be exaggerated adding to extrasynaptic glutamate concentrations [46]. The physical connections between the glial processes possess specialized gap junctions known as "connexin channels," which allow transfer of several neuroactive chemicals (such as glutamate) and ions (such as Na^+ and Ca^+) between cells. Disruption of these connexin channel proteins by inflammatory processes has been demonstrated in patients with mood disorders [50–52].

Increased Release through x_c-System Activation

Neurometabolic activity and cellular stress leads to the generation of large quantities of reactive oxygen species (ROS) and/or reactive nitrogen species (RNS) [53, 54]. Increases in ROS/RNS can lead to a breakdown in cellular bioenergetics and integrity (by deactivating enzymes involved in energy regulation and respiration), while activating enzymes that lead to cell death [55]. Neural and glial cells use antioxidant molecules such as glutathione (GSH) to buffer ROS/RNS and prevent cell death. GSH is synthesized by combining cysteine with glutamate in astroglial and neural cells. Cysteine is an unstable molecule and is synthesized from cystine obtained from the extracellular space using dedicated surface transporters known as x_c-systems. x_c-transporters permit entry of 1 cystine molecule in exchange for extrusion of 1 glutamate molecule from the glial cell to enable synthesis of 1 GSH molecule [56, 57]. Consequently, activation of x_c-system transporters leads to net increases in the extracellular/extrasynaptic concentration of glutamate [56]. Exaggerated neural activity (stress) and increased immune activity (inflammation) can lead to increased demand for GSH leading to consequent x_c-system activation [56, 57]. In addition to oxidative stress and inflammation, increased glutamate by itself can increase the activity of the x_c-system, further amplifying its own release [56–58].

Increased Activation of Purine Receptors

Activation of purine receptors (such as P2X7-type receptors) can trigger the release of glutamate through ion channels under toxic conditions resulting in profound neurotoxicity [46]. Once considered mostly relevant to neurological disorders, recent data indicate that purine signaling might also mediate glutamate release during psychological stress [59]. Of note, purines also activate the intracellular inflammasome system of proteins leading to the release of the cytokine IL-1b, which in turn can further amplify the release of other cytokines [32].

Release of Glutamate by Activated Immune Cells

Astro- and oligodendroglial cells express glutamate binding sites and uptake transporters on their surface regions and mount demonstrable electrophysiological responses consequent to the binding of glutamate to these target sites. Microglia, which are the primary immune effector cells of the brain, also express glutamatergic activity, but only when activated by immune signaling molecules [60, 61]. In addition, activated resident microglia and trafficking macrophages release copious amounts of glutamate into the extrasynaptic space adding to the ambient glutamatergic surge [60, 62]. Most of the microglial release is believed to occur via activation of x_c-transporters due to a high intracellular antioxidant demand resulting from inflammatory activation [63]. Another mechanism of release might involve the release of glutamate through connexin hemichannels, whereby the released glutamate is passed on to other glial cells, especially astrocytes and oligodendroglial cells [60, 62, 64]. The precise function of this release is still a matter of investigation, but increases in intracellular glutamate in astrocytes and oligodendrocytes cause a further decrease in EAAT activity and an increase in extracellular glutamate [64–66]. Accordingly, the blockade of microglial glutamate release using microglial inhibitors has been associated with neuroprotective effects [64, 67]. The recent awareness of the role played by microglia in synaptic pruning has led many investigators to propose a "quadripartite synapse," which includes microglia along with astrocytes in the neuronal elements [66]. It would be intriguing to reframe the currently known facts on glutamate metabolism in the context of the quardripartite synapse. Taken together, it could be stated that microglial activation associated with increases in glutamate-mediated neurotoxicity and microglial inhibition has been associated with neuroprotective functions [67, 68].

Effects of Inflammatory Activation on the Kynurenine Pathway

The kynurenine (KYN) pathway is an alternate tryptophan-metabolizing pathway. This pathway is an enzymatic cascade that sequentially degrades tryptophan into multiple neuroactive intermediaries with glutamate-like and prooxidant properties. Some of the glutamate-like products, including KYN acid (KYNA), 3-hydroxy-KYN, and quinolinic acid (QUIN), have been examined in depth recently [69]. The synthesis of KYN metabolites and shunting of tryptophan away from other pathways such as serotonin, melatonin, and protein synthesis is tightly regulated by the immune system [69, 70]. The pathway is initiated when immune molecules act upon the rate-limiting enzymes – indoleamine 3,4-dioxygenase in the brain and periphery and tryptophan 3,4-dioxygenase in the liver – to liberate metabolites that act as substrates to other enzymes of the pathway [70–72]. A detailed discussion of the regulation of this pathway and its by-products can be found in some of the references cited later [69, 70, 73, 74]. Only end products of relevance to glutamate metabolism will be discussed here. QUIN is a powerful NMDA receptor agonist and a potent neurotoxin [75]. QUIN is known to increase glutamate release by both neurons and glial cells, and induces direct toxicity to neurons and glial cells by binding to NMDA receptors-independent and in addition of its effects on glutamate release. QUIN has been associated with the severity of depressive symptoms in patients treated with IFN-α and suicidality among patients with mood disorders [75–78]. It has also been associated with neuroprogressive (kindling) effects in seizure disorders and is associated with acceleration of neurodegenerative changes in patients with Huntington disease and acquired immune deficiency [75]. KYNA, on the other hand, is an NMDA antagonist and has been ascribed putative neuroprotective functions [63, 79, 80]. KYNA is believed to exert its NMDA antagonist properties by binding to the glycine subunit of the NMDA receptor [80]. KYNA also decreases extracellular glutamate concentrations in every brain region studied so far, stemming from its blockade of α_7 nicotinic acetyl choline receptors (nAChR7) located on most glutamatergic neurons [80]. Thus, better understanding of the role played by KYNA in neuroprogression might be complicated by its dual role on glutamate: antagonism at the NMDA site and decreased release of intrasynaptic glutamate. QUIN and KYNA represent two of the multiple trajectories in the KYN pathway, with the former being controlled by the enzyme KYN monooxygenase and the latter being controlled by KYN aminotransferase. Intriguingly, the former is located exclusively on microglia, while the latter is exclusively localized on astroglial cells [71]. Consequently, microglial activation would be expected to increase QUIN, while astroglial hyperactivity would increase KYNA. Other molecules generated by this pathway, such as 3H-KYN, induce oxidative stress that contributes indirectly to increases in brain glutamate. Thus, under inflammatory conditions, metabolites of the KYN pathway can significantly contribute to glutamate increases.

Empirical Evidence of the Association between Inflammatory Activation and Glutamate Dysregulation

Immune Stimulation, Brain Activity, and Behavioral Changes

Our preliminary experiments in this area were conducted using an experimental model of inflammatory stimulation resulting from the therapeutic administration of IFN-α [81, 82]. IFN-α is an antiviral and antiproliferative cytokine used to treat viral infections such as hepatitis C and cancer (e.g., malignant melanoma and renal cell carcinoma) and induces profound inflammatory responses [83]. At high doses, IFN-α is also known to cause severe behavioral toxicity in the form of major depression-like symptoms in more than 50% of patients during the first 4–12 weeks of administration [81, 84, 85]. Similar behavioral changes are seen in nonhuman primates following 4 weeks of IFN-α treatment [86]. IFN-α-induced increases in depressive symptoms are paralleled by concurrent changes in neurobiological markers characteristic of depression, including increases in brain (CSF) and plasma inflammatory markers, and hypothalamic-pituitary-adrenal axis activation, and decreases in dopaminergic and serotonin turnover in the brain [78, 82, 87]. IFN-α administration is also associated with increases in neurometabolic activity in the basal ganglia (indexed by increased uptake of FDG during PET) and dorsal anterior cingulate cortical (dACC) activity in response to cognitive challenge (measured using BOLD-fMRI), both of which also predicted the severity of depressive symptoms and cognitive impairment following IFN-α [88–90]. Susbequent studies indicated that the effect of IFN-α on basal ganglion function and the expression of anhedonic symptoms might be mediated by its adverse impact on dopamine synthesis pathways in the striatum and elsewhere [91–94]. These data support the idea that inflammatory activation (induced by IFN-α) leads to behavioral symptoms such as depression, anhedonia, psychomotor slowing, and fatigue by selectively targeting brain regions and neural circuits involved in the processing of these complex behavioral functions.

Impact of Interferon-α on Brain Glutamate

Based on this information, we examined if 4-week treatment with IFN-α for hepatitis C virus disease was associated with glutamate increases in the basal ganglia and dACC, and if any observed increases were associated with the development of depression and related symptoms [95]. We used proton magnetic resonance spectroscopy to measure changes in glutamate concentrations over the 4-week study period in the regions of interest – namely bilateral basal ganglia and dACC [95]. As shown in Figure 1a, 4-week treatment with IFN-α led to significant increases in glutamate normalized to creatinine in the left basal ganglion region of interest, and this increase was in turn associated with reduced motivation [95]. These findings helped confirm our hypothesis that inflammatory activation might lead to behavioral changes by increasing brain glutamate concentrations in basal ganglia regions that serve as hubs that connect affective, motivational, and cognitive circuitry [95].

Impact of Aging on Inflammation and CNS Glutamate

As a follow-up to the above study, we examined whether older age may interact with inflammation to exaggerate the effects of IFN-α on CNS glutamate and behavior. Intriguingly, older patients treated with IFN-α exhibited a significantly greater increase in glutamate in the left basal ganglia over the 4-week study period compared to older controls and younger IFN-α-treated and untreated subjects [96] (Fig. 1b). In addition, increased glutamate in older but not younger IFN-α-treated and untreated patients was associated with increased tumor necrosis factor, reduced motivation, and prolonged reaction time [96]. This exaggerated response to immune stimula-

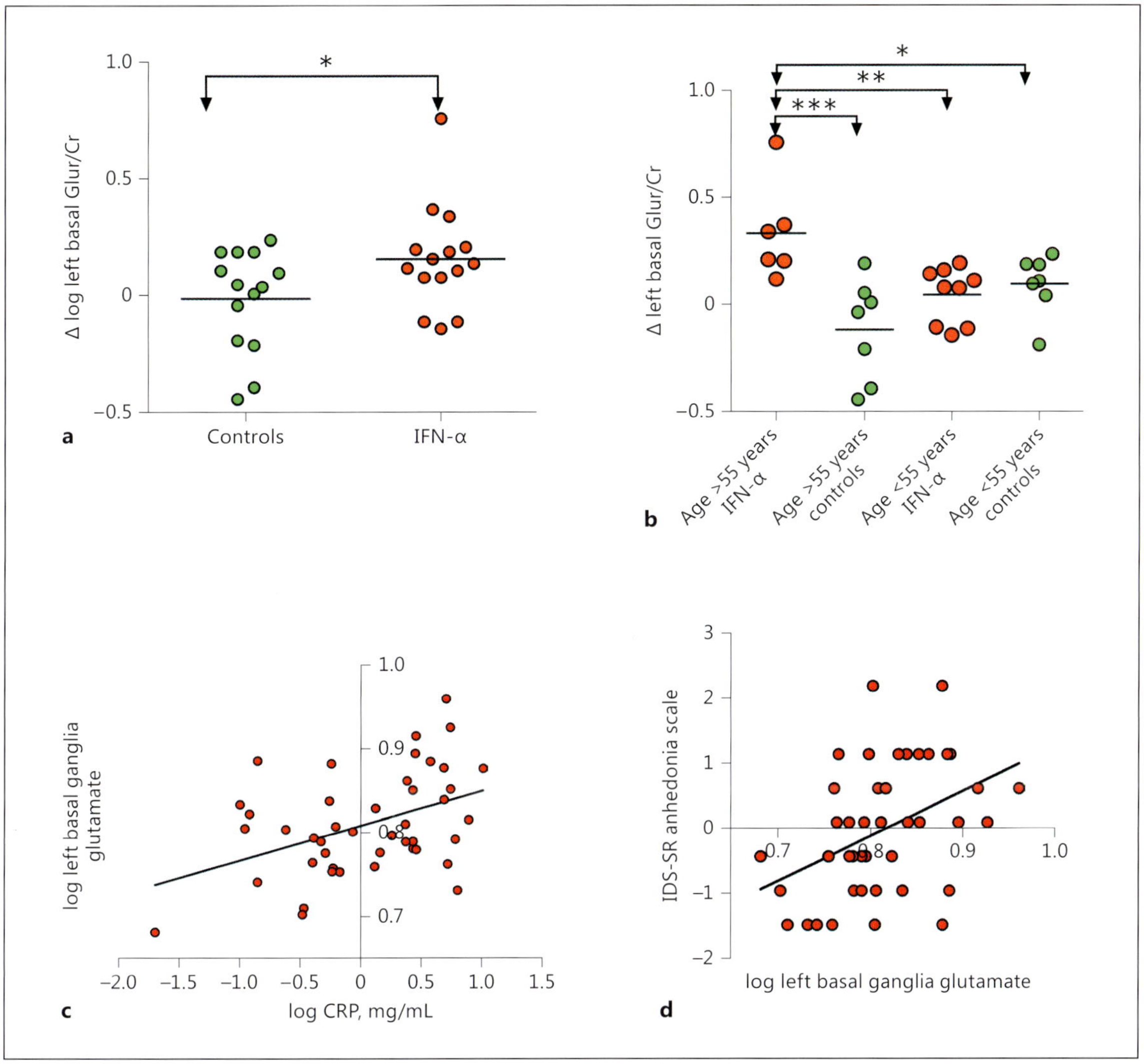

Fig. 1. a Interferon (IFN)-α increases glutamate (Glu) normalized to creatine (Cr) in the dorsal anterior cingulate cortex and left basal ganglion region: increased Glu/Cr in the left basal ganglion region following 4 doses of IFN-α (t = 2.10, df = 27, p = 0.046) in nondepressed, medically ill (hepatitis C) patients. **b** Increased left basal ganglion Glu in older IFN-α-treated subjects: 4-week treatment with IFN-α increased left log basal ganglion Glu/Cr to a greater extent in older (>55 years) IFN-α-treated subjects than controls and younger (<55 years) IFN-α-treated and untreated subjects with hepatitis C infection ($F_{1,22}$ = 10.4, p = 0.004. * $p < 0.05$, * $p < 0.01$, *** $p < 0.001$). **c** Association between plasma C-reactive protein (CRP) and the left basal ganglion Glu (mg/mL) in depressed patients: increased log plasma CRP correlated with log left basal ganglion absolute Glu concentrations in depressed subjects after controlling for age, sex, race, smoking status, body mass index, and 17-item Hamilton depression rating scale scores (β = 0.36, t = 2.57, p = 0.014) in a sample of medically healthy, depressed subjects. **d** Correlations between the left basal ganglion Glu and anhedonia in depressed patients: log values of absolute Glu concentrations in the left basal ganglia were correlated with the Inventory of Depressive Symptoms (IDS)-Self-Rated (SR) anhedonia subscale (β = 0.42, t = 3.03, p = 0.004) in a sample of medically healthy, depressed subjects. [Adapted with permission from 95 (**a**), 96 (**b**), and 103 (**c**, **d**).]

tion seen among older individuals has been attributed to senescent changes in brain immune cells such as microglia [97–99]. Effects of senescence upon microglia and other immune cells lead to the development of a "primed" phenotype – a partially activated state in which microglial cells demonstrate exaggerated basal and evoked responses to immune stimulation [100]. Based on our results, we concluded that the primed microglial phenotype might also have led to glutamate increases and associated behavioral impairments in older subjects [96]. Of note, priming effects have been noted in microglial cells exposed to psychological stress [101] and in microglia extracted from postmortem brain of suicide victims [102].

Inflammation-Induced Glutamate Changes in Major Depression

Given our data linking inflammatory activation, depression, and glutamate following IFN-α administration, we examined if the association would hold in a group of subjects with major depression. We hypothesized that increased inflammation in major depression may lead to increased glutamate in the basal ganglia in association with glial dysfunction leading to behavioral symptom dimensions of depression such as anhedonia and psychomotor slowing [103]. We measured CRP levels in plasma and CSF to index inflammatory activity in our depressed patients [9–12]. We classified depressed patients into high- (>3 mg/L), medium- (1–2 mg/L), and low-inflammation groups (<1 mg/L) in accordance with recommendations provided by the American Heart Association based on cardiovascular atherosclerosis risk studies [104, 105]. Increased plasma CRP was significantly associated with increased left basal ganglion glutamate controlling for age, sex, race, body mass index, smoking status, and depression severity (Fig. 1c). Left basal ganglion glutamate was in turn associated with anhedonia (Fig. 1d), psychomotor slowing, and information processing delays [data not shown, see 103 for details]. Plasma and CSF CRP was also associated with measures of basal ganglion glutamate and the glial marker myoinositol using chemical shift imaging.

Impact of Inflammation-Induced Glutamate Changes on Behavioral Neural Circuits

Depression is associated with altered regional neural activity and whole brain connectivity [106–109]. More specifically, alterations in the connectivity of the corticostriatal and corticolimbic circuitry have been associated with neuroprogressive symptoms of depression, such as anhedonia, psychomotor slowing, and executive dysfunction [107, 110, 111]. Inflammatory activation has been demonstrated to lead to behavioral changes by disrupting the connectivity of corticostriatal circuits [112–114]. The ability of inflammatory molecules to selectively target and negatively impact brain circuits involved in mood and cognition processing has been discussed in previous sections. This selective targeting by inflammatory molecules can be profiled with fMRI-based connectivity analysis, where the level of synchrony and overlap between low-frequency BOLD fluctuations in an originating "seed" and destination "target" regions is estimated and used for statistical analysis. Using a variant of this approach (seed-whole brain connectivity analysis), our group recently discovered that decreases in resting BOLD fMRI-based functional connectivity between ventral striatal seeds and ventromedial prefrontal cortex targets and dorsal striatal seeds and ventromedial prefrontal cortex/presupplementary motor targets were associated with inflammation-induced anhedonia and motor slowing, respectively [113]. Given our earlier findings associating inflammatory activation and glutamate increases in hubs associated with corticostriatal circuits such as basal ganglion and dACC regions, it remains to be determined whether increased glutamate contributes to these changes in the neurocircuitry, in particular by altering synaptic fidelity through glutamate spillover.

Immune Regulation of Glutamate-Modulating Agents

Findings using other paradigms have cross-validated the association between inflammatory activation and glutamate changes in the brain. Laboratory animals administered bacterial endotoxin develop well-validated, time-bound, self-limiting constellation of depression-like behavioral changes, including lethargy, anorexia, and apathy known as "sickness behavior." Pretreatment with the noncompetitive NMDA antagonist and antidepressant agent ketamine prior to immune stimulation with endotoxin prevented the development of sickness behavior without changing the underlying inflammatory activation [115]. Glutamate dysfunction induced by stress also has been associated with activation of the intracellular inflammasome proteins in glial cells [59, 116]. Glial cell pathology and glutamate gene dysregulation resulting from chronic immune stimulation has been one of the most replicated histopathological findings in postmortem studies using brain specimens from individuals diagnosed with major depression [117–121]. Taken together, these data indicate the intimate association between chronic nonresolving inflammation and glutamate dysfunction in mood disorders.

Immune-Glutamate Dysfunction as a Pathway to Neuroprogression – A Working Model and Hypothesis

Concept of Neuroprogression

Neuroprogression is a term used to describe the progressive neural changes and behavioral impairments associated with repeated relapses of mental illnesses [122]. Neuroprogression has been consistently associated with earlier age of onset, multiple relapses with progressively poorer response, partial response, and lack of remission, increasing number or length of episodes (kindling), longer duration of illness, and treatment resistance, ultimately resulting in exaggerated personality and functional and cognitive decline [54, 123–125]. Neuronal loss and structural changes in the hippocampus, amygdala, orbitofrontal cortex, anterior cingulate cortex, basal ganglia, and pituitary glands have consistently been associated with progressive pathology in mood disorders [124, 126]. The long-term consequences of repeated and uncontrolled mood dysfunction have been of concern with chronically depressed patients, carrying about 2–5 times higher risk of developing dementia in the later years of their life [127]. It has also been stated that each episode of major depression results in a 2–3% drop in memory test performance, and each episode increases the risk of developing dementia by 13% [18, 128, 129].

Synthesis and Working Model

Increased glutamate levels can subsequently lead to all the classic features of neuroprogression, including oxidative stress, mitochondrial dysfunction, toxic alterations in KYN pathway metabolites (decreased KYNA and increased QUIN levels) and aberrant extrasynaptic glutamate signaling. Persistent increases in brain glutamate and inflammatory activity have been reported in all phases of mood disorder pathology, including the phase of symptomatic remission [17, 130–133]. Dysregulation of glutamate by glial cells in the context of overactive inflammatory stimulation forms the basis of our working model and hypothesis.

Early Changes

Given the key role played by glial cells in regulating glutamate and their sensitivity to immune modulation, we propose that glial elements form a key focal and unifying point of integration of this pathophysiology [2, 32]. It is by now well established that psychological, physiological, and cellular stress induces immune activity in the periphery and in the brain with prominent effects on glial function, glutamate regulation, and increased trafficking of nonglial immune cells such as peripheral macrophages into the brain (Fig. 2) [101, 134, 135]. Stressed cells communicate

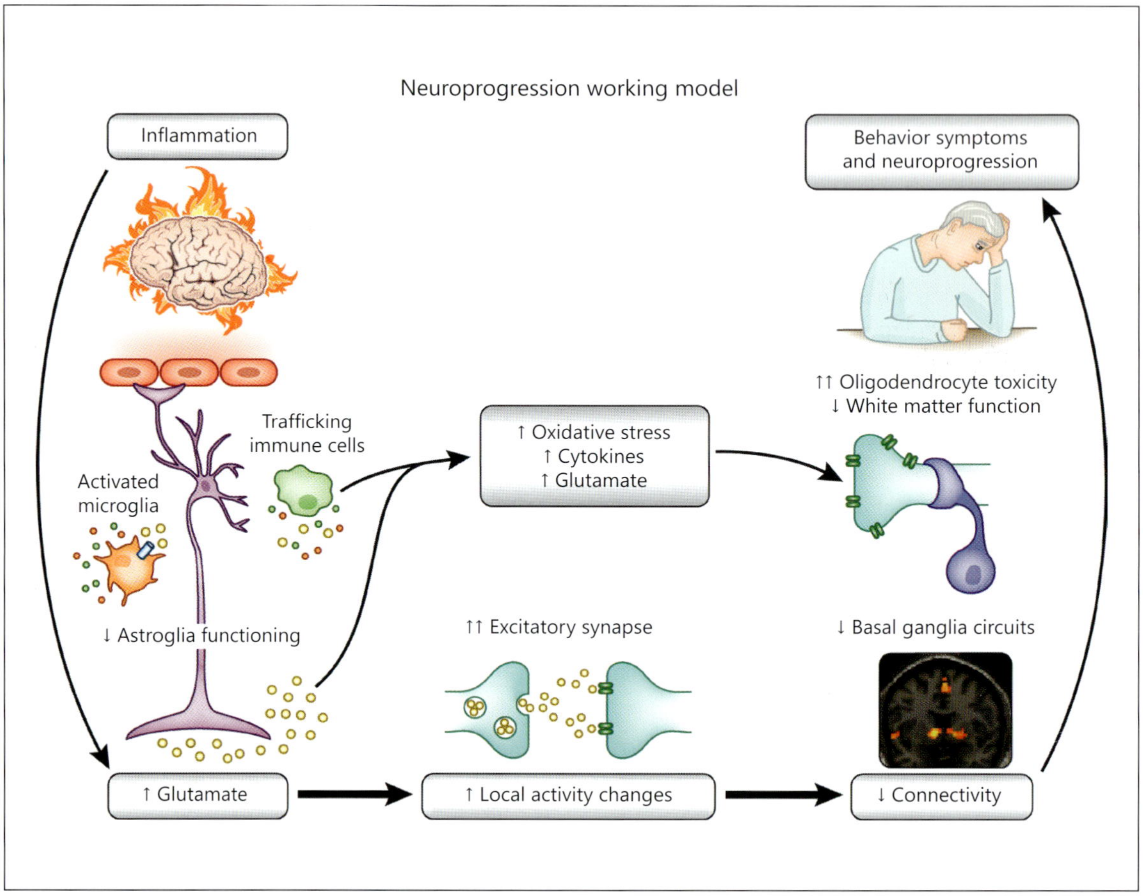

Fig. 2. Neuroprogression in mood disorders – working model and hypothesis: this figure illustrates our working model and hypothesizes that inflammatory pathology mediates transformation of mood disorder pathology into a neuroprogressive state with its behavioral consequences. We propose that the core pathology in neuroprogression revolves around inflammation-induced impairments in glutamate handling and loss of neurotrophic support by glial cells. Promotion of a proinflammatory environment in the brain milieu results from increases in the release of immune molecules by activated microglia and by trafficking immune cells. Inflammatory molecules selectively target functioning of mood- and cognition-regulating neural regions by altering the ability of astro- and oligodendroglial networks to regulate tonic and pulse phasic neural activity at the local level. During acute inflammatory contexts, this adaptive sculpting of neural and glial plasticity by the immune system enables neuroprotection and helps to reorient behavioral priorities to promote healing and recovery. Persisting immune effects on astrocytes/oligodendroglial cells lead to loss of key regulatory functions, oxidative stress, and dysregulation of glutamate clearance, resulting in glutamate spillover into the extrasynaptic space. Release of glutamate by microglia and trafficking blood cells during conditions of inflammatory activation further adds to the overall glutamate binding to extrasynaptic signaling targets. The excess in extrasynaptic glutamate stimulation leads to chaotic, disarrayed neural signaling resulting in neuropsychiatric symptoms and behavioral disorganization. In contrast, sustained loss of intrasynaptic glutamate signaling leads to loss of synaptic fidelity, learning, plasticity, and declining responses to pharmacological and psychological remediation techniques. In addition, chronic loss of intrasynaptic glutamate signaling ("go pathway") in the presence of excessive extrasynaptic signaling ("stop pathway") leads to a pervasive synaptic dysfunction, synaptic loss, and neuroprogression. Finally, loss of oligodendrocytes resulting from oxidative, immune, and glutamate toxicity can lead to myelin loss and white matter dysfunction leading to decreased connectivity and coordination between large-scale brain networks and to treatment resistance and neuroprogression.

with the immune cells in the brain and elsewhere using specialized molecules known as danger-associated molecular pattern molecules [15], e.g., high-mobility group protein HMGB-1, heat shock protein HSP-72, and adenosine triphosphate, which signal and activate the intracellular cytokine-secreting machinery known as the "inflammasome" in glia and other trafficking immune cells, leading to the secretion of cytokines such as IL-1β [136]. Irrespective of the source, inflammatory activation can lead to progressive increases in intra- and extrasynaptic glutamate through the mechanisms described earlier. Defective cradling and containment of glutamate in the intrasynaptic space by immune-activated astrocytic cells lead to increased glutamate spillover into extrasynaptic regions [2]. The glutamate spillover in consonance with glutamate released by activated microglia/macrophages can bind to and overactivate extrasynaptic glutamate receptors, leading to signaling disarray, loss of synaptic fidelity, and ultimately neuropsychiatric symptoms.

Late Changes

Over the longer term, we propose that a different type of pathology might set in. The persistent activation of the extrasynaptic NMDA receptors might lead to depletion of neurotrophic factors such as brain-derived neurotrophic factor, resulting in synaptic death, neuronal loss, and loss of antidepressant efficacy [33]. The high concentrations of extrasynaptic glutamate might also provide feedback inhibition to presynaptic mGluRs, which block presynaptic release of glutamate, leading to decreased intrasynaptic stimulation, desensitization, and atrophic loss of intrasynaptic inotropic receptors [2]. The loss of intrasynaptic ionotropic stimulation in the presence of excessive extrasynaptic inotropic stimulation is believed to lead to further reduction in neurotrophic support and synaptic loss [33]. Furthermore, recent data have demonstrated that intrasynaptic NMDA receptor stimulation and activity is coupled with GSH synthesis to cope with increased antioxidant demand during increased neuronal activity [137]. Activated resident immune cells, such as microglia and trafficking immune cells, release large quantities of glutamate, which further adds to the extrasynaptic load, which, acting in concert with increases in glutamate-like molecules such as QUIN, serve to further accentuate neurotoxicity and neuroprogression [75]. The persistent elevations in glutamate, KYNs, and ROS/RNS also result in toxic death of oligodendrocytes leading to loss of regional brain connectivity [138, 139]. In addition to loss of connectivity, the death of oligodendrocytes further compromises glutamate clearance as these cells are key regulators of extrasynaptic glutamate in the white matter tissue [140]. Taken together, persisting inflammation and chronic glutamate dysregulation might lead to neurotoxicity and neuroprogression through a confluence of pathological influences, including an imbalance in ionotropic physiology (decreased intrasynaptic versus increased extrasynaptic activation), loss of neurotrophic support, dysregulated neurotoxic metabolite release, failing antioxidant buffering, and declining glial plasticity.

Translational and Treatment Implications

Anti-Inflammatory Therapies

A better understanding of the above degenerative cascades can lead to selective targeting of inflammation and/or glutamate systems in the management of chronicity and neuroprogression in mood disorders. Limited data support the use of anti-inflammatory agents to augment and improve the efficacy of currently available antidepressants and mood stabilizers [7]. Administering immune-altering agents to a group of chronic mentally ill is an endeavor best approached with caution, and the caveats and pitfalls have been discussed extensively [7]. Notwithstanding, increasing support from recent publications makes this an attractive avenue to pursue. Minocycline, a tetracycline group antibiotic with a well-documented ability to downregulate microglial activation, has been used as adjunctive therapeutic agent in the treatment of refractory mood and psychotic disorders [141]. Again, the current data on minocycline are limited but promising. Data from studies using nonsteroidal anti-inflammatory agents are also limited, but results of larger-scale studies are awaited [14, 142, 143].

Glutamatergic Antidepressants

The rapid antidepressant actions and longer-term efficacy of the noncompetitive NMDA antagonist ketamine is under extensive investigation, and the reader is referred to expert reviews on this topic elsewhere [144, 145]. However, the utility of immune markers to predict treatment response to ketamine is just beginning to be studied. For example, the antidepressant efficacy of ketamine was predicted by measures of the baseline inflammatory status, including CRP, IL-6, and adipokines [146–149]. A similar approach (stratifying subjects by inflammation status) has yet to be employed to study antidepressant effects of glutamate-modulating agents such as riluzole or memantine – both of which have shown mixed therapeutic benefits in large-scale studies [150–154]. Similarly, attempts to correlate acute antidepressant response to ketamine with brain changes in glutamate have met with mixed success [155]. A major limitation in this regard has been the difficulty in associating in vivo glutamate measurements with antidepressant response to glutamate-modulating agents, and recent studies have attempted to overcome these methodological shortcomings [156]. Ketamine also has multiple off-target effects extending beyond the glutamate system, including effects on the glucose synthase kinase (GSK-3) system, PSD-95, and AMPA receptor trafficking mechanisms – targets originally associated with mood-stabilizing and anticonvulsant medications [157]. Thus, some caution is advised in considering ketamine exclusively as a glutamate-modulating agent [158].

Mood Stabilizers

Both glutamate increases and inflammatory activation have consistently been documented in bipolar disorders [130, 159]. The hypothesis that mood stabilizers and anticonvulsants might impact glutamate has not yet been well studied. In this regard, lithium is known to downregulate both inflammation and glutamate neurotransmission via inhibition of GSK-3 and hence might enable concurrent targeting of both processes [160]. Furthermore, lithium also enhances glial reuptake of glutamate from the synapses and protects neurons from NMDA-mediated calcium influx and excitotoxicity through its effects on GSK-3 and PSD-95 [161, 162]. Other mood stabilizers that impact glutamate neurotransmission include lamotrigine, which is believed to reduce and stabilize presynaptic release of glutamate [162, 163]. Agents such as riluzole that stabilize synaptic glutamate concentrations and glial functions might offer some promise in the future, though current studies of this agent have been disappointing [152].

Concluding Remarks

In this chapter, we have highlighted the key areas of interaction between inflammatory molecules and glutamate neurotransmission, and how chronic activation of these systems can lead to deleterious pathologies, including neuroprogression. We have also provided an overview of the promises and pitfalls of using drugs targeting glutamate and the immune system to address this pathophysiologic cascade. In conclusion, it might be stated that this composite targeting of inflammation, glial dysfunction, and glutamate dysregulation offers promising avenues to understand and treat neuroprogression.

References

1 Berk M, Conus P, Kapczinski F, et al: From neuroprogression to neuroprotection: implications for clinical care. Med J Aust 2010;193:S36–S40.

2 Haroon E, Miller AH, Sanacora G: Inflammation, glutamate, and glia: a trio of trouble in mood disorders. Neuropsychopharmacology 2017;42:193–215.

3 Moussavi S, Chatterji S, Verdes E, Tandon A, Patel V, Ustun B: Depression, chronic diseases, and decrements in health: results from the World Health Surveys. Lancet 2007;370:851–858.

4 National Center for Injury Prevention and Control: Suicide – Facts at a Glance 2015. Atlanta, Centers for Disease Control, 2015.

5 Brundin L, Bryleva EY, Thirtamara Rajamani K: Role of inflammation in suicide: from mechanisms to treatment. Neuropsychopharmacology 2017;42: 271–283.

6 Nemeroff CB: Prevalence and management of treatment-resistant depression. J Clin Psychiatry 2007;68(suppl 8):17–25.

7 Miller AH, Haroon E, Felger JC: Therapeutic implications of brain-immune interactions: treatment in translation. Neuropsychopharmacology 2017;42: 334–359.

8 Haroon E, Raison CL, Miller AH: Psychoneuroimmunology meets neuropsychopharmacology: translational implications of the impact of inflammation on behavior. Neuropsychopharmacology 2012;37:137–162.

9 Dowlati Y, Herrmann N, Swardfager W, et al: A meta-analysis of cytokines in major depression. Biol Psychiatry 2010; 67:446–457.

10 Howren MB, Lamkin DM, Suls J: Associations of depression with C-reactive protein, IL-1, and IL-6: a meta-analysis. Psychosom Med 2009;71: 171–186.

11 Munkholm K, Brauner JV, Kessing LV, Vinberg M: Cytokines in bipolar disorder vs. healthy control subjects: a systematic review and meta-analysis. J Psychiatr Res 2013;47:1119–1133.

12 Valkanova V, Ebmeier KP, Allan CL: CRP, IL-6 and depression: a systematic review and meta-analysis of longitudinal studies. J Affect Disord 2013;150:736–744.

13 Raison CL, Rutherford RE, Woolwine BJ, et al: A randomized controlled trial of the tumor necrosis factor antagonist infliximab for treatment-resistant depression: the role of baseline inflammatory biomarkers. JAMA Psychiatry 2013; 70:31–41.

14 Halaris A, Meresh E, Sharma A, Alvi N, Sinacore J: 90. Inflammation control augments antidepressant response in bipolar depression. Brain Behav Immun 2013;32(suppl):e26.

15 Miller AH, Raison CL: The role of inflammation in depression: from evolutionary imperative to modern treatment target. Nat Rev Immunol 2016;16:22–34.

16 Raison CL: Inflammatory depression: a trifecta of trouble. J Clin Psychiatry 2014;75:663–664.

17 Berk M, Kapczinski F, Andreazza AC, et al: Pathways underlying neuroprogression in bipolar disorder: focus on inflammation, oxidative stress and neurotrophic factors. Neurosci Biobehav Rev 2011;35:804–817.

18 Moylan S, Maes M, Wray NR, Berk M: The neuroprogressive nature of major depressive disorder: pathways to disease evolution and resistance, and therapeutic implications. Mol Psychiatry 2013; 18:595–606.

19 Duman RS, Aghajanian GK, Sanacora G, Krystal JH: Synaptic plasticity and depression: new insights from stress and rapid-acting antidepressants. Nat Med 2016;22:238–249.

20 Popoli M, Yan Z, McEwen BS, Sanacora G: The stressed synapse: the impact of stress and glucocorticoids on glutamate transmission. Nat Rev Neurosci 2011;13: 22–37.

21 Braitenberg V, Schüz A: Cortex: Statistics and Geometry of Neuronal Connectivity. Berlin, Springer, 2013.

22 Hall CN, Howarth C, Kurth-Nelson Z, Mishra A: Interpreting BOLD: towards a dialogue between cognitive and cellular neuroscience. Philos Trans R Soc Lond B Biol Sci 2016;371: 20150348.

23 Chesselet MF: Presynaptic regulation of neurotransmitter release in the brain: facts and hypothesis. Neuroscience 1984;12:347–375.

24 Nedergaard M, Takano T, Hansen AJ: Beyond the role of glutamate as a neurotransmitter. Nat Rev Neurosci 2002;3: 748–755.

25 Ajilore O, Haroon E, Kumaran S, et al: Measurement of brain metabolites in patients with type 2 diabetes and major depression using proton magnetic resonance spectroscopy. Neuropsychopharmacology 2007;32:1224–1231.

26 Nemeth CL, Haroon E, Neigh GN: Heartsick: psychiatric and inflammatory implications of cerebromicrovascular disease. Int J Geriatr Psychiatry 2014;29: 577–585.

27 Zhou Y, Danbolt NC: Glutamate as a neurotransmitter in the healthy brain. J Neural Transm (Vienna) 2014;121: 799–817.

28 Danbolt NC: Glutamate uptake. Prog Neurobiol 2001;65:1–105.

29 Danbolt NC, Chaudhry FA, Dehnes Y, et al: Properties and localization of glutamate transporters. Prog Brain Res 1998; 116:23–43.

30 McCullumsmith RE, Sanacora G: Regulation of extrasynaptic glutamate levels as a pathophysiological mechanism in disorders of motivation and addiction. Neuropsychopharmacology 2015;40: 254–255.

31 Niciu MJ, Kelmendi B, Sanacora G: Overview of glutamatergic neurotransmission in the nervous system. Pharmacol Biochem Behav 2012;100:656–664.

32 Haroon E, Miller AH: Inflammation effects on brain glutamate in depression: mechanistic considerations and treatment implications. Curr Top Behav Neurosci 2017;31:173–198.

33 Hardingham GE, Bading H: Synaptic versus extrasynaptic NMDA receptor signalling: implications for neurodegenerative disorders. Nat Rev Neurosci 2010;11:682–696.

34 Okubo Y, Iino M: Visualization of glutamate as a volume transmitter. J Physiol 2011;589:481–488.

35 Sykova E: Extrasynaptic volume transmission and diffusion parameters of the extracellular space. Neuroscience 2004; 129:861–876.

36 Agnati LF, Guidolin D, Guescini M, Genedani S, Fuxe K: Understanding wiring and volume transmission. Brain Res Rev 2010;64:137–159.

37 Garcia-Munoz M, Lopez-Huerta VG, Carrillo-Reid L, Arbuthnott GW: Extrasynaptic glutamate NMDA receptors: key players in striatal function. Neuropharmacology 2015;89:54–63.
38 Haydon PG, Carmignoto G: Astrocyte control of synaptic transmission and neurovascular coupling. Physiol Rev 2006;86:1009–1031.
39 Haydon PG, Nedergaard M: How do astrocytes participate in neural plasticity? Cold Spring Harb Perspect Biol 2014;7:a020438.
40 Duman RS, Aghajanian GK: Synaptic dysfunction in depression: potential therapeutic targets. Science 2012;338: 68–72.
41 Duman RS: Pathophysiology of depression and innovative treatments: remodeling glutamatergic synaptic connections. Dialogues Clin Neurosci 2014;16: 11–27.
42 Nagy C, Suderman M, Yang J, et al: Astrocytic abnormalities and global DNA methylation patterns in depression and suicide. Mol Psychiatry 2015;20:320–328.
43 Tzingounis AV, Wadiche JI: Glutamate transporters: confining runaway excitation by shaping synaptic transmission. Nat Rev Neurosci 2007;8: 935–947.
44 Dorsett CR, McGuire JL, DePasquale EA, Gardner AE, Floyd CL, McCullumsmith RE: Glutamate neurotransmission in rodent models of traumatic brain injury. J Neurotrauma 2017;34:263–272.
45 Ida T, Hara M, Nakamura Y, Kozaki S, Tsunoda S, Ihara H: Cytokine-induced enhancement of calcium-dependent glutamate release from astrocytes mediated by nitric oxide. Neurosci Lett 2008; 432:232–236.
46 Malarkey EB, Parpura V: Mechanisms of glutamate release from astrocytes. Neurochem Int 2008;52:142–154.
47 Tilleux S, Hermans E: Neuroinflammation and regulation of glial glutamate uptake in neurological disorders. J Neurosci Res 2007;85:2059–2070.
48 Zou JY, Crews FT: TNFα potentiates glutamate neurotoxicity by inhibiting glutamate uptake in organotypic brain slice cultures: neuroprotection by NFκB inhibition. Brain Res 2005;1034: 11–24.
49 Verkhratsky A, Nedergaard M, Hertz L: Why are astrocytes important? Neurochem Res 2015;40:389–401.
50 Takeuchi H, Suzumura A: Gap junctions and hemichannels composed of connexins: potential therapeutic targets for neurodegenerative diseases. Front Cell Neurosci 2014;8:189.
51 Sun JD, Liu Y, Yuan YH, Li J, Chen NH: Gap junction dysfunction in the prefrontal cortex induces depressive-like behaviors in rats. Neuropsychopharmacology 2012;37:1305–1320.
52 Miguel-Hidalgo JJ, Wilson BA, Hussain S, Meshram A, Rajkowska G, Stockmeier CA: Reduced connexin 43 immunolabeling in the orbitofrontal cortex in alcohol dependence and depression. J Psychiatr Res 2014;55:101–109.
53 Maes M, Fisar Z, Medina M, Scapagnini G, Nowak G, Berk M: New drug targets in depression: inflammatory, cell-mediated immune, oxidative and nitrosative stress, mitochondrial, antioxidant, and neuroprogressive pathways. And new drug candidates – Nrf2 activators and GSK-3 inhibitors. Inflammopharmacology 2012;20:127–150.
54 Bakunina N, Pariante CM, Zunszain PA: Immune mechanisms linked to depression via oxidative stress and neuroprogression. Immunology DOI: 10.1111/imm.12443, Epub ahead of print.
55 Chong ZZ, Li F, Maiese K: Oxidative stress in the brain: novel cellular targets that govern survival during neurodegenerative disease. Prog Neurobiol 2005;75: 207–246.
56 Massie A, Boillee S, Hewett S, Knackstedt L, Lewerenz J: Main path and byways: non-vesicular glutamate release by system x_c^- as an important modifier of glutamatergic neurotransmission. J Neurochem 2015;135:1062–1079.
57 Lewerenz J, Hewett SJ, Huang Y, et al: The cystine/glutamate antiporter system x_c^- in health and disease: from molecular mechanisms to novel therapeutic opportunities. Antioxid Redox Signal 2013;18:522–555.
58 Jensen CJ, Massie A, De Keyser J: Immune players in the CNS: the astrocyte. J Neuroimmune Pharmacol 2013;8:824–839.
59 Iwata M, Ota KT, Li XY, et al: Psychological stress activates the inflammasome via release of adenosine triphosphate and stimulation of the purinergic type 2X7 receptor. Biol Psychiatry 2016; 80:12–22.
60 Bezzi P, Domercq M, Brambilla L, et al: CXCR4-activated astrocyte glutamate release via TNFα: amplification by microglia triggers neurotoxicity. Nat Neurosci 2001;4:702–710.
61 Persson M, Brantefjord M, Hansson E, Ronnback L: Lipopolysaccharide increases microglial GLT-1 expression and glutamate uptake capacity in vitro by a mechanism dependent on TNF-α. Glia 2005;51:111–120.
62 Piani D, Fontana A: Involvement of the cystine transport system x_c^- in the macrophage-induced glutamate-dependent cytotoxicity to neurons. J Immunol 1994;152:3578–3585.
63 Dantzer R, Walker AK: Is there a role for glutamate-mediated excitotoxicity in inflammation-induced depression? J Neural Transm (Vienna) 2014;121:925–932.
64 Takeuchi H, Jin S, Wang J, et al: Tumor necrosis factor-α induces neurotoxicity via glutamate release from hemichannels of activated microglia in an autocrine manner. J Biol Chem 2006;281: 21362–21368.
65 Riazi K, Galic MA, Kentner AC, Reid AY, Sharkey KA, Pittman QJ: Microglia-dependent alteration of glutamatergic synaptic transmission and plasticity in the hippocampus during peripheral inflammation. J Neurosci 2015;35:4942–4952.
66 Schafer DP, Stevens B: Microglia function in central nervous system development and plasticity. Cold Spring Harb Perspect Biol 2015;7:a020545.
67 Takeuchi H, Jin S, Suzuki H, et al: Blockade of microglial glutamate release protects against ischemic brain injury. Exp Neurol 2008;214:144–146.
68 Brown GC, Vilalta A: How microglia kill neurons. Brain Res 2015;1628:288–297.
69 Schwarcz R, Bruno JP, Muchowski PJ, Wu HQ: Kynurenines in the mammalian brain: when physiology meets pathology. Nat Rev Neurosci 2012;13:465–477.
70 Dantzer R, O'Connor JC, Lawson MA, Kelley KW: Inflammation-associated depression: from serotonin to kynurenine. Psychoneuroendocrinology 2011;36: 426–436.
71 Guillemin GJ, Smythe G, Takikawa O, Brew BJ: Expression of indoleamine 2,3-dioxygenase and production of quinolinic acid by human microglia, astrocytes, and neurons. Glia 2005;49: 15–23.

72 Halaris A: Inflammation, heart disease, and depression. Curr Psychiatry Rep 2013;15:400.
73 Savitz J: Role of kynurenine metabolism pathway activation in major depressive disorders. Curr Top Behav Neurosci 2017;31:249–267.
74 Stone TW, Forrest CM, Darlington LG: Kynurenine pathway inhibition as a therapeutic strategy for neuroprotection. FEBS J 2012;279:1386–1397.
75 Guillemin GJ: Quinolinic acid, the inescapable neurotoxin. FEBS J 2012;279: 1356–1365.
76 Erhardt S, Lim CK, Linderholm KR, et al: Connecting inflammation with glutamate agonism in suicidality. Neuropsychopharmacology 2013;38:743–752.
77 Steiner J, Walter M, Gos T, et al: Severe depression is associated with increased microglial quinolinic acid in subregions of the anterior cingulate gyrus: evidence for an immune-modulated glutamatergic neurotransmission? J Neuroinflammation 2011;8:94.
78 Raison CL, Dantzer R, Kelley KW, et al: CSF concentrations of brain tryptophan and kynurenines during immune stimulation with IFN-α: relationship to CNS immune responses and depression. Mol Psychiatry 2010;15:393–403.
79 Savitz J, Drevets WC, Smith CM, et al: Putative neuroprotective and neurotoxic kynurenine pathway metabolites are associated with hippocampal and amygdalar volumes in subjects with major depressive disorder. Neuropsychopharmacology 2015;40:463–471.
80 Schwarcz R: Kynurenines and glutamate: multiple links and therapeutic implications. Adv Pharmacol 2016;76: 13–37.
81 Capuron L, Neurauter G, Musselman DL, et al: Interferon-α-induced changes in tryptophan metabolism. Relationship to depression and paroxetine treatment. Biol Psychiatry 2003;54:906–914.
82 Raison CL, Borisov AS, Woolwine BJ, Massung B, Vogt G, Miller AH: Interferon-α effects on diurnal hypothalamic-pituitary-adrenal axis activity: relationship with proinflammatory cytokines and behavior. Mol Psychiatry 2010;15:535–547.
83 Raison CL, Broadwell SD, Borisov AS, et al: Depressive symptoms and viral clearance in patients receiving interferon-α and ribavirin for hepatitis C. Brain Behav Immun 2005;19:23–27.
84 Capuron L, Gumnick JF, Musselman DL, et al: Neurobehavioral effects of interferon-α in cancer patients: phenomenology and paroxetine responsiveness of symptom dimensions. Neuropsychopharmacology 2002;26:643–652.
85 Musselman DL, Lawson DH, Gumnick JF, et al: Paroxetine for the prevention of depression induced by high-dose interferon alfa. N Engl J Med 2001;344:961–966.
86 Felger JC, Alagbe O, Hu F, et al: Effects of interferon-α on rhesus monkeys: a nonhuman primate model of cytokine-induced depression. Biol Psychiatry 2007;62:1324–1333.
87 Raison CL, Borisov AS, Majer M, et al: Activation of central nervous system inflammatory pathways by interferon-α: relationship to monoamines and depression. Biol Psychiatry 2009;65:296–303.
88 Juengling FD, Ebert D, Gut O, et al: Prefrontal cortical hypometabolism during low-dose interferon α treatment. Psychopharmacology (Berl) 2000;152:383–389.
89 Capuron L, Pagnoni G, Demetrashvili MF, et al: Basal ganglia hypermetabolism and symptoms of fatigue during interferon-α therapy. Neuropsychopharmacology 2007;32:2384–2392.
90 Capuron L, Pagnoni G, Demetrashvili M, et al: Anterior cingulate activation and error processing during interferon-α treatment. Biol Psychiatry 2005;58:190–196.
91 Felger JC, Mun J, Kimmel HL, et al: Chronic interferon-α decreases dopamine 2 receptor binding and striatal dopamine release in association with anhedonia-like behavior in nonhuman primates. Neuropsychopharmacology 2013;38:2179–2187.
92 Felger JC, Miller AH: Cytokine effects on the basal ganglia and dopamine function: the subcortical source of inflammatory malaise. Front Neuroendocrinol 2012;33:315–327.
93 Felger JC, Li L, Marvar PJ, et al: Tyrosine metabolism during interferon-α administration: association with fatigue and CSF dopamine concentrations. Brain Behav Immun 2013;31:153–160.
94 Capuron L, Pagnoni G, Drake DF, et al: Dopaminergic mechanisms of reduced basal ganglia responses to hedonic reward during interferon alfa administration. Arch Gen Psychiatry 2012;69: 1044–1053.
95 Haroon E, Woolwine BJ, Chen X, et al: IFN-α-induced cortical and subcortical glutamate changes assessed by magnetic resonance spectroscopy. Neuropsychopharmacology 2014;39:1777–1785.
96 Haroon E, Felger JC, Woolwine BJ, et al: Age-related increases in basal ganglia glutamate are associated with TNF, reduced motivation and decreased psychomotor speed during IFN-α treatment: preliminary findings. Brain Behav Immun 2015;46:17–22.
97 Norden DM, Muccigrosso MM, Godbout JP: Microglial priming and enhanced reactivity to secondary insult in aging, and traumatic CNS injury, and neurodegenerative disease. Neuropharmacology 2015;96:29–41.
98 Norden DM, Godbout JP: Review: microglia of the aged brain: primed to be activated and resistant to regulation. Neuropathol Appl Neurobiol 2013;39: 19–34.
99 Dilger RN, Johnson RW: Aging, microglial cell priming, and the discordant central inflammatory response to signals from the peripheral immune system. J Leukoc Biol 2008;84:932–939.
100 Godbout JP, Johnson RW: Age and neuroinflammation: a lifetime of psychoneuroimmune consequences. Immunol Allergy Clin North Am 2009;29: 321–337.
101 Niraula A, Sheridan JF, Godbout JP: Microglia priming with aging and stress. Neuropsychopharmacology 2017;42:318–333.
102 Torres-Platas SG, Cruceanu C, Chen GG, Turecki G, Mechawar N: Evidence for increased microglial priming and macrophage recruitment in the dorsal anterior cingulate white matter of depressed suicides. Brain Behav Immun 2014;42:50–59.
103 Haroon E, Fleischer CC, Felger JC, et al: Conceptual convergence: increased inflammation is associated with increased basal ganglia glutamate in patients with major depression. Mol Psychiatry 2016;21:1351–1357.
104 Ridker PM: Clinical application of C-reactive protein for cardiovascular disease detection and prevention. Circulation 2003;107:363–369.

105 Ridker PM: From C-reactive protein to interleukin-6 to interleukin-1: moving upstream to identify novel targets for atheroprotection. Circ Res 2016;118: 145–156.

106 Fox MD, Buckner RL, Liu H, Chakravarty MM, Lozano AM, Pascual-Leone A: Resting-state networks link invasive and noninvasive brain stimulation across diverse psychiatric and neurological diseases. Proc Natl Acad Sci USA 2014;111:E4367–E4375.

107 Greicius MD, Flores BH, Menon V, et al: Resting-state functional connectivity in major depression: abnormally increased contributions from subgenual cingulate cortex and thalamus. Biol Psychiatry 2007;62:429–437.

108 Iwabuchi SJ, Krishnadas R, Li C, Auer DP, Radua J, Palaniyappan L: Localized connectivity in depression: a meta-analysis of resting state functional imaging studies. Neurosci Biobehav Rev 2015;51:77–86.

109 Kaiser RH, Andrews-Hanna JR, Wager TD, Pizzagalli DA: Large-scale network dysfunction in major depressive disorder: a meta-analysis of resting-state functional connectivity. JAMA Psychiatry 2015;72:603–611.

110 Dombrovski AY, Szanto K, Clark L, et al: Corticostriatothalamic reward prediction error signals and executive control in late-life depression. Psychol Med 2015;45:1413–1424.

111 Taylor WD, Aizenstein HJ, Alexopoulos GS: The vascular depression hypothesis: mechanisms linking vascular disease with depression. Mol Psychiatry 2013;18:963–974.

112 Miller AH, Haroon E, Raison CL, Felger JC: Cytokine targets in the brain: impact on neurotransmitters and neurocircuits. Depress Anxiety 2013;30: 297–306.

113 Felger JC, Li Z, Haroon E, et al: Inflammation is associated with decreased functional connectivity within corticostriatal reward circuitry in depression. Mol Psychiatry 2016;21:1358–1365.

114 Felger JC, Treadway MT: inflammation effects on motivation and motor activity: role of dopamine. Neuropsychopharmacology 2017;42:216–241.

115 Walker AK, Budac DP, Bisulco S, et al: NMDA receptor blockade by ketamine abrogates lipopolysaccharide-induced depressive-like behavior in C57BL/6J mice. Neuropsychopharmacology 2013;38:1609–1616.

116 Iwata M, Ota KT, Duman RS: The inflammasome: pathways linking psychological stress, depression, and systemic illnesses. Brain Behav Immun 2013;31:105–114.

117 Rajkowska G, Stockmeier CA: Astrocyte pathology in major depressive disorder: insights from human postmortem brain tissue. Curr Drug Targets 2013;14:1225–1236.

118 Ongur D, Bechtholt AJ, Carlezon WA Jr, Cohen BM: Glial abnormalities in mood disorders. Harv Rev Psychiatry 2014;22:334–337.

119 Mechawar N, Savitz J: Neuropathology of mood disorders: do we see the stigmata of inflammation? Transl Psychiatry 2016;6:e946.

120 Banasr M, Duman RS: Glial loss in the prefrontal cortex is sufficient to induce depressive-like behaviors. Biol Psychiatry 2008;64:863–870.

121 Bernard R, Kerman IA, Thompson RC, et al: Altered expression of glutamate signaling, growth factor, and glia genes in the locus coeruleus of patients with major depression. Mol Psychiatry 2011;16:634–646.

122 Berk M: Neuroprogression: pathways to progressive brain changes in bipolar disorder. Int J Neuropsychopharmacol 2009;12:441–445.

123 Post RM, Speer AM, Hough CJ, Xing G: Neurobiology of bipolar illness: implications for future study and therapeutics. Ann Clin Psychiatry 2003;15: 85–94.

124 Post RM, Fleming J, Kapczinski F: Neurobiological correlates of illness progression in the recurrent affective disorders. J Psychiatr Res 2012;46: 561–573.

125 Gama CS, Kunz M, Magalhaes PV, Kapczinski F: Staging and neuroprogression in bipolar disorder: a systematic review of the literature. Rev Bras Psiquiatr 2013;35:70–74.

126 Price JL, Drevets WC: Neurocircuitry of mood disorders. Neuropsychopharmacology 2010;35:192–216.

127 Byers AL, Yaffe K: Depression and risk of developing dementia. Nat Rev Neurol 2011;7:323–331.

128 Diniz BS, Butters MA, Albert SM, Dew MA, Reynolds CF 3rd: Late-life depression and risk of vascular dementia and Alzheimer's disease: systematic review and meta-analysis of community-based cohort studies. Br J Psychiatry 2013;202:329–335.

129 da Costa SC, Passos IC, Lowri C, Soares JC, Kapczinski F: Refractory bipolar disorder and neuroprogression. Prog Neuropsychopharmacol Biol Psychiatry 2016;70:103–110.

130 Yuksel C, Ongur D: Magnetic resonance spectroscopy studies of glutamate-related abnormalities in mood disorders. Biol Psychiatry 2010;68: 785–794.

131 Bhagwagar Z, Wylezinska M, Jezzard P, et al: Reduction in occipital cortex γ-aminobutyric acid concentrations in medication-free recovered unipolar depressed and bipolar subjects. Biol Psychiatry 2007;61:806–812.

132 Bhattacharya A, Derecki NC, Lovenberg TW, Drevets WC: Role of neuro-immunological factors in the pathophysiology of mood disorders. Psychopharmacology (Berl) 2016;233: 1623–1636.

133 Reus GZ, Fries GR, Stertz L, et al: The role of inflammation and microglial activation in the pathophysiology of psychiatric disorders. Neuroscience 2015;300:141–154.

134 Wohleb ES, Powell ND, Godbout JP, Sheridan JF: Stress-induced recruitment of bone marrow-derived monocytes to the brain promotes anxiety-like behavior. J Neurosci 2013;33: 13820–13833.

135 Wohleb ES, McKim DB, Sheridan JF, Godbout JP: Monocyte trafficking to the brain with stress and inflammation: a novel axis of immune-to-brain communication that influences mood and behavior. Front Neurosci 2014;8: 447.

136 Fleshner M, Frank M, Maier SF: Danger signals and inflammasomes: stress-evoked sterile inflammation in mood disorders. Neuropsychopharmacology 2017;42:36–45.

137 Baxter PS, Bell KF, Hasel P, et al: Synaptic NMDA receptor activity is coupled to the transcriptional control of the glutathione system. Nat Commun 2015;6:6761.

138 Matute C: Oligodendrocyte NMDA receptors: a novel therapeutic target. Trends Mol Med 2006;12:289–292.

139 Matute C: Interaction between glutamate signalling and immune attack in damaging oligodendrocytes. Neuron Glia Biol 2007;3:281–285.
140 Matute C, Domercq M, Sanchez-Gomez MV: Glutamate-mediated glial injury: mechanisms and clinical importance. Glia 2006;53:212–224.
141 Miller BJ, Goldsmith DR: Towards an immunophenotype of schizophrenia: progress, potential mechanisms, and future directions. Neuropsychopharmacology 2017;42:299–317.
142 Muller N, Schwarz MJ, Dehning S, et al: The cyclooxygenase-2 inhibitor celecoxib has therapeutic effects in major depression: results of a double-blind, randomized, placebo controlled, add-on pilot study to reboxetine. Mol Psychiatry 2006;11:680–684.
143 Na KS, Lee KJ, Lee JS, Cho YS, Jung HY: Efficacy of adjunctive celecoxib treatment for patients with major depressive disorder: a meta-analysis. Prog Neuropsychopharmacol Biol Psychiatry 2014;48:79–85.
144 Newport DJ, Carpenter LL, McDonald WM, et al: Ketamine and other NMDA antagonists: early clinical trials and possible mechanisms in depression. Am J Psychiatry 2015;172:950–966.
145 Zarate CA Jr, Singh JB, Carlson PJ, et al: A randomized trial of an N-methyl-D-aspartate antagonist in treatment-resistant major depression. Arch Gen Psychiatry 2006;63:856–864.
146 Machado-Vieira R, Gold PW, Luckenbaugh DA, et al: The role of adipokines in the rapid antidepressant effects of ketamine. Mol Psychiatry 2017;22:127–133.
147 Hashimoto K: Inflammatory biomarkers as differential predictors of antidepressant response. Int J Mol Sci 2015;16:7796–7801.
148 Walker AJ, Kim Y, Price JB, et al: Stress, inflammation, and cellular vulnerability during early stages of affective disorders: biomarker strategies and opportunities for prevention and intervention. Front Psychiatry 2014;5:34.
149 Walker AJ, Foley BM, Sutor SL, McGillivray JA, Frye MA, Tye SJ: Peripheral proinflammatory markers associated with ketamine response in a preclinical model of antidepressant-resistance. Behav Brain Res 2015;293:198–202.
150 Mathew S, Fava M, Guerguieva R, Sanacora G: A randomized placebo-controlled adjunctive trial of riluzole in treatment-resistant major depressive disorder – NCT01204918 – efficacy and tolerability of riluzole in treatment resistant depression. ASCP Annual Meeting, Miami, 2015.
151 Niciu MJ, Luckenbaugh DA, Ionescu DF, et al: Riluzole likely lacks antidepressant efficacy in ketamine non-responders. J Psychiatr Res 2014;58:197–199.
152 Pittenger C, Coric V, Banasr M, Bloch M, Krystal JH, Sanacora G: Riluzole in the treatment of mood and anxiety disorders. CNS Drugs 2008;22:761–786.
153 Sanacora G, Zarate CA, Krystal JH, Manji HK: Targeting the glutamatergic system to develop novel, improved therapeutics for mood disorders. Nat Rev Drug Discov 2008;7:426–437.
154 Salardini E, Zeinoddini A, Mohammadinejad P, et al: Riluzole combination therapy for moderate-to-severe major depressive disorder: a randomized, double-blind, placebo-controlled trial. J Psychiatr Res 2016;75:24–30.
155 DeWilde KE, Levitch CF, Murrough JW, Mathew SJ, Iosifescu DV: The promise of ketamine for treatment-resistant depression: current evidence and future directions. Ann NY Acad Sci 2015;1345:47–58.
156 Milak MS, Proper CJ, Mulhern ST, et al: A pilot in vivo proton magnetic resonance spectroscopy study of amino acid neurotransmitter response to ketamine treatment of major depressive disorder. Mol Psychiatry 2016;21:320–327.
157 Beurel E, Grieco SF, Amadei C, Downey K, Jope RS: Ketamine-induced inhibition of glycogen synthase kinase-3 contributes to the augmentation of α-amino-3-hydroxy-5-methylisoxazole-4-propionic acid (AMPA) receptor signaling. Bipolar Disord 2016;18:473–480.
158 Zanos P, Moaddel R, Morris PJ, et al: NMDAR inhibition-independent antidepressant actions of ketamine metabolites. Nature 2016;533:481–486.
159 Gigante AD, Bond DJ, Lafer B, Lam RW, Young LT, Yatham LN: Brain glutamate levels measured by magnetic resonance spectroscopy in patients with bipolar disorder: a meta-analysis. Bipolar Disord 2012;14:478–487.
160 Jope RS: Glycogen synthase kinase-3 in the etiology and treatment of mood disorders. Front Mol Neurosci 2011;4:16.
161 Machado-Vieira R, Manji HK, Zarate CA Jr: The role of lithium in the treatment of bipolar disorder: convergent evidence for neurotrophic effects as a unifying hypothesis. Bipolar Disord 2009;11(suppl 2):92–109.
162 Krystal JH, Sanacora G, Blumberg H, et al: Glutamate and GABA systems as targets for novel antidepressant and mood-stabilizing treatments. Mol Psychiatry 2002;7(suppl 1):S71–S80.
163 Frye MA, Watzl J, Banakar S, et al: Increased anterior cingulate/medial prefrontal cortical glutamate and creatine in bipolar depression. Neuropsychopharmacology 2007;32:2490–2499.

Ebrahim Haroon, MD
Emory Behavioral Immunology Program
Department of Psychiatry and Behavioral Sciences
Winship Cancer Institute
Emory University School of Medicine
101 Woodruff Circle, WMRB Suite #6003
Atlanta, GA 30322 (USA)
E-Mail eharoon@emory.edu

Halaris A, Leonard BE (eds): Neuroprogression in Psychiatric Disorders.
Mod Trends Pharmacopsychiatry. Basel, Karger, 2017, vol 31, pp 56–66 (DOI: 10.1159/000470807)

Major Depression as a Neuroprogressive Prelude to Dementia: What Is the Evidence?

Brian E. Leonard

National University of Ireland, Galway, Ireland

Abstract

Epidemiological studies implicate chronic depression as a predisposing factor for dementia in later life. However, the link is incompletely understood and controversial. The aim of this review is to consider some of the biological factors that contribute to neuroprogressive brain dysfunction in late life as a consequence of prolonged, low-grade inflammation in the course of depressive episodes. As chronic inflammation is known to precipitate increased apoptosis of neurons and astrocytes, this could be a contributing factor to brain dysfunction. In addition, certain proinflammatory cytokines activate the neurotoxic derivatives of the tryptophan-kynurenine pathway. This results in the synthesis of the NMDA glutamate agonist, quinolinic acid, and kynurenine metabolites which initiate oxidative stress and insulin receptor resistance. As a consequence of these changes, combined with a structural and functional defect in brain mitochondria, glucose transport into the brain is affected. Due to the ensuing reduction in the metabolic energy needed to sustain brain function, brain cells die prematurely. These changes could provide a link between chronic inflammation and dementia, at least in some patients with recurrent and chronic depression. This outcome may be particularly true in poor responders and treatment-resistant depression.

Introduction

Epidemiological studies indicate that chronic depression is an important predisposing factor for dementia in late life. Depression has been shown to be a frequent antecedent of late-onset Alzheimer disease, and there is now evidence that depression may be an early manifestation of dementia before the cognitive symptoms become apparent [1, 2]. In support of these observations, patients with depression who later develop dementia often have a poorer baseline performance in cognitive tasks [3]. Further, it has been estimated that mild cognitive impairment in depression increases the risk of developing dementia at least twofold [3]. This suggests that depression might be a prodrome for dementia [4].

However, the link between major depression and dementia is complex. For example, depressive symptoms are prominently associated with current Alzheimer disease [5], and those with a premorbid history of major depression are more likely to develop Alzheimer disease [6], which opens up the possibility that depression may be a predictor or a risk factor for Alzheimer disease. Thus, the current consensus is that major depression in earlier life may be a risk factor for Alzheimer disease, the odds ratio being 1.5–2.0 [7]. Crews and Masliah [8] have summarized the possible molecular mechanisms of neurodegeneration in Alzheimer disease, many of which overlap with those reported to occur in major depression.

Depression and Alzheimer disease are associated with neuroinflammatory changes in the brain [4]. Clinical studies have demonstrated that proinflammatory cytokines are elevated in the blood of many depressed patients [9, 10] while the increase in oxidative and nitrosative stress pathways by inflammation plays a significant part in the consequent neurodegeneration [11]. The diagnostic features of Alzheimer disease include the accumulation of hyperphosphorylated tau protein, which destroys the microfilamentous structure of neurons, and extracellular β-amyloid plaques [12]. The deposition of β-amyloid in the brain activates microglia which release proinflammatory cytokines and chemokines [13]. Thus, current evidence suggests that brain inflammation is a pathological feature of Alzheimer disease that results from the interaction between activated microglia, neurons, and β-amyloid plaques [14, 15].

In addition, elevated concentrations of cortisol in both depression and Alzheimer disease not only cause direct damage to neurons but also potentiate the neurotoxic effects of β-amyloid [16, 17]. Such effects contribute to the impairment of cognitive function in both disorders [18]. Furthermore, as there is evidence that neuronal apoptosis is increased in both disorders, resulting in structural changes in the hippocampus, frontal cortex, amygdala, and in other brain regions [16, 17, 19], it is concluded that inflammatory mediators play a central role in the pathology of these conditions.

The purpose of this review is to consider the primarily clinical evidence to support the hypothesis that neuroinflammation in major depression is an important neuroprogressive risk factor for late-onset Alzheimer disease.

The Importance of the Immune System and Inflammation in Major Depression and Alzheimer Disease

Neuroinflammation in chronic depression and Alzheimer disease, which is mediated by proinflammatory cytokines, occurs as a result of the interaction between stress and/or β-amyloid, activated microglia and astrocytes [12]. In addition to the proinflammatory cytokines, such as interleukin (IL)-6, tumor necrosis factor (TNF)-α, and interferon (IFN)-γ, activated neuroglia also produce chemokines, prostaglandins, leukotrienes, thromboxanes, nitic oxide, and oxygen free radicals [20].

Microglia are the principal cells in the brain which are involved in innate immunity and are mediators of neuroinflammation. The structure of microglia changes according to their functional role. Thus, in the healthy normal brain, they are round with numerous branching processes. This form supports and partially protects neuronal growth and synaptic remodeling in the brain and spinal cord. In addition, in this state (termed the M2 phenotype), microglia exhibit a scavenger expression and by phagocytosis remove cell debris, release neurotrophic factors, and thereby preserve neuronal function [21]. When microglia are activated by a bacterial or viral infection, traumatic injury, ischemia, or, in the case of depression, by stress, they are transformed into an amyloid, neurotoxic state which facilitates their movement, termed the M1 phenotype.

This M1 phenotype releases predominantly proinflammatory cytokines such as IL-1, IL-6, TNF-α and IFN-γ in addition to free radicals and nitric oxide. In the brain, this neurotoxic combination can damage neurons and astrocytes, but it is also a mechanism whereby invading microorganisms are destroyed [22–24]. It is suggested that dysfunctional microglia-neuronal cross talk may be the primary cause of stress-induced microglial activation [25]. In major depression, the astrocytes are vulnerable to the neurotoxic effects of quinolinic acid and are, therefore, unlikely to contribute to the inflammatory changes in the brain. This will be discussed further in this review.

In Alzheimer disease, microglial activation is linked to the presence of β-amyloid which enhances the inflammatory response by activating nuclear factor-κB and consequently the synthesis of proinflammatory cytokines [26]. In major depression, stress initiates the mitogen-activated protein kinase pathway leading to the synthesis of proinflammatory cytokines and chemokines [27].

In addition to microglia, astrocytes also play a crucial role in neuroprotection. An important function of astrocytes involves the regulation of synaptogenesis and angiogenesis, processes that promote neuronal survival and maturation [28]. However, there is evidence that in Alzheimer disease astrocytes are also activated by β-amyloid to produce proinflammatory cytokines and chemokines thereby contributing to brain inflammation [29]. This contrasts with the evidence from animal models of Alzheimer disease where astrocytes undergo degeneration and atrophy at early stages of the pathological process, which is thought to contribute to the early cognitive deficits in these models. In later stages, degenerated astrocytes are associated with neuritic plaques in animal models.

In Alzheimer disease, these inflammatory mediators enhance the synthesis of β-amyloid and thereby increase the amyloidogenic cascade, which is considered to be the pathological hallmark of this disease [20]. As there is evidence that amyloid plaques also accumulate in geriatric patients with chronic depression [30, 31], it is reasonable to postulate that Alzheimer disease could be a possible outcome in these patients [32].

An important event which results in reductions in hippocampal volume and other brain regions in major depression involves the loss of synapses and an increase in neuronal apoptosis. It is well established that the increased vulnerability to stress in depression plays an important role in the reduction in brain repair which is caused by the reduction in the synthesis of neurotrophic factors, such as brain-derived neurotrophic factor (BDNF) and vascular endothelial growth factor. In normal brain, the development of neuronal networks is largely dependent on neuronal activity. Active neurons release neurotrophic factors, and there is evidence that specific neurons, which form part of the network, are targeted. These neurons release neurotrophic factors at a lower level of neuronal activity than normally occurs in single neurons [33]. By contrast, neurons that are largely inactive are eliminated. This might explain how functioning neuronal networks are formed, and the consequences to brain structure and function if there is a reduction in the synthesis of BDNF and related neurotrophic factors, as occurs in depression and Alzheimer disease.

In chronic depression, bipolar disorder, and Alzheimer disease, the BDNF concentration is reduced [34–36], an event associated with the activation of central glucocorticoid receptors by glucocorticoids. Thus, the combined effects of the neurotoxic consequences of elevated proinflammatory cytokines, the activation of N-methyl-D-aspartate (NMDA) receptors by nitric oxide and quinolinic acid (discussed later in this review), together with oxidative stress, provide a toxic environment for neurons in the brain of patients with depression or Alzheimer disease. This has been discussed in detail elsewhere [37].

The Hypothalamic-Pituitary-Adrenal Axis and Its Importance in Neuroprogression

The impact of the stress-induced activation of the hypothalamic-pituitary-adrenal (HPA) axis is well established in most cases of major depression [38, 39]. The chronic elevation of glucocorticoids, as distinct from the transitory changes which occur following acute stress, is linked to neuronal damage [40]. As mentioned previously, the reduction in hippocampal volume is associated with hypercortisolemia and the suppression of neurogenesis, events which are reversed by chronic antidepressant treatments in experimental studies [41, 42]. However, it must be emphasized that the reduction in neurogenesis by stress is largely based on rodent studies, and the evidence from human studies is, at best, equivocal [43, 44]. Nevertheless, the elevation of plasma cortisol in depression is positively associated with cognitive dysfunction [45], but this does not necessarily mean that the mechanism whereby glucocorticoids cause cognitive dysfunction is due to the suppression of hippocampal neurogenesis [46].

In addition to the hippocampus, the prefrontal cortex and amygdala are also affected by chronic stress. For example, the changes in the medial prefrontal cortex are similar to those seen in the hippocampus following chronic stress [47, 48]. In both major depression and Alzheimer disease, the raised glucocorticoid concentrations present concurrently with the raised proinflammatory cytokines. As glucocorticoids are considered to suppress inflammation, this appears to be an anomaly. There are two possible explanations for this phenomenon. Firstly, the glucocorticoid receptors on immune cells are desensitized by chronic low-grade inflammation and, therefore, do not respond to the glucocorticoids. Secondly, there is a qualitative difference between central and peripheral effects, and there is evidence that glucocorticoids exert a proinflammatory action in the brain [49].

In Alzheimer disease, raised plasma and CSF cortisol levels are commonly (like depression) resistant to dexamethasone suppression, thereby suggesting that (like depression) the glucocorticoid receptors are resistant to feedback inhibition [50]. However, there is a difference between the circadian regulation of cortisol in these disorders. Unlike depression, the circadian rhythm in Alzheimer disease is relatively normal despite the rise in the glucocorticoid concentration, whereas in depression it is partly accentuated and fairly flat [51]. However, in both conditions, cognitive dysfunction is associated with raised cortical levels and also with normal ageing.

The apolipoprotein E4 genotype is considered to be a major risk factor for late-onset Alzheimer disease, but it is also associated with raised CSF cortisol levels in nondepressed geriatric individuals [52]. The link between elevated glucocorticoid levels and the pathophysiology of Alzheimer disease is provided by the increased synthesis of β-amyloid and phosphorylated tau protein shown to occur in a genetically expressed mouse model of the disease; stress was shown to increase the deposition of these neurotoxic proteins [53]. As there is evidence that amyloid plaques also occur in the brains of elderly patients with chronic depression, this is further evidence that major depression and subsequent dementia are linked.

Proinflammatory Cytokines, Glucocorticoids, and Neurotoxicity in Major Depression and Alzheimer Disease

Over 40 years ago, Lapin and Oxenkrug [54] and Lapin [55] published a series of experimental studies demonstrating that depressive-like behavior in rodents could be caused by the metabolic end products of the tryptophan-kynurenine pathway. These authors demonstrated that quinolinic acid and 3-hydroxykynurenine, formed from kynurenine as components of the neurodegenerative arm of the pathway, caused anxiety and stress-like changes in the animals.

In recent years, there has been renewed interest in the role of the tryptophan-kynurenine pathway in depression, which has been stimulated by evidence that the plasma free tryptophan concentration is decreased and that kynurenine metabolites are increased. Under no-stress conditions, tryptophan is primarily metabolized in the liver to kynurenine by tryptophan 2,3-dioxygenase under the influence of circulating glucocorticoids [56–58]. However, in stressful and inflammatory conditions, tryptophan is metabolized by indoleamine 2,3-dioxygenase (IDO), which is widely distributed in immune cells, lung, kidney, and the brain [59, 60]. Anti-inflammatory cytokines reduce the activity of IDO.

Approximately 60% of brain kynurenine arises from the blood [61], where it is further metabolized to either the excitotoxic metabolite quinolinic acid, which acts as an agonist at NMDA glutamate receptors, or to the neuroprotective metabolite, kynurenic acid, that acts as an antagonist at the NMDA receptor [62]. Under noninflammatory/no-stress conditions, there is a balance between the end products of the neurodegenerative and neuroprotective pathways [63].

The distribution of the tryptophan-kynurenine pathway in the brain is concentrated mainly in astrocytes and microglia [64, 65]. In astrocytes, kynurenine is mainly synthesized to kynurenic acid by kynurenine amino transferase, whereas it is primarily converted to 3-hydroxykynurenine and quinolinic acid in the microglia [66, 67]. Under noninflammatory conditions, the end product of kynurenine metabolism is kynurenic acid, but following the activation of microglia by stress or inflammation the neurodegenerative pathway predominates [67]. This is further enhanced under inflammatory conditions by the apoptosis of astrocytes induced by quinolinic acid [67]. This results in the exposure of neurons to excitotoxic kynurenine metabolites thereby contributing to the decrease in brain volume reported to occur in patients with chronic schizophrenia or depression [68, 69].

The regions of the brain involved in the metabolism of kynurenine may differ according to the pathophysiology of the psychiatric disorder. In major depression, Steiner et al. [57] have shown that the quinolinic acid concentration in microglia from subregions of the anterior cingulated gyrus is increased. This evidence that the excitotoxic pathway is increased in depression is further supported by the increase in plasma 3-hydroxykynurenine [70].

Thus, as a consequence of immune activation, the changes in the tryptophan-kynurenine pathway play a major role in the dysfunctional neurotransmitter systems in the brain and, in addition, contribute to the changes in the brain structure and function which characterize depression.

A summary of the main metabolic stages in the tryptophan-kynurenine pathway is shown in Figure 1 in the chapter by Tang et al. [this vol., p. 110].

In recent years, attention has been focused on the neurotoxic consequences of the increase in quinolinic acid and the intermediates formed from kynurenine in the tryptophan-kynurenine pathway. While such neurotoxins undoubtedly play a critical role in the neurodegenerative changes associated with chronic psychiatric disorders, such as depression and schizophrenia, it is often overlooked that quinolinic acid is also an important substrate for the formation of nicotinamide adenine dinucleotide (NAD+). As NAD+ is a key component of the respiratory chain, chronic pathological changes that reduce its formation are liable to have adverse consequences for intermediary metabolism particularly in neurons that are critically dependent on high-energy sources.

It is estimated that approximately 99% of tryptophan that is not used for protein and serotonin synthesis is metabolized to NAD+ via the tryptophan-kynurenine pathway, and, therefore, this pathway is important for the synthesis of this vital

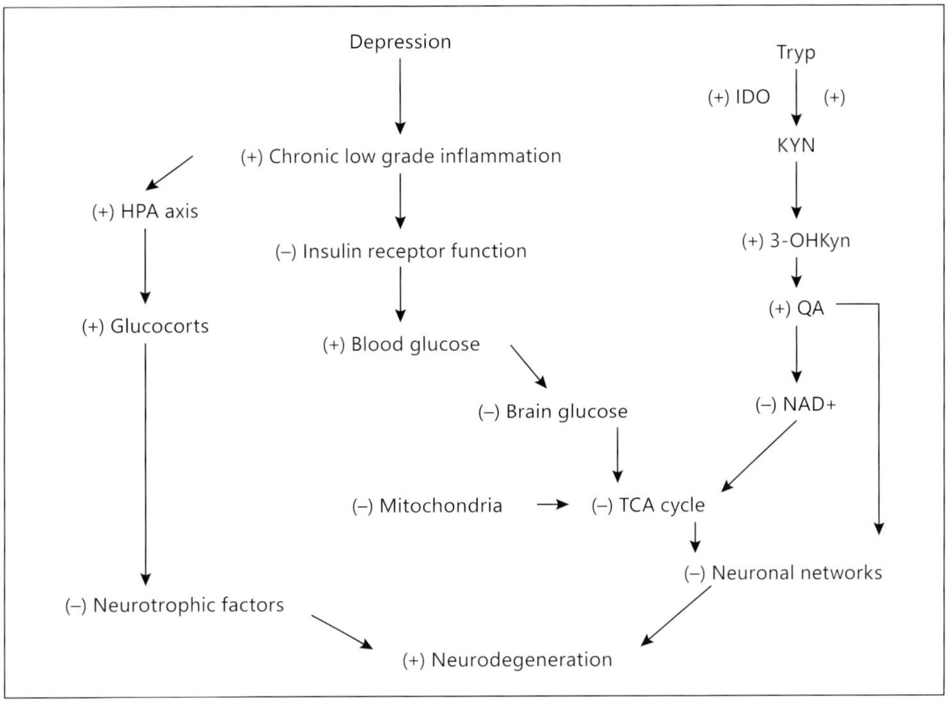

Fig. 1. Dysfunctional brain glucose metabolism caused by inflammation: a possible link to dementia? HPA, hypothalamic-pituitary-adrenal axis; KYN, kynurenine; QUIN, quinolinic acid; TCA, tricarboxylic acid; TRP, tryptophan.

cofactor [70, 71]. This situation would be compounded by a reduction in the availability of insulin, a key factor in the transport of glucose into neurons [72]. As there is evidence that insulin receptor resistance is a frequent feature of depression, other major psychiatric disorders, and age-related pathology associated with dementia [73], it is reasonable to conclude that a chronic decrease in high-energy substrates resulting from a deficit in glucose and essential cofactors may be of crucial importance in understanding the causes of increased neuronal apoptosis (Fig. 1) [73].

This situation is further complicated by mitochondrial dysfunction in depression which results in decreased synthesis of adenosine triphosphate (ATP) and related high-energy molecules, combined with an increase in oxidative damage. In addition, the synthesis of superoxide radicals results from a decrease in the respiratory chain and causes increased damage to mitochondrial membranes by opening the permeability transition pores [74]. The synthesis of oxygen free radicals is further enhanced by xanthurenic acid and 3-hydroxykynurenine, which are formed in the brain as a result of the inflammation-enhanced tryptophan-kynurenine pathway.

Thus, the combined impact of chronic inflammation, endogenous neurotoxins, and oxidative stress on the adverse changes in brain energy metabolism may provide the pathological link between the degenerative changes in the brain of the elderly depressed patient and the onset of dementia, particularly Alzheimer disease.

The tryptophan-kynurenine pathway is also activated in Alzheimer disease, and there is evidence that β-amyloid induces this pathway and also increases the concentration of quinolinic acid by activating microglia [65]. Such changes are consistent with the neurodegenerative role of quinolinic acid, but it is somewhat surprising that there is evidence that the neurotoxin is unchanged in the brain [75, 76]. Whether there is an increase in intracellular compartments and/or localized regions of the brain is unknown. However, studies in rodent models have indicated that specific areas of the brain, such as the basal ganglia, are particularly vulnerable to the toxic effects of quinolinic acid [77]. As this is the main cholinergic region of the brain which is thought to be dysfunctional in Alzheimer disease, this could provide a link between the increased activity of the tryptophan-kynurenine pathway and the pathophysiology of the condition.

Proinflammatory cytokines, such as IFN-γ, are known to activate IDO, and more recently it has been proposed that IL-18 also plays a significant role in Alzheimer disease and possibly also in major depression [78]. This cytokine is induced by stress and, following its cleavage by caspase-1, mediates the increase in IFN-γ [79]. This could be an important mechanism in the pathology of both depression and Alzheimer disease. An increase in IL-18 has been demonstrated in the brain of Alzheimer disease patients and in the CSF of those with cognitive impairment. IL-18 also shows a synergic interaction with the apolipoprotein E4 allele associated with the increased susceptibility of patients with Alzheimer disease [80, 81]. This mechanism provides another link between the pathological changes initiated by inflammation and dementia.

A Working Hypothesis

Dysfunctional brain glucose metabolism as a cause of neurodegeneration in depression. Is this the important link to Alzheimer disease?

The adult human brain constitutes approximately 2% of the body weight but consumes 20% of the available glucose. Glucose is the primary source of energy for the brain. In order to provide neurons and neuroglia with energy, glucose must be transported across the blood-brain barrier and into the cells by specific glucose transporters. The GLUT-1 transporter operates at the blood-brain barrier and the GLUT-3 undertakes the same function across the neuropil, axons, and dendrites. Insulin functions by translocating some of the glucose transporters from the cytosol to the plasma membrane and, in this way, contributes to the availability of glucose to support cellular function [82]. Clearly, a reduction in the availability of insulin will have a major detrimental effect on brain energy metabolism.

Astrocytes play an important part in maintaining the energy availability of neurons. The GLUT-1 transporter moves glucose into astrocytes where it is converted to lactate and passed on to neurons. This is the main source of energy to sustain neural function rather than the direct transport of glucose into neurons by the GLUT-3 transporter [83]. Thus, any deficiency in GLUT-1 transporter proteins, or apoptotic changes in the astrocytes, will have a major impact on brain energy metabolism [84]. These changes are implicated to occur in depression.

On entering brain cells, glucose is converted to pyruvate by aerobic metabolism and via the tricarboxylic acid cycle to carbon dioxide and water. The energy created by this metabolic process generates high-energy phosphates, such as ATP, which act as a major source of energy for synthetic processes in the brain. Of the many cofactors which are essential for the efficient function of the tricarboxylic acid cycle, NAD+ and its phosphorylated derivative NADP+ are of primary importance [85]. Thus, any major disturbance in glucose and brain energy metabolism will profoundly affect neuronal survival and thereby brain function, such as disruption of neurotransmitter metabolism, cerebral blood flow, and the integrity of the blood-brain barrier, which was indicated by the changes observed in patients with uncontrolled diabetes [86].

Physical illnesses, such as coronary heart disease, type 2 diabetes, cancer, and autoimmune diseases, are known to be significantly increased in patients with chronic major depression [87]. While a minority of depressed patients develop diabetes, metabolically adverse changes occur in depression which profoundly affect brain glucose metabolism. In this respect, there is already evidence from the serum of bipolar patients that there are critical changes in insulin, ghrelin, glycogen, and glycogen-like peptide-1 [88], while in medication-resistant depressed patents glucose metabolism in the prefrontal cortex is abnormal [89].

Recently, much attention has been directed towards the tryptophan-kynurenine pathway and, as already discussed above, the diabetogenic kynurenine metabolites xanthurenic acid and 3-hydroxykynurenine, which have an antagonistic effect on the availability of insulin and glucose metabolism. In addition to these changes, the reduction in the synthesis of NAD+ and the dysfunctional changes in the mitochondria of depressed patients combine to impair brain glucose metabolism and neuronal function [90].

Conclusions

The hypothesis linking a dysfunctional brain glucose metabolism to the neurodegenerative changes found in the brains of chronically depressed patients, which may trigger the onset of Alzheimer disease and other neurodegenerative diseases, may be summarized as follows:

1 Chronic low-grade inflammation and stress-induced increase in glucocorticoid levels cause insulin receptor insensitivity and reduce the availability of glucose for optimal neuronal function.

2 Activation of the tryptophan-kynurenine pathway by proinflammatory cytokines and the increase in circulating glucocorticoids in depression lead to increased synthesis of diabetogenic kynurenine metabolites which inhibit insulin activity and reduce the synthesis of NAD+.

3 Glucose transport into astrocytes and neurons is disrupted thereby contributing to the defective metabolic network.

4 Structural disruption of mitochondria in depression reduces the synthesis of high-energy phosphates which support and maintain neuronal function.

Such long-term changes in the brain would contribute to the neurodegenerative changes apparent in the chronically depressed patient and possibly contribute to the increase in the prevalence of Alzheimer disease in those patients.

References

1 Geerlings MI, Schoevers RA, Beckman AT, et al: Depression and risk of cognitive decline and Alzheimer's disease. Results of two prospective community-based studies in the Netherlands. Br J Psychiatry 2000;176:568–575.

2 Visser PJ, Verhey FR, Ponds RW, et al: Distinction between preclinical Alzheimer's disease and depression. J Am Geriatr Soc 2000;48:479–484.

3 Modrego PJ, Fernandez J: Depression in patients with mild cognitive impairment increases the risk of developing dementia of the Alzheimer type. Arch Neurol 2004;61:1290–1293.

4 Steffens DC, Plassman BL, Helms MJ, et al: A twin study of late-onset depression and apolipoprotein E ε4 as risk factors for Alzheimer's disease. Biol Psychiatry 1997;41:851–856.

5 Benoit M, Berrut G, Doussaint S, et al: Apathy and depression in mild Alzheimer's disease: a cross-sectional study using diagnostic criteria. J Alzheimers Dis 2012;31:325–334.

6 Harwood DG, Barker WW, Ownby RL, Duara R: Association between premorbid history of depression and current depression in Alzheimer's disease. J Geriatr Psychiatry Neurol 1999;12:72–75.

7 Green RC, Cupples LA, Kurz A, et al: Depression as a risk factor for Alzheimer's disease: the MIRAGE study. Arch Neurol 2003;60:753–759.

8 Crews L, Masliah E: Molecular mechanisms of neurodegeneration in Alzheimer's disease. Hum Mol Genetics 2010;19:R12–R20.

9 Maes M, Bosmans E, De Jongh R, et al: Increased serum IL-6 and IL-1 receptor antagonist concentrations in major depression and treatment resistant depression. Cytokine 1997;9:853–858.

10 Kang J, Lemaire HG, Unterbeck A, et al: The precursor of Alzheimer's disease, amyloid A4 protein, resembles cell surface receptor. Nature 1987;325:733–736.
11 Maes M, Galecki P, Chang YS, Berk M: A review on the oxidative and nitrosative stress pathways in major depression and their possible contribution to the neurodegenerative process in that illness. Prog Neuropsychopharmacol Biol Psychiatry 2011;35:676–692.
12 Finch CE, Morgan TE: Systematic inflammation, infection, ApoE alleles, and Alzheimer disease: a position paper. Curr Alzheimer Res 2007;4:185–189.
13 Bloom GS: Amyloid-β and tau: the trigger and bullet in Alzheimer disease pathogenesis. JAMA Neurol 2014;71: 505–508.
14 Sapolsky RM, Uno H, Rebert CS, Finch CE: Hippocampal damage associated with prolonged glucocorticoid exposure in primates. J Neurosci 1990;10:2897–2902.
15 Margariños AM, McEwen BS, Flügge G, Fuchs E: Chronic psychosocial stress causes apical dendritic atrophy of hippocampal CA3 pyramidal neurons in subordinate tree shrews. J Neurosci 1996;16:3534–3540.
16 Ragnarsson O, Berglund P, Eder DN, Johannsson G: Long-term cognitive impairments and attentional deficits in patients with Cushing's disease and cortisol-producing adrenal adenoma in remission. J Clin Endocrinol Metab 2012;97:E1640–E1648.
17 McEwen BS: Possible mechanisms for atrophied human hippocampus. Mol Psychiatry 1997;2:255–262.
18 Sapolsky RM: Stress, glucocorticoids and damage to the nervous system: the current state of confusion. Stress 1996;1: 1–19.
19 Gilbertson HW, Sherton ME, Ciszewski A, et al: Smaller hippocampal volume predicts pathological vulnerability to psychological trauma. Nat Neurosci 2002;5:1242–1247.
20 Attwood CS, Obrenovich ME, Liu T, et al: Amyloid-β: a chameleon walking in two worlds: a review of the trophic and toxic properties of amyloid-β. Brain Res Brain Res Rev 2003;43:1–16.
21 Delpech J-C, Madoc C, Nadjar A, et al: Microglia in neuronal plasticity: influence of stress. Neuropharmacology 2015;96(pt A):19–28.
22 Kettenmann H, Hanisch UK, Noda M, Verkhratsky A: Physiology of microglia. Physiol Rev 2011;91:461–553.
23 Dheen ST, Kaur C, Ling EA: Microglial activation and its implications in the brain diseases. Curr Med Chem 2007;14: 1189–1197.
24 Czeh M, Gressens P, Kaindl AM: The yin and yang of microglia. Dev Neurosci 2011;33:199–209.
25 Jurgens HA, Johnson RW: Dysregulated neuronal-microglial cross-talk during aging, stress and inflammation. Exp Neurol 2010;233:40–48.
26 Combs CK, Karlo JC, Kao SC, Landreth GE: β-Amyloid stimulation of microglia and monocytes results in TNFα-dependent expression of inducible nitric oxide synthase and neuronal apoptosis. J Neurosci 2001;231:1179–1188.
27 Ho GJ, Drego R, Hakimian E, Masliah E: Mechanisms of cell signaling and inflammation in Alzheimer's disease. Curr Drug Targets Inflamm Allergy 2005;4: 247–256.
28 Schousboe A, Westergaard N, Sonnewald U, et al: Regulatory role of astrocytes for neuronal biosynthesis and homeostasis of glutamate and GABA. Prog Brain Res 1992;94:199–211.
29 Aloisi F: Cytokine production; in Kettenmann H, Ransom B (eds): Neuroglia. Oxford, Oxford University Press, 2005.
30 Rapp MA, Dahlman K, Sano M, et al: Neuropsychological differences between late-onset and recurrent geriatric major depression. Am J Psychiatry 2005;162: 691–698.
31 Rapp MA, Schnaider-Beeri M, Grossman HAT, et al: Increased hippocampal plaques and tangles in patients with Alzheimer disease with a lifetime history of major depression. Arch Gen Psychiatry 2006;63:161–167.
32 Borsini A, Zunszain PA, Thuret S, Pariante CM: The role of inflammatory cytokines as key modulators of neurogenesis. Trends Neurosci 2015;38:145–157.
33 Song H, Poo M: The cell biology of neuronal navigation. Nat Cell Biol 2001; 3:E81–E88.
34 Chou YH, Ling JF, Hsieh WC, et al: Neither cortisol nor brain-derived neurotrophic factor is associated with serotonin transport in bipolar disorder. Eur Neuropsychopharmacol 2016;26:280–287.
35 Kapczinski F, Frey BN, Kauer-Sant'Anna M, Grassi-Oliveira B: Brain-derived neurotrophic factors and neuroplasticity in bipolar disorder. Expert Rev Neurother 2008;8:1101–1113.
36 Castren E, Rautamaki T: The role of BDNF and its receptors in depression and antidepressant drug action: reactivation of developmental plasticity. Dev Neurobiol 2010;70:289–297.
37 Leonard BE, Myint A-M: Inflammation and depression: is there a causal connection with dementia? Neurotox Res 2006; 10:149–160.
38 Cowen PJ: Not fade away: the HPA axis and depression. Psychol Med 2010;40: 1–4.
39 Stetler C, Miller GE: Depression and HPA activation: a quantitative summary of 4 decades of research. Psychosom Med 2011;73:114–126.
40 McEwen BS: Protective and damaging effects of stress mediators: central role of the brain. Dialogues Clin Neurosci 2006;8:367–381.
41 Duman RS: Depression: a case of a neuronal life and death? Biol Psychiatry 2004;56:140–145.
42 Sahay A, Hen R: Adult hippocampal neurogenesis in depression. Nat Neurosci 2007;10:1110–1115.
43 Reif A, Fritzen S, Finger M, et al: Neural stem cell proliferation is decreased in schizophrenia but not in depression. Mol Psychiatry 2006;11:514–522.
44 Boldrini M, Underwood D, Hen R, et al: Antidepressants increase neural progenitor cells in the human hippocampus. Neuropsychopharmacology 2009;34: 2376–2389.
45 Hinkelmann K, Morit S, Botzenhardt J, et al: Cognitive impairment in major depression: association with salivary cortisol. Biol Psychiatry 2009;66:879–885.
46 Kempermann G, Krebs J, Fabel K: The contribution of failing adult hippocampal neurogenesis to psychiatric disorders. Curr Opin Psychiatry 2008;21: 290–295.
47 Czeh B, Lucassen PJ: What causes the hippocampal volume decrease in depression? Are neurogenesis, glial changes and apoptosis implicated? Eur Arch Psychiatry Clin Neurosci 2007;257:250–260.

48 Banasr M, Valentine GW, Li YY, et al: Chronic unpredictable stress decreases cell proliferation in the cerebral cortex of the adult rat. Biol Psychiatry 2007;62: 496–504.
49 Sorells SF, Caso JR, Munhoz CD, Sapolsky RM: The stressed CNS: when glucocorticoids aggravate inflammation. Neuron 2009;64:33–39.
50 Davis KL, Davis BM, Greenwald BS, et al: Cortisol and Alzheimer's disease: 1. Basal studies. Am J Psychiatry 1986;143: 300–305.
51 Hartmann A, Veldhuis JD, Deuschle M, et al: Twenty-four hour cortisol release profiles in patients with Alzheimer's disease and Parkinson's disease compared to normal controls: ultradian secretory pulsatility and diurnal variation. Neurobiol Aging 1997;18:285–289.
52 Peskind ER, Wilkinson CW, Petrie EC, et al: Increased CSF cortisol is a function of ApoE genotype. Neurology 2001;56: 1094–1098.
53 Marlatt MW, Lucassen PJ: Neurogenesis and Alzheimer's disease: biology and pathophysiology in mice and men. Curr Alzheimer Res 2010;7:113–125.
54 Lapin IP, Oxenkrug GF: Intensification of the central serotoninergic processes as a possible determinant of the thymoleptic affect. Lancet 1969;1:132–136.
55 Lapin IP: Kynurenines as probable participants of depression. Pharmakopsychiatr Neuropsychopharmakol 1973;6: 273–279.
56 Myint A-M: Inflammation, neurotoxins and psychiatric disorders; in Halaris A, Leonard BE (eds): Inflammation in Psychiatry. Mod Trends Pharmacopsychiatry. Basel, Karger, 2013, vol 28, pp 61–74.
57 Steiner J, Bielau H, Brisch R, et al: Immunological aspects in the neurobiology of suicide: elevated microglial density in schizophrenia and depression is associated with suicide. J Psychiatr Res 2008; 42:151–157.
58 Satyanarayana U, Rao BS: Dietary tryptophan level and the enzymes of tryptophan NAD pathway. Br J Nutr 1980;43: 107–113.
59 Carlin JM, Borden EC, Sondel PM, Byrne GI: Interferon-induced indoleamine 2,3-dioxygenase activity in human mononuclear phagocytes. J Leukoc Biol 1989;45:29–34.
60 Taylor MS, Feng GS: Relationship between interferon gamma, indoleamine 2,3-dioxygenase and tryptophan catabolism. FASEB J 1991;5:2516–2522.
61 Gál EM, Sherman AD: L-Kynurenine: its synthesis and possible regulatory function in brain. Neurochem Res 1980;5: 223–239.
62 Myint AM, Kim YK: Cytokine-serotonin interaction through IDO: a neurodegeneration hypothesis of depression. Med Hypotheses 2003;61:519–525.
63 Perkins MN, Stone TW: An iontophoretic investigation of the actions of convulsant kynurenines and their interaction with the endogenous excitant quinolinic acid. Brain Res 1982;247: 184–187.
64 Grant RS, Kapoor V: Murine glial cells regenerate NAD, after peroxide-induced depletion, using either nicotinic acid, nicotinamide, or quinolinic acid as substrates. J Neurochem 1998;70:1759–1763.
65 Guillemin GJ, Brew BJ: Implications of the kynurenine pathway and quinolinic acid in Alzheimer's disease. Redox Report 2002;7:199–206.
66 Guillemin GJ, Smythe G, Takikawa O, Brew BJ: Expression of indoleamine 2,3-dioxygenase and production of quinolinic acid by human microglia, astrocytes, and neurons. Glia 2005;49: 15–23.
67 Guillemin GJ, Wang L, Brew BJ: Quinolinic acid selectively induces apoptosis in human astrocytes: potential in AIDS dementia complex. J Neuroinflammation 2005;2:16.
68 Sheline YI: Neuroimaging studies of mood disorder: effects on the brain. Biol Psychiatry 2002;54:338–352.
69 Bremmer JD, Narayan M, Anderson ER, et al: Hippocampal volume reduction in major depression. Am J Psychiatry 2000; 157:115–118.
70 Han Q, Tao DA, Li J: Structure, expression, and function of kynurenine aminotransferases in human and rodent brains. Cell Mol Life Sci 2010;67:353–368.
71 Oxenkrug GF: Insulin resistance and dysregulation of the tryptophan-kynurenine-nicotinamide adenine dinucleotide metabolic pathways. Mol Neurobiol 2013;48:294–301.
72 Lee YJ, Han SB, Nam SY, et al: Inflammation and Alzheimer's disease. Arch Pharm Res 2010;33:1539–1556.
73 Sas K, Robotka H, Toldi J, Vécsei L: Mitochondria, metabolic disturbance, oxidative stress and the kynurenine system, with focus on neurodegenerative disorders. J Neurol Sci 2009;257:221–239.
74 Moudrian MM, Heyes MP, Pan JB, et al: No changes in central quinolinic acid levels in Alzheimer's disease. Neurosci Lett 1989;105:233–238.
75 Sofic E, Halket J, Przyborowska A, et al: Brain quinolinic acid in Alzheimer's dementia. Eur Arch Psychiatry Neurol Sci 1989;239:171–179.
76 Schwarcz R, Köhler C: Differential vulnerability of central neurons of the rat to quinolinic acid. Neurosci Lett 1983;58: 85–90.
77 Sekiyama A, Ueda H, Kashiwamura S, et al: IL-18: a cytokine translates a stress into medical science. J Med Invest 2005; 52(suppl):236–239.
78 Gillespie MT, Horwood NJ: IL-18: perspectives on the newest interleukin. Cytokine Growth Factor Rev 1998;9:109–116.
79 Ojala J, Alafuzoff I, Herukka SK, et al: Expression of IL-18 is increased in the brains of Alzheimer's disease patients. Neurobiol Aging 2009;30:198–209.
80 Yu JT, Tan L, Song JH, et al: Interleukin-18 promoter polymorphisms and risk of late onset Alzheimer's disease. Brain Res 2009;1253:169–175.
81 Jurcovicova J: Glucose transport in brain – effect of inflammation. Endocr Regul 2014;48:35–48.
82 Benarroch EE: Brain glucose transporters: implications for neurologic diseases. Neurology 2014;15:1374–1379.
83 Mergenthaler P, Lindauer L, Dienel GA, Meisel A: Sugar for the brain: the role of glucose in physiological and pathological brain function. Trends Neurosci 2013;36:587–597.
84 Van Linden M, Skoge H, Ziegler M: Discovery, metabolism and future of NAD and NADP. Biochemist 2015;37: 9–13.
85 McCall AL: Cerebral glucose metabolism in diabetes mellitus. Eur J Pharmacol 2004;490:147–158.

86 Black PH: The inflammatory consequences of psychologic stress: relationship to insulin resistance, obesity, atherosclerosis and diabetes mellitus type II. Med Hypotheses 2006;67:879–891.
87 Rosso G, Cattanco A, Zarardini R, et al: Glucose metabolism alterations in patients with bipolar disorder. J Affect Disord 2015;184:293–298.
88 Li CT, Su TP, Wang SJ, Tu PC, Hsieh JC: Prefrontal glucose metabolism in medication-resistant major depression. Br J Psychiatry 2015;206:316–323.
89 Myint A-M: Kynurenines: from the perspective of major psychiatric disorders. FEBS J 2012;279:1375–1385.
90 Kotake Y, Ueda T, Mori T, et al: Abnormal tryptophan metabolism and experimental diabetes by xanthurenic acid. Acta Vitaminol Enzymol 1975;29:236–239.

Prof. em. Brian E. Leonard, PhD, DSc
National University of Ireland
Galway H91 TK33 (Ireland)
E-Mail brian@psycholeonard.com

Halaris A, Leonard BE (eds): Neuroprogression in Psychiatric Disorders.
Mod Trends Pharmacopsychiatry. Basel, Karger, 2017, vol 31, pp 67–78 (DOI: 10.1159/000470808)

Innate Immune Memory: Implications for Microglial Function and Neuroprogression

Alex P. Salam • Carmine M. Pariante • Patricia Zunszain

Stress, Psychiatry, and Immunology Laboratory, Institute of Psychiatry, Psychology, and Neuroscience, King's College London, London, UK

Abstract

Immunostimulatory insults such as stress and infection are risk factors for the development of several neuropsychiatric disorders characterized by neuroprogression. Inflammatory and neurotoxic molecules in the brain can cause disruptions in neurogenesis, neuronal excitability, synaptic transmission, synaptic plasticity, and neuronal survival – changes that characterize neuroprogression. We draw on recent findings in the immunology literature that peripheral innate immune cells are capable of retaining long-term memory of infectious insults and displaying long-lasting upregulated proinflammatory function in response to repeated infectious insults – a concept known as "innate immune memory." In turn, we hypothesize that microglia, the resident innate immune cells of the brain, are also capable of retaining long-term memory of infectious and noninfectious insults, including stress. Microglia are capable of producing a variety of proinflammatory neurotoxic cytokines and chemokines. Persistent upregulation of microglial proinflammatory function as a result of memory for immunostimulatory insults may therefore contribute to persistent and progressive inflammation in neuropsychiatric illnesses and be an important driver of neuroprogression.

Introduction

Inflammation is widely implicated in the etiology and progression of several neuropsychiatric disorders characterized by neuroprogression, including depression [1, 2] and schizophrenia [3]. Early-life insults such as stress and infection are common inflammatory risk factors for the development of both depression and schizophrenia [4]. Such early-life insults appear to lead to a proinflammatory phenotype that persists into adulthood [5, 6]. Critically, such insults may sensitize peripheral and central nervous system (CNS) innate immune responses to subsequent immunostimulatory insults, resulting in neuroinflammation and the enhanced release of neurotoxic molecules [7]. This is one mechanism by which an immunostimulatory insult in early life might render an individual more vulnerable to the development of neuropsychiatric illness in adulthood following a second immunostimulatory insult – a concept known in general as the "double-hit" model [7]. This sensitization also appears to occur in adulthood. Prior episodes of depression can sensitize immune responses to subsequent depressive episodes, resulting in increased levels of peripheral proinflammatory cytokines and chemokines with sequential depressive episodes [8]. This observed sensitization cannot be explained by adaptive immune memory given its nonspecificity and is therefore likely to be mediated by innate immunity.

This raises the obvious question as to how an event early in life, or indeed at any stage of life, can lead to a change in innate immune function that persists well beyond the inciting event. An increasing body of evidence suggests that certain immune challenges are capable of reprogramming the function of peripheral innate immune cells over the long term, such that the production of proinflammatory molecules is enhanced upon a secondary and distant stimulation. This is known as "innate immune memory" or "trained innate immunity," and is mechanistically characterized by stable epigenetic changes in immune and metabolic pathways [9].

Below, we present the evidence for innate immune memory and then draw on some suggestions in the literature that microglia, the resident innate immune cells of the CNS, are also susceptible to long-term reprogramming. In response to brain injury or inflammation, or systemic inflammation, microglia can develop an activated proinflammatory phenotype, characterized by morphological changes, upregulation of cell surface receptors, and the release of various inflammatory molecules, including cytokines (e.g., interleukin (IL)-1β, IL-6, interferon (IFN)-γ, tumor necrosis factor (TNF)-α) and reactive oxygen species (ROS) and reactive nitrogen species (RNS) [10]. These can cause disruptions in neuroplasticity and neurogenesis, including growth, differentiation, myelination, apoptosis, synaptic branching, and neurotrophin regulation [10], ultimately contributing to and driving neuroprogression [11]. Thus, persistent upregulation of microglial proinflammatory function as a result of memory for immunostimulatory insults may act as a significant contributor to neuroprogression. We also hypothesize that reprogramming of microglial function is not specific to early life, but that it can occur at any stage of life. If this is case, then therapeutic agents that reverse or prevent training of microglia following immunostimulatory insults may have a role in halting neuroprogression.

Peripheral Innate Immune Memory

The Innate Immune System

The vertebrate immune system is classically divided into the adaptive and the innate immune system. Historically, the major difference between them was considered to be the ability of the adaptive immune system to retain memory of specific pathogens through the generation of antibodies and specific T-cell clones towards specific antigens, often resulting in lifelong protection to reinfection with the same pathogen. In contrast, for many years, the innate immune system was considered being capable of displaying neither specificity nor memory, offering instead a first-line, rapid, and generalized response to invading microorganisms. In the last decade however, it has become clear that cells of the innate immune system do display a certain degree of specificity for pathogens. Innate recognition of pathogens involves specific recognition of conserved microbial molecules known as pathogen-associated molecular patterns (PAMPs). This occurs through pattern recognition receptors (PRRs), that are present on the cell surface, in the endosome, or in the cytoplasm of innate immune cells. PRRs include Toll-like receptors (TLRs), C-type lectin receptors, nucleotide-binding oligomerization domain-like receptors, and retinoic acid-inducible gene helicases [12]. PRRs recognize a diverse array of microbial lipids, proteins, nucleic acids and carbohydrates as danger signals. Microorganisms may stimulate one or multiple PRRs. As a result, PRRs enable cells of the innate immune system to recognize different kinds of microorganisms – though not individual species – offering a degree of partial specificity. The innate immune response to different microorganisms may vary depending on the type of PAMP-PRR interaction. Often, TLR-PRR binding results in the downstream production of proinflammatory, chemokines, ROS, and NOS [12].

Innate Immune Memory

More recently, the innate immune system has also been shown to be capable of retaining long-term memory for infectious insults through epigenetic reprogramming. This has been termed "innate immune memory" or "trained innate immunity" [9]. It is well described that innate immunity in plants and invertebrates can exhibit memory characteristics after the first encounter with a pathogen. Increasingly, evidence suggests that innate immune memory also exists in vertebrates, including humans. This represents a paradigm shift in our understanding of the innate immune system [13]. In essence, an initial activation of a PRR can sometimes lead to an altered cellular immune response upon reactivation of the same PRR at a distant time, or the reactivation of a different PRR, but one which feeds into the same or similar signaling pathways [14]. As a result of certain infections or stimuli, innate immune responses can be primed to respond more strongly upon challenge with a second stimulus (either identical or different to the initial stimulus) [15]. Often this exaggerated immune response often declines and resolves rapidly after the initial priming insult (hours to days). However, it does not always return to baseline and sometimes can remain altered for weeks/months/years [15]. When the response remains enhanced and above baseline after the initial short-term priming phase, this memory effect is known as "training." This can lead to both enhanced innate cellular immune responses and a lower threshold for innate cellular activation [15]. However, the response can instead be downregulated and can remain below baseline after the initial priming phase, which is known as "tolerance" [16]. Indeed, short-term innate immune tolerance has been known for some time in the context of inhibited innate immune cell function following repeated lipopolysaccharide (LPS) exposure, known as LPS-induced tolerance. This is responsible for the immunoparalysis that occurs after gram-negative sepsis [16]. Stimulation of innate immune cells can therefore result in short-term priming alone, priming followed by training, or priming followed by tolerance. Whether tolerance or training develops depends on a multitude of factors, including the organism species and strain, the dose, and the duration of exposure [15]. An important observation, and one that is potentially highly relevant to the pathogenesis of neuropsychiatric disorders, is that trained immunity lacks specificity. Protection is provided not only against the original microorganism but also against unrelated microorganisms, known as heterogeneity or cross protection [15]. This likely occurs because different organisms can activate the same PRRs, and because different PRRs share many of the same downstream signaling pathways [14]. Trained immunity, rather than immune tolerance, is most relevant to neuropsychiatric disorders given the contribution of neuroinflammation to disease pathology [11]. It is therefore the primary focus of this discussion. However, given the critical and sensitive role that microglia play in neurodevelopment [10], any medium or longterm dysregulation in microglial function (whether up- or downregulation) may have an impact on neurohomeostasis and pathological processes.

Training of Macrophages and Monocytes

Monocytes, macrophages, natural killer cells and myeloid progenitor cells have lifespans which range from weeks to years, and are the cells of the innate immune system that have been shown to display training [15]. We will describe the evidence and mechanisms for trained immunity in monocytes/macrophages since they share many functional similarities with microglia [10]. This serves as a precedent for the possibility that microglia may also display training.

Human monocytes stimulated in vitro with a low dose of *Candida albicans* or β-glucan (a fungal cell wall component) demonstrate increased production of TNF-α and IL-6 mRNA and protein after secondary stimulation 2 weeks later with either *C. albicans*, various PRRs ligands, or bacteria [17]. Monocytes that have not received a primary stimulation do not. Inhibiting dectin-1, the PRR for β-glucan, during primary stimulation prevents this, as does inhibiting one of its downstream signaling pathways, Raf-1. This training effect is accompanied by dramatic and lasting changes in gene expression, with more than 5,000 upregulated genes and 1,000 downregulated genes. Genes for various TLRs, chemokines, nitric oxide synthases, and histone methyl-transferases are up-regulated following primary stimulation with β-glucan. Similarly, monocytes isolated from humans 2 weeks after BCG vaccination and then exposed to BCG, *Staphylococcus aureus, C. albicans* or LPS show increased production of the proinflammatory cytokines IFN-γ, TNF-α, and IL-1β compared to monocytes isolated from humans prior to BCG vaccination [18]. This effect lasts for months and is accompanied by increased cell surface expression of multiple activation markers and PRRs, such as CD11b and TLR4. Monocytes recovered from humans 1 year after BCG vaccination still have increased expression of various activation markers and PRRs (e.g., CD11b, CD14, TLR4) and increased TNF-α and IL-1β production after secondary stimulation with LPS [19]. This training effect is prevented by inhibiting the PRR for mycobacteria (NOD2) or its downstream signaling pathway (Rip2 kinase) during primary stimulation. Further, the NOD2-specific ligand MDP mimics the effects of primary stimulation with BCG. Similarly, and of interest given the age difference, monocytes isolated 4 weeks after neonatal human BCG vaccination show increased production of IL-1β, IL-6, IFN-γ, and TNF-α upon secondary stimulation with various TLR agonists [20].

As alluded to earlier, the dose and duration of the primary stimulation may influence whether cells display training or tolerance. Primary stimulation of human monocytes with high doses of Pam3CSK4, LPS, flagellin, or polyinosinic:polycytidylic acid (which all engage TLRs) induces a tolerant state in which monocytes produce less TNF-α and IL-6 upon secondary stimulation with either Pam3CKS4 or LPS 5 days later [21]. However, primary stimulation with low doses enhances production of TNF-α and IL-6 and expression of CD11b upon secondary stimulation with Pam3CSK4 or LPS. In contrast, different doses of Tri-DAP and MDP (which engage NODs) train the cells toward an increased production of TNF-α and IL-6.

Together, these studies demonstrate that monocytes/macrophages, which share many functional characteristics and developmental origins with microglia [10, 22], can retain long-term homogeneous and heterogeneous memory of a challenge with a PAMP. Whether tolerance and downregulation of proinflammatory function occures, or training and upregulation of proinflammatory function occurs, is dependent on both the nature of the PAMP as well as its dose.

Training of Macrophages/Monocytes by Nonmicrobial Stimuli

Training has also been shown to be induced by nonmicrobial stimuli. The fact that monocytes/macrophages (and therefore possibly microglia) can be trained by nonmicrobial stimuli is important given that a number of nonmicrobial immune stimuli (e.g., stress) are implicated in the etiology of neuropsychiatric illnesses, neuroinflammation, and neuroprogression [23]. Danger-associated molecular patterns (DAMPs) are endogenous molecules that are typically released by host cells following cellular damage or stress. They activate many of the same PRRs as PAMPs and initiate an inflammatory response [24]. Examples of DAMPs include high-mobility group box-1 (HMGB-1), S-100 proteins, heat shock proteins (HSPs), hyaluronan, surfactant protein, uric acid, fibronectin, β-defensin, and cardiolipin [24]. Many DAMPs either include oxidation-specific epitopes present on apoptotic cells and cellular debris (produced as a result of ROS generation during cellular damage) [25], or involve hydrophobic portions of molecules exposed as a result of cell death or damage [24]. Many microorganisms have PAMPs which have large hydrophobic portions that are exposed when microorganisms are damaged [24] or PAMPs that have molecular mimicry with oxidation-specific epitopes [25]. We would therefore expect DAMPs to also induce training of innate immune cells. Indeed, human monocytes exposed to oxidized low-density lipoprotein, a DAMP, show increased production of TNF-α, IL-6, IL-18, IL-8, and MCP-1 upon restimulation with different TLR4 or TLR2 agonists 7 days later [26]. To date, this is the only study that we are aware of that has directly investigated the ability of DAMPs to re-program monocytes/macrophages over the medium term. There exists, however, a significant body of evidence showing that peripheral innate immune cells can be primed over the short term by DAMPs to display enhanced pro-inflammatory cytokine production in response to an immediate or near-immediate (within hours) secondary stimulation [27]. As a result of this and the fact that DAMPs engage many of the same PRRs and downstream signaling pathways as PAMPs, we hypothesize that many DAMPs will be able to reprogram cells of the innate immune system heterogeneously.

Mechanisms Underlying Training of Monocytes/ Macrophages

A number of studies have demonstrated that a single episode of PRR stimulation can result in profound alterations in epigenetic architecture lasting for months and possibly longer, with resultant changes in cellular metabolic and immune pathways. β-Glucan, *Candida*, and BCG training of human monocytes results in increased levels of H3K4me3 (a histone modification associated with the regulation of immune-related genes) at the promoters of various important target genes, including TNF-α, IL-6, IL-18, TLR4, and dectin-1 (amongst others) [17, 18]. Blocking histone methylation using the methyltransferase inhibitor methyl-thioadenosine prevents these training-induced modifications of the epigenome. Other studies have found that β-glucan training in human monocytes reinforces specific gene expression programs involved in metabolic pathway remodeling, revealing that trained phenotypes involve marked changes in cellular metabolism [28, 29]. Further, training with LPS in mice induces phosphorylation of the stress response transcription factor ATF7 in tissue macrophages [30]. ATF7 normally suppresses a group of genes encoding factors involved in innate immunity in macrophages by recruiting the histone H3K9 dimethyltransferase G9a. Training leads to the release of ATF7 from chromatin and a decrease in repressive histone H3K9me2 marks. Interestingly, ATF7 is also phosphorylated as a result of social isolation stress and other stressors [31]. Finally, training of monocytes by oxidized low-density lipoprotein is also dependent on enrichment of H3K4me3 at the promoters of various immune-related genes [26].

In summary, multiple studies have shown that PAMPs can train monocytes/macrophages homogeneously and heterogeneously, resulting in increased production of pro-inflammatory cytokines. The effects are underpinned by epigenetic reprogramming and appear to last for extended periods of time. Whilst, to our knowledge, no one has investigated the long-term training of functions such as respiratory burst and phagocytosis, ample evidence exists that such functions in monocytes/macrophages are enhanced by immediate or near-immediate secondary stimulation [27], suggesting that they may also be capable of being trained over the long term. Given that microglia share developmental origins and many functional characteristics with monocytes/macrophages [10, 22], it would not be surprising if microglia are also capable of being trained.

Central Innate Immune Memory

Training of Microglia by Microbial Stimuli

β-Amyloid accumulation in Alzheimer disease has been shown to prime microglia to have exaggerated proinflammatory cytokine production in response to concurrent heterogeneous insults such as systemic infections or exposure to endotoxins [32]. Similarly, concurrent or recent stress also primes microglia to have exaggerated proinflammatory cytokine production in response to heterogeneous insults [33]. Though these findings may not specifically represent innate immune memory per se, given the simultaneous or very recent presence of DAMPs or other immunostimulatory molecules at the time of the additional insult, they demonstrate that microglia are capable of being heterogeneously primed. Since adult microglia arise from a founding population of bone marrow-derived primitive macrophages during development and are maintained throughout life [22], they are prime candidates for long-term training.

A single intraperitoneal injection of attenuated *Salmonella typhimurium* (which contains LPS) in mice results in increased microglial immunoreactivity for the activation markers CD11c and MHCII in response to a single intrahippocampal injection of LPS 4 weeks later [34]. Importantly, there is no increase in CD11 expression following *Salmonella* infection alone. This suggests that the increase in CD11 following secondary stimulation is due to an increase in CD11 expression on the surface of microglia, i.e., a functional change, rather than simply an increase in the number of microglia. Further, there are no differences in CD11 or MHCII immunoreactivity on microglia between mice with severe combined immunodeficiency (SCID) (which have absent/dysfunctional T and B cells [35]) and control mice shortly after infection with *Salmonella* alone. Although secondary LPS stimulation in mice with SCID was not investigated, this still suggests that microglia rather than adaptive immune cells are responsible for the memory effect.

Similar effects occur as a result of early-life immune stimulation. Neonatal rats injected with live *Escherichia coli* (which contains LPS) have an increase in microglial CD11b gene expression, which persists into adulthood and is most pronounced in the hippocampus [36]. Subsequent peripheral injection of LPS in adulthood results in a greater increase in hippocampal CD11b mRNA relative to control adults, as well as faster and prolonged increases in the pro-inflammatory and neurotoxic cytokine IL-1β in the hippocampus and cortex. These exaggerated IL-1β responses to LPS in adulthood are specific to the brain and do not occur in the periphery. As before, the frequency of adult hippocampal microglia does not differ as consequence of neonatal infection [37]. Instead, microglial morphology is altered. Microglia have larger cell volumes and shorter, thicker, processes [37]. Importantly, these effects are not due interactions between the adaptive and the innate immune system. Microglia isolated from adult rats which have been injected postnatally with *E. coli* and then stimulated ex vivo with LPS have greater CD11b cell surface expression on a per-cell basis as well as increased IL-1β expression [38]. This demonstrates that the memory effect is intrinsic to microglia and does not require interactions with other cell types.

Maternal immune activation has also been shown to induce memory effects in microglia. Maternal immune activation and immune responses in utero can have profound consequences for the development of neuropsychiatric disorders in adulthood as a result of alterations in the adult neuroinflammatory environment [7]. Cao et al. [39] administered either LPS or saline to near-term fetal sheep, isolated microglia from the fetuses, maintained them in vitro for 19 days, and then exposed them to LPS. Compared to in vivo saline-exposed microglia, in vivo LPS-exposed microglia had significantly higher IL-1β production in response to LPS administration ex vivo. Even in the absence of secondary LPS stimulation, in vivo LPS-exposed microglia secreted more IL-1β ex vivo. There were dramatic changes in microglial gene regulation 19 days after in vivo LPS exposure, with 4,000 upregulated and 2,000 downregulated genes. Further, compared to microglia exposed to LPS in vitro only, in vivo LPS-pre-exposed microglia showed a total of 6 differentially expressed genes upon re-exposure to LPS in vitro. In particular, heme oxygenase decycling 1 (HMOX1), thought to play an anti-inflammatory role [40], was strongly downregulated.

Although limited in number and scope, these studies suggest that microglia, much like peripheral monocytes and macrophages, may be capable of being trained both during early life and adult life.

Training of Microglia by Nonmicrobial Stimuli

It is possible that microglia, like monocytes/macrophages, may be capable of being trained by noninfectious stimuli. Exposure to psychological and physical stressors results in the release of multiple DAMPs. Predatory fear increases heat shock protein 70 kDa protein 1 (HSP72) in the blood for example [41], while tail shock increases uric acid and HSP72 in the blood [42], and HMGB-1 in the hippocampus [43]. Furthermore, stressor exposure also releases microbial-associated molecular patterns (MAMPs) from nonpathogenic symbiotic gut microbiota into the blood [44]. Such MAMPs have been shown to prime the innate immune response to subsequent pathogenic challenges such as *E. coli* and LPS over the short term through several signals, including catecholamines and glucocorticoids [44]. Given the association between early-life stress and inflammation and neuropsychiatric illness in adulthood [7], the possibility that D/MAMPs released as a result of stress may train microglia over the medium to long term in a homogeneous or heterogeneous manner via epigenetic reprogramming is an intriguing avenue of investigation. Microglia isolated from mice injected with LPS 1 month after traumatic brain injury have increased IL-1β and TNF-α mRNA expression relative to control/LPS mice [45]. These changes are accompanied by depressive-like behaviors. Traumatic brain injury (TBI) releases DAMPs, results in increased microglial activation that persists decades after injury in humans, and is a risk factor for depression [46]. TBI is a complex and multifaceted phenomenon, making it difficult to disentangle the contribution of microglial training to heightened microglial responses upon secondary stimulation. One study suggests that stress can indeed result in heterogeneous training of microglia. Microglia isolated from mice 24 days after repeated social defeat stress have higher gene expression of IL-1β, IL-6, and TNF-α and increased IL-1β in supernatant in response to ex vivo LPS stimulation relative to microglia from non-stressed controls [47].

Epigenetic Reprogramming of Microglia

To our knowledge, there exists only one study investigating epigenetic changes in microglia as a result of a single immune stimulus. Schaafsma et al. [48] demonstrated that 6 days following a primary high-dose LPS exposure, microglia displayed proinflammatory cytokine tolerance but enhanced phagocytic activity and RNS production in response to a secondary LPS stimulation in vivo and in vitro. These changes were mediated by epigenetic alterations. Although only a single study, this highlights 3 important possibilities: (1) that the memory behavior of microglia, like peripheral monocytes/macrophages, is likely to be dependent on the type, dose, and duration of the primary stimulation; (2) that microglia are capable of undergoing stable medium-term epigenetic modifications as a result of a single stimulus; and (3) that the dichotomy of tolerance versus training is possibly simplistic, and that some innate immune cell functions may be downregulated while others are simultaneously upregulated.

Innate Immune Memory and Neuroprogression

Stress is a potent risk factor for psychiatric illness, in particular depression. It can be an inducer or suppressor of immune function, depending on the severity and duration of stress, as well as additional environmental factors [49]. One of the principal mechanisms by which stress is hypothesized to contribute to the development of depression is immune activation, particularly activation of the innate immune system and microglia [33, 50]. Depression is itself, however, a stressful state, and depression may, theoretically, itself sensitize innate immune responses, leading to a positive feedback loop and refractory disease. Enhanced innate immunity is observed in some patients with depression. Monocytes from depressed and chronically stressed patients showed increased IL-1β and IL-6 gene expression upon in vitro stimulation with LPS relative to monocytes from controls [51–53]. Increased monocyte IL-6 gene expression and activation of signal transducer and activation of transcription 1 and nuclear factor-κB is observed in response to acute stress in adults with depression or a history of traumatic stress [54, 55]. Methylation profiles in peripheral immune cells from individuals with depression or a history of traumatic stress suggest upregulation in immune system-related genes [56–58]. The innate immune system, therefore, appears to be shifted towards a proinflammatory state in patients with depression or in individuals suffering from chronic stress, a shift that is possibly underlined by epigenetic modifications. Stress-induced systemic and central release of DAMPs, resulting in training of peripheral and central innate immune cells, including microglia, is a possible and plausible explanation for this observed phenomenon. The initiation of a vicious cycle in which stress trains innate immune cells to respond more aggressively to subsequent inflammatory stimuli, including stress, is a theoretical model that may account for the heterogeneity in life events that can trigger depression and depressive relapses with sequentially increasing severity and treatment refractoriness. Pathological sensitization and cross sensitization to stressors, including relapses, have been demonstrated in both animal models and humans [59]. Such heterologous sensitization manifests as increased severity and duration of episodes and vulnerability to recurrence, characteristic of clinical neuroprogression [59]. Interestingly, lithium, one of the only medical therapies that prevents or halts neuroprogression, has significant anti-inflammatory effects [60]. It would be interesting to see whether lithium is able to prevent training of innate immune cells in vitro and in vivo.

General Conclusions

There are suggestions that microglia are capable of being trained to display enhanced proinflammatory cytokine production, and potentially even enhanced RNS production and phagocytosis. Microglia appear to be capable of being trained during different stages of life, including in utero, postnatally, and in adulthood. To what degree microglia can be trained in a heterogeneous manner is unclear, as is the degree to which epigenetic reprogramming underlies microglial training. If microglia are capable of being trained towards a proinflammatory phenotype, this may partly explain the mechanism by which early-life neuroimmunostimulatory insults are risk factors for the development of neuropsychiatric illnesses in adulthood, in particular when accompanied by a second neuroimmunostimulatory insult later in life [7]. A single or series of neuroimmunostimulatory insults early in life of a particular origin, dose, and duration may train microglia towards an exaggerated production of proinflammatory cytokines, ROS, nitric oxide, and/or phagocytic activity in response to subsequent homogenous or heterogeneous neuroimmunostimulatory insults at a distant time. Many of the species released by activated proinflammatory microglia, such as IL-1β, ROS, and RNS, result in impairments in neuroplasticity and neurogenesis, including growth, differentiation, myelination, apoptosis, synaptic branching, and neurotrophin regulation [10, 61], and are proposed to be drivers of neuroprogression [11]. Thus, trained microglia may render an individual more susceptible to neurological injury and the development of neuropsychiatric illness through enhanced production of these neurotoxins. It is also possible that neuroimmunostimulatory insults in adulthood may train microglia. While a number of sources of inflammation have been proposed for depression [23], it is not clear why some individuals might develop depression in response to one of these sources, while others do not, even when taking into account genetic risk. Microglial training as a result of previous events may be one mechanism that accounts for interindividual variation in susceptibility to depression and other neuropsychiatric illnesses.

Once microglia are trained, this may lead to a vicious cycle whereby neuroinflammation begets neuroinflammation. In response to neuroinflammation and cell damage and death, a host of DAMPs are released [11]. These DAMPs may, depending on conditions, train microglia, causing an exaggerated neurotoxic response towards subsequent stimulation with DAMPs or PAMPs. This could lead to a cycle of increasing neuroinflammation with each subsequent neuroimmunostimulatory event. Indeed, this may explain how prior episodes of depression sensitize immune responses to subsequent depressive episodes, with levels of pro-inflammatory cytokines and chemokines increasing with sequential depressive episodes [8].

The finding that monocytes, macrophages, and natural killer cells can retain long-term memory of infectious insults, resulting in the upregulation of proinflammatory cytokines and chemokines, represents a paradigm shift in our understanding of the innate immune system, with potentially profound implications for the study of autoimmune diseases and the design of immunotherapeutics such as vaccines. It is possible that microglia, which are extensively implicated in the pathogenesis of several psychiatric disorders, may also retain long-term memory of infectious and noninfectious insults. Additional studies are needed to elucidate whether, when, and how microglia display training. If confirmed, microglial training may also have significant implications for our understanding and treatment of neuropsychiatric diseases and neuroprogression.

References

1 Zunszain PA, Hepgul N, Pariante CM: Inflammation and depression; in Cowen PJ, Sharp T, Lau JYF (eds): Behavioral Neurobiology of Depression and Its Treatment. Curr Top Behav Neurosci. Berlin, Springer, 2012, vol 14, pp 135–151.

2 Zunszain PA, Anacker C, Cattaneo A, Carvalho LA, Pariante CM: Glucocorticoids, cytokines and brain abnormalities in depression. Prog Neuropsychopharmacol Biol Psychiatry 2011;35:722–729.

3 Anderson G, Berk M, Dodd S, Bechter K, Altamura AC, Dell'Osso B, Kanba S, Monji A, Fatemi SH, Buckley P, Debnath M, Das UN, Meyer U, Müller N, Kanchanatawan B, Maes M: Immuno-inflammatory, oxidative and nitrosative stress, and neuroprogressive pathways in the etiology, course and treatment of schizophrenia. Prog Neuropsychopharmacol Biol Psychiatry 2013;42:1–4.

4 Bale TL, Baram TZ, Brown AS, Goldstein JM, Insel TR, McCarthy MM, Nemeroff CB, Reyes TM, Simerly RB, Susser ES, Nestler EJ: Early life programming and neurodevelopmental disorders. Biol Psychiatry 2010;68:314–319.

5 Cattaneo A, Macchi F, Plazzotta G, Veronica B, Bocchio-Chiavetto L, Riva MA, Pariante CM: Inflammation and neuronal plasticity: a link between childhood trauma and depression pathogenesis. Front Cell Neurosci 2015;9:40.

6 McDade TW: Early environments and the ecology of inflammation. Proc Natl Acad Sci 2012;109:17281–17288.

7 Bilbo SD: Early-life programming of later-life brain and behavior: a critical role for the immune system. Front Behav Neurosci 2009;3:14.

8 Moylan S, Maes M, Wray NR, Berk M: The neuroprogressive nature of major depressive disorder: pathways to disease evolution and resistance, and therapeutic implications. Mol Psychiatry 2012;18:595–606.

9 Netea MG, Latz E, Mills KHG, O'Neill LAJ: Innate immune memory: a paradigm shift in understanding host defense. Nat Immunol 2015;16:675–679.

10 Tremblay ME, Sierra A: Microglia in Health and Disease. New York, Springer, 2014.

11 Bakunina N, Pariante CM, Zunszain PA: Immune mechanisms linked to depression via oxidative stress and neuroprogression. Immunology 2015, Epub ahead of print.

12 Akira S, Uematsu S, Takeuchi O: Pathogen recognition and innate immunity. Cell 2006;124:783–801.

13 Netea MG, Quintin J, van der Meer JWM: Trained immunity: a memory for innate host defense. Cell Host Microbe 2011;9:355–361.

14 Martin SF: Adaptation in the innate immune system and heterologous innate immunity. Cell Mol Life Sci 2014;71:4115–4130.

15 Netea MG: Training innate immunity: the changing concept of immunological memory in innate host defence. Eur J Clin Invest 2013;43:881–884.

16 Biswas SK, Lopez-Collazo E: Endotoxin tolerance: new mechanisms, molecules and clinical significance. Trends Immunol 2009;30:475–487.

17 Quintin J, Saeed S, Martens JHA, Giamarellos-Bourboulis EJ, Ifrim DC, Logie C, Jacobs L, Jansen T, Kullberg B-J, Wijmenga C, Joosten LAB, Xavier RJ, van der Meer JWM, Stunnenberg HG, Netea MG: *Candida albicans* infection affords protection against reinfection via functional reprogramming of monocytes. Cell Host Microbe 2012;12:223–232.

18 Kleinnijenhuis J, Quintin J, Preijers F, Joosten LAB, Ifrim DC, Saeed S, Jacobs C, van Loenhout J, de Jong D, Stunnenberg HG, Xavier RJ, van der Meer JWM, van Crevel R, Netea MG: Bacille Calmette-Guérin induces NOD2-dependent nonspecific protection from reinfection via epigenetic reprogramming of monocytes. Proc Natl Acad Sci USA 2012;109:17537–17542.

19 Kleinnijenhuis J, Quintin J, Preijers F, Benn CS, Joosten LAB, Jacobs C, van Loenhout J, Xavier RJ, Aaby P, van der Meer JWM, van Crevel R, Netea MG: Long-lasting effects of BCG vaccination on both heterologous Th1/ Th17 responses and innate trained immunity. J Innate Immun 2014;6:152–158.

20 Jensen KJ, Larsen N, Biering-Sørensen S, Andersen A, Eriksen HB, Monteiro I, Hougaard D, Aaby P, Netea MG, Flanagan KL, Benn CS: Heterologous immunological effects of early BCG vaccination in low-birth-weight infants in Guinea-Bissau: a randomized-controlled trial. J Infect Dis 2015;211:956–967.

21 Ifrim DC, Quintin J, Joosten LAB, Jacobs C, Jansen T, Jacobs L, Gow NAR, Williams DL, van der Meer JWM, Netea MG: Trained immunity or tolerance: opposing functional programs induced in human monocytes after engagement of various pattern recognition receptors. Clin Vaccine Immunol 2014;21:534–545.

22 Ginhoux F, Greter M, Leboeuf M, Nandi S, See P, Gokhan S, Mehler MF, Conway SJ, Ng LG, Stanley ER, Samokhvalov IM, Merad M: Fate mapping analysis reveals that adult microglia derive from primitive macrophages. Science 2010;330:841–845.

23 Berk M, Williams LJ, Jacka FN, Neil AO, Pasco JA, Moylan S, Allen NB, Stuart AL, Hayley AC, Byrne ML, Maes M: So depression is an inflammatory disease, but where does the inflammation come from? BMC Med 2013;11:200.

24 Seong SY, Matzinger P: Hydrophobicity: an ancient damage-associated molecular pattern that initiates innate immune responses. Nat Rev Immunol 2004;4:469–478.

25 Matt U, Sharif O, Martins R, Knapp S: Accumulating evidence for a role of oxidized phospholipids in infectious diseases. Cell Mol Life Sci 2014;72:1059–1071.

26 Bekkering S, Quintin J, Joosten LAB, van der Meer JWM, Netea MG, Riksen NP: Oxidized low-density lipoprotein induces long-term proinflammatory cytokine production and foam cell formation via epigenetic reprogramming of monocytes. Arterioscler Thromb Vasc Biol 2014;34:1731–1738.

27 Edwards SW: Biochemistry and Physiology of the Neutrophil. Cambridge, Cambridge University Press, 2005.

28 Saeed S, Quintin J, Kerstens HHD, Rao NA, Aghajanirefah A, Matarese F, Cheng SC, Ratter J, Berentsen K, van der Ent MA, Sharifi N, Janssen-Megens EM, Ter Huurne M, Mandoli A, van Schaik T, Ng A, Burden F, Downes K, Frontini M, Kumar V, Giamarellos-Bourboulis EJ, Ouwehand WH, van der Meer JWM, Joosten LAB, Wijmenga C, Martens JHA, Xavier RJ, Logie C, Netea MG, Stunnenberg HG: Epigenetic programming of monocyte-to-macrophage differentiation and trained innate immunity. Science 2014;345:1251086.
29 Cheng SC, Quintin J, Cramer RA, Shepardson KM, Saeed S, Kumar V, Giamarellos-Bourboulis EJ, Martens JHA, Rao NA, Aghajanirefah A, Manjeri GR, Li Y, Ifrim DC, Arts RJW, van der Veer BMJW, Deen PMT, Logie C, O'Neill LA, Willems P, van de Veerdonk FL, van der Meer JWM, Ng A, Joosten LAB, Wijmenga C, Stunnenberg HG, Xavier RJ, Netea MG: mTOR- and HIF-1-mediated aerobic glycolysis as metabolic basis for trained immunity. Science 2014;345: 1250684.
30 Yoshida K, Maekawa T, Zhu Y, Renard-Guillet C, Chatton B, Inoue K, Uchiyama T, Ishibashi K-I, Yamada T, Ohno N, Shirahige K, Okada-Hatakeyama M, Ishii S: The transcription factor ATF7 mediates lipopolysaccharide-induced epigenetic changes in macrophages involved in innate immunological memory. Nat Immunol 2015;16:1034–1043.
31 Maekawa T, Kim S, Nakai D, Makino C, Takagi T, Ogura H, Yamada K, Chatton B, Ishii S: Social isolation stress induces ATF-7 phosphorylation and impairs silencing of the 5-HT 5B receptor gene. EMBO J 2009;29:196–208.
32 Perry VH, Holmes C: Microglial priming in neurodegenerative disease. Nat Rev Neurol 2014;10:217–224.
33 Walkera FR, Nilsson M, Jones K: Acute and chronic stress-induced disturbances of microglial plasticity, phenotype and function. Curr Drug Targets 2013;14: 1262–1276.
34 Püntener U, Booth SG, Perry VH, Teeling JL: Long-term impact of systemic bacterial infection on the cerebral vasculature and microglia. J Neuroinflammation 2012;9:146.
35 Bosma GC, Custer RP, Bosma MJ: A severe combined immunodeficiency mutation in the mouse. Nature 1983; 301:527–530.
36 Bilbo SD: Neonatal infection-induced memory impairment after lipopolysaccharide in adulthood is prevented via caspase-1 inhibition. J Neurosci 2005; 25:8000–8009.
37 Bland ST, Beckley JT, Watkins LR, Maier SF, Bilbo SD: Neonatal *Escherichia coli* infection alters glial, cytokine, and neuronal gene expression in response to acute amphetamine in adolescent rats. Neurosci Lett 2010;474:52–57.
38 Williamson LL, Sholar PW, Mistry RS, Smith SH, Bilbo SD: Microglia and memory: modulation by early-life infection. J Neurosci 2011;31:15511–15521.
39 Cao M, Cortes M, Moore CS, Leong SY, Durosier LD, Burns P, Fecteau G, Desrochers A, Auer RN, Barreiro LB, Antel JP, Frasch MG: Fetal microglial phenotype in vitro carries memory of prior in vivo exposure to inflammation. Front Cell Neurosci 2015;9:12.
40 Ye M, Wang Q, Zhang W, Li Z, Wang Y, Hu R: Oroxylin A exerts anti-inflammatory activity on lipopolysaccharide-induced mouse macrophage via Nrf2/ARE activation. Biochem Cell Biol 2014;92: 337–348.
41 Fleshner M, Campisi J, Amiri L, Diamond DM: Cat exposure induces both intra- and extracellular Hsp72: the role of adrenal hormones. Psychoneuroendocrinology 2004;29:1142–1152.
42 Maslanik T, Mahaffey L, Tannura K, Beninson L, Greenwood BN, Fleshner M: The inflammasome and danger associated molecular patterns (DAMPs) are implicated in cytokine and chemokine responses following stressor exposure. Brain Behav Immun 2013;28:54–62.
43 Weber MD, Frank MG, Tracey KJ, Watkins LR, Maier SF: Stress induces the danger-associated molecular pattern HMGB-1 in the hippocampus of male Sprague Dawley rats: a priming stimulus of microglia and the NLRP3 inflammasome. J Neurosci 2015;35:316–324.
44 Fleshner M: Stress-evoked sterile inflammation, danger associated molecular patterns (DAMPs), microbial associated molecular patterns (MAMPs) and the inflammasome. Brain Behav Immun 2013;27:1–7.
45 Fenn AM, Gensel JC, Huang Y, Popovich PG: Immune activation promotes depression 1 month after diffuse brain injury: a role for primed microglia. Biol Psychiatry 2014;76:575–584.
46 Ramlackhansingh AF, Brooks DJ, Greenwood RJ, Bose SK, Turkheimer FE, Kinnunen KM, Gentleman S, Heckemann RA, Gunanayagam K, Gelosa G, Sharp DJ: Inflammation after trauma: microglial activation and traumatic brain injury. Ann Neurol 2011;70:374–383.
47 Ramirez K, Shea DT, McKim DB, Reader BF, Sheridan JF: Imipramine attenuates neuroinflammatory signaling and reverses stress-induced social avoidance. Brain Behav Immun 2015;46:212–220.
48 Schaafsma W, Zhang X, van Zomeren KC, Jacobs S, Georgieva PB, Wolf SA, Kettenmann H, Janova H, Saiepour N, Hanisch UK, Meerlo P, van den Elsen PJ, Brouwer N, Boddeke HWGM, Eggen BJL: Long-lasting pro-inflammatory suppression of microglia by LPS-preconditioning is mediated by RelB-dependent epigenetic silencing. Brain Behav Immun 2015;48:205–221.
49 Iwata M, Ota KT, Duman RS: The inflammasome: pathways linking psychological stress, depression, and systemic illnesses. Brain Behav Immun 2013;31: 105–114.
50 Yirmiya R, Rimmerman N, Reshef R: Depression as a microglial disease. Trends Neurosci 2015;38:637–658.
51 Miller GE, Murphy MLM, Cashman R, Ma R, Ma J, Arevalo JMG, Kobor MS, Cole SW: Greater inflammatory activity and blunted glucocorticoid signaling in monocytes of chronically stressed caregivers. Brain Behav Immun 2014;41: 191–199.
52 Lisi L, Camardese G, Treglia M, Tringali G, Carrozza C, Janiri L, Russo CD, Navarra P: Monocytes from depressed patients display an altered pattern of response to endotoxin challenge. PLoS One 2013;8:e52585–e52587.
53 Rohleder N, Joksimovic L, Wolf JM, Kirschbaum C: Hypocortisolism and increased glucocorticoid sensitivity of pro-inflammatory cytokine production in Bosnian war refugees with posttraumatic stress disorder. Biol Psychiatry 2004;55:745–751.
54 Miller GE, Rohleder N, Stetler C, Kirschbaum C: Clinical depression and regulation of the inflammatory response during acute stress. Psychosom Med 2005; 67:679–687.

55 Schwaiger M, Grinberg M, Moser D, Zang JCS, Heinrichs M, Hengstler JG, Rahnenführer J, Cole S, Kumsta R: Altered stress-induced regulation of genes in monocytes in adults with a history of childhood adversity. Neuropsychopharmacology 2016;41:2530–2540.
56 Uddin M, Koenen KC, Aiello AE, Wildman DE, de los Santos R, Galea S: Epigenetic and inflammatory marker profiles associated with depression in a community-based epidemiologic sample. Psychol Med 2010;41:997–1007.
57 Uddin M, Aiello AE, Wildman DE, Koenen KC, Pawelec G, de los Santos R, Goldmann E, Galea S: Epigenetic and immune function profiles associated with posttraumatic stress disorder. Proc Natl Acad Sci USA 2010;107:9470–9475.
58 Smith AK, Conneely KN, Kilaru V, Mercer KB, Weiss TE, Bradley B, Tang Y, Gillespie CF, Cubells JF, Ressler KJ: Differential immune system DNA methylation and cytokine regulation in posttraumatic stress disorder. Am J Med Genet B Neuropsychiatr Genet 2011; 156:700–708.
59 da Costa SC, Passos IC, Lowri C, Soares JC, Kapczinski F: Refractory bipolar disorder and neuroprogression. Prog Neuropsychopharmacol Biol Psychiatry 2016;70:103–110.
60 Nassar A, Azab AN: Effects of lithium on inflammation. ACS Chem Neurosci 2014;5451–458.
61 Zunszain PA, Anacker C, Cattaneo A, Choudhury S, Musaelyan K, Myint AM, Thuret S, Price J, Pariante CM: Interleukin-1β: a new regulator of the kynurenine pathway affecting human hippocampal neurogenesis. Neuropsychopharmacology 2012;37:939–949.

Alex P. Salam
Stress, Psychiatry, and Immunology Laboratory
Institute of Psychiatry, Psychology, and Neuroscience
King's College London, London SE5 8AF (UK)
E-Mail alexsalam@doctors.org.uk

Halaris A, Leonard BE (eds): Neuroprogression in Psychiatric Disorders.
Mod Trends Pharmacopsychiatry. Basel, Karger, 2017, vol 31, pp 79–95 (DOI: 10.1159/000470809)

Inflammatory and Innate Immune Markers of Neuroprogression in Depressed and Teenage Suicide Brain

Ghanshyam N. Pandey

Department of Psychiatry, University of Illinois at Chicago, Chicago, IL, USA

Abstract

Several studies suggest that major depressive disorder (MDD) and bipolar disorder (BPD) are neuroprogressive illnesses. Besides clinical features, neurobiological mechanisms have been suggested to contribute to the neuroprogression of mood disorders. Biological factors that have been shown to contribute significantly toward the neuroprogressive course of these disorders are inflammatory markers, such as cytokines. Cytokines have been extensively investigated, primarily in the serum of MDD and BPD patients, and these studies show cytokine abnormalities in both adolescent and adult patients with mood disorders. However, cytokine abnormalities in the brain may also contribute toward neuroprogression, but brain cytokines have not been adequately investigated. To examine the role of cytokines in neuroprogression, we have studied the markers of adaptive and innate immunity in postmortem brain obtained from teenage and adult suicide victims and gene expression of cytokines and their membrane-bound receptors in lymphocytes of MDD and BPD patients. Cytokines and Toll-like receptors (TLRs) were studied in 24 teenage suicide victims and 24 normal control (NC) subjects, and also in 22 adult depressed suicide victims and 20 adult NC subjects. We found that the protein and mRNA expression of the proinflammatory cytokines tumor necrosis factor (TNF)-α, interleukin (IL)-1β, and IL-6 were significantly higher in the prefrontal cortex (PFC). We also found that the protein and mRNA expression of TLRs, which are major mediators of innate immunity, is increased in the PFC of adult depressed suicide victims and NC subjects. In patients, mRNA and protein expression of TNF-α, IL-1β, and IL-6 was significantly increased in both MDD and BPD patients. Similarly, mRNA expression of some specific membrane-bound receptors, such as IL1R1, TNFR1, IL1RA, were significantly increased in lymphocytes of MDD and BPD patients. These studies indicate the existence of abnormal cytokines and TLRs in the brain of teenage and adult suicide victims. Future studies, including both teenage and adult postmortem samples, will be needed to further clarify the role of cytokines and TLRs in neuroprogression.

Inflammation and Neuroprogression in Mood Disorders

Several studies suggest that major depressive disorder (MDD) and bipolar disorder (BPD) are neuroprogressive illnesses. Clinical studies and experience confirm that mood disorders are chronic and progressive illnesses associated with recurrence and relapse, and, therefore, are neuroprogressive in nature. Main factors associated with neuroprogression are the chronic and repetitive nature of depressive and manic episodes, gender, age at onset, episode frequency, partial response, and treatment resistance. The higher the frequency of depressive and manic episodes and the longer the course of the illness, the more severe the neuroprogressive course resulting in an unfavorable prognosis.

Besides the clinical features, neurobiological mechanisms have also been suggested to contribute to the neuroprogression of mood disorders. These biological factors contribute to the pathophysiology of neuroprogression and include neurotransmitter systems, inflammation, oxidative and nitrosative stress, neurotrophins, neurogenesis, hypothalamic-pituitary-adrenal axis abnormalities, and mitochondrial dysfunction, for example. Among the main biological mechanisms, inflammatory markers appear to contribute significantly to neuroprogression, as evidenced by several studies. Several meta-analyses of studies of cytokines suggest that levels of some of the proinflammatory cytokines, such as interleukin (IL)-1β, IL-6, and tumor necrosis factor (TNF)-α, are increased in the serum of MDD and BPD patients. The changes in inflammatory markers appear to be related to the number of depressive episodes. IL-1β and TNF-α are significantly higher in patients with more than 3 depressive episodes, suggesting that prior depressive episodes potentially increase vulnerability to further depressive episodes. In addition, some studies suggest that a lifetime history of MDD is associated with increased IL-6 and soluble IL-1 receptor antagonist (IL1RA). These cytokines may influence neuroprogression through different mechanisms, e.g., TNF-α-mediated cell signaling causes apoptosis associated with caspase-dependent mechanisms. Thus, inflammation likely contributes to the recurrence of MDD episodes.

We have examined the contribution and abnormality of cytokines in MDD and BPD in the following manner.

(1) Most of the studies of cytokines in mood disorders are based on the determination of protein expression of cytokines and their soluble receptors in the serum of MDD and BPD patients and normal control (NC) subjects. The biological and behavioral effects of cytokines are mediated through the membrane-bound cytokine receptors. Soluble receptors, which are degraded products of the membrane-bound receptors, also alter the function of these receptors by competing with cytokines for the membrane-bound receptors.

(2) Whereas most of the studies of cytokines have determined their protein expression in the serum of MDD and BPD patients, relatively very few studies have determined gene (mRNA) expression of inflammatory cytokines. Also, whereas several studies have determined soluble cytokine receptors, the membrane-bound receptors have not been adequately investigated.

(3) Since most of the studies of cytokines are carried out in peripheral cells or in serum, it is not clear whether these abnormalities are also present in the brain. Bearing in mind all these factors, we have determined the gene and protein expression of cytokines and their membrane-bound receptors in lymphocytes obtained from MDD, BPD, and NC subjects. To examine the status of proinflammatory cytokines in the brain, we have determined the gene and protein expression of these cytokines in postmortem brain of teenage suicide victims. To investigate the reasons for increased cytokine levels in the brain, we also determined the innate immunity receptors, known as Toll-like receptors (TLRs), which mediate the func-

tions of cytokines and also stimulate their production. These studies are described in the following pages.

Cytokines and Depression

There are many interactions between neural, immune, and neuroendocrine systems, and this has raised the question of whether the immune system may also be involved in some of the brain-related disorders, such as depression [1–4]. In recent years, it has been suggested that depression, which is one of the major psychiatric disorders known to be related to changes in the neuroendocrine system, may also be related to or caused by changes in the immune system.

Cytokines are a diverse group of proteins that can be considered as hormones of the immune system. These small molecules are secreted by various cells and act as signals between the cells to regulate the immune responses to injury and infection. The responses of cytokines are mediated through cytokine receptors. As is the case with other receptors, specific cytokine receptors respond to the presence of specific cytokines and thus produce their physiological responses. Cytokine receptors are present both in soluble forms and membrane bound.

The classification and nomenclature of cytokines is complex. Cytokines have been classified into families of IL, TNF, interferons (IFN), chemokines, hematoproteins, and colony-stimulating factors. In terms of biological activity, cytokines have been classified as pro- or anti-inflammatory. The main proinflammatory cytokines, which are involved in inflammatory processes, include IL-1, IL-6, IFN-α, and TNF-α. The anti-inflammatory cytokines include IL-4, IL-10, and IL-13. Cytokines have been shown to regulate growth, differentiation, and function of many cell types. Several of the cytokines are present in more than one form, and these are termed as α, β, or γ.

One of the two major lines of evidence suggesting that cytokines play an important role in the pathophysiology of depression is derived from the observation that administration of some cytokines, such as IFN-α, to patients with hepatitis or melanoma produced a syndrome known as "sickness behavior" that is very similar to depression [5–16]. The other major piece of evidence for a role of cytokines in the pathophysiology of depression is derived from the observations that levels of pro- and/or anti-inflammatory cytokines are altered in the serum [17–31] or cerebrospinal fluid (CSF) [32, 33] of depressed patients. Other indirect evidence suggesting a role for cytokines in depression is the observation that stress, which is a major risk factor for depressive illness, alters not only the immune system but also the levels of several cytokines [33–38].

Several observations suggest that altered immune states may be present in patients suffering from depressive disorders. For example, Kronfol et al. [39] measured lymphocyte response to mitogens in patients diagnosed with melancholic depression, and those patients were found to have a decreased lymphocyte response compared with controls. The findings of lower lymphocyte responsiveness have been replicated in several studies [39–43]. Several investigators have also observed that natural killer cell activity was decreased in patients with depression compared with NC subjects [39, 42–45]. The other immune function studied in depressed patients – neutrophil activity – was found to be lower among depressed patients than control subjects. These studies do indicate that the immune process may be reduced in patients with depression (for review, see Weisse [31]).

Cytokines in Suicide

The role of cytokines and immune dysregulation has been studied in great detail in patients with mood disorders and schizophrenia. However, their role in suicide is less clear. Since both depression and stress are major risk factors for suicide, it is quite likely that abnormalities in proinflammatory cytokines may be associated with the pathophysiology of suicide. There is also direct and indirect evidence suggesting a relationship between immune dysregulation and suicide. In a study by Steiner et al. [46], microgliosis was increased in postmortem brain of suicide victims with affective disorders and schizophrenia compared with NC subjects. Goodwin and Eaton [47] found a significant association between asthma and increased suicidal ideation and suicide attempts among adults in the community. Goodwin et al. [48] reported that youth who are hospitalized for asthma have higher than expected levels of suicidal ideation. That an abnormality in cytokines may be associated with suicidal behavior is supported by a recent report by Tonelli et al. [49] who revealed increased mRNA expression of IL-4 and IL-3 in the prefrontal cortex (PFC) of female suicide victims and IL-13 in male suicide victims compared with NC subjects. Lindqvist et al. [50] have observed increased levels of IL-6 in the CSF of suicide attempters. In sum, these studies suggest that cytokines are abnormally regulated in suicide.

Although abnormal levels of cytokines are observed in the serum of patients with depression, it is not clear if there are also abnormal levels of cytokines in the brain. The immunological aspects of the neurobiology of suicide have been reported by Steiner et al. [46], but cytokines in the brain of suicide victims or subjects with depression have not been systematically studied.

Inflammatory Cytokines in the Brain as Risk Factors for Depression and Suicide

The major evidence suggesting that cytokine abnormalities may be associated with suicide is derived from studies of inflammatory cytokines in the serum of patients with suicidal behavior and nonsuicidal patients. For example, a meta-analysis of studies in patients with suicidal behavior by Black and Miller [51] suggested that some of the cytokines may be abnormally expressed in the serum of patients with suicidal behavior. Although there are several studies of cytokines in patients with suicidal behavior, it is not clearly known if similar cytokine abnormalities are also present in the brain. The presence of cytokine immune responses in the brain is now well established. Besides cytokines, the receptors for IL-1, TNFα and IL-6 have also been measured in a number of brain regions, and they are found on both glia and neurons [2, 52]. Many of the effects of infection, such as fever, disturbed sleep, and decreased food and water intake, are also altered after IFN-α administration. These proinflammatory cytokines can produce behavioral effects through various mechanisms, and it has been suggested that cytokines interact with monoaminergic systems, the serotonergic system in particular.

Although the presence of cytokines in the brain has been demonstrated, how are these cytokines produced in the brain? Brain cytokines and their receptors can be synthesized in the brain itself, or they can be derived from peripheral sources. Although, under normal conditions, cytokines cannot cross the blood-brain barrier, their transport to the brain may occur through several mechanisms and pathways. For example, cytokines can enter the brain through areas with a weaker blood-brain barrier. Cytokines could also enter the brain via an active transport system. Although these data have been primarily derived from studies in animals, they appear to be consistent with human studies, in that it has been shown that the administration of proinflammatory cytokines, such as IFN-α, to patients with hepatitis is associated with the activation of inflammatory responses in the brain. Increased CSF concentrations of IL-6 and other cytokines have been observed in depressed and suicidal patients. Activation of central inflammatory responses has also been shown to correlate with changes in neurotransmitter metabolism and stimulation of the kynurenine pathway, which may lead to changes in the CSF concentration of kynurenic acid, a metabolite of this pathway.

Although the presence of cytokines in the brain has been demonstrated, it also suggests that cytokines may be transported to the brain through various mechanisms. However, the studies of cytokines and their receptors in human postmortem brain are limited. Therefore, to examine if cytokines are present in human brain and are also abnormally expressed in depression and/or suicide, we conducted studies of proinflammatory cytokines in the PFC obtained from 24 teenage suicide victims and 24 teenage NC subjects [53].

Cytokine Studies in the Brain of Teenage Subjects

Characteristics of Postmortem Brain Samples

Postmortem brain samples were obtained from the Brain Collection Program of the Maryland Psychiatric Research Center, Baltimore, MD, USA. The samples were obtained from 24 teenage suicide victims and 24 teenage NC subjects. The samples were screened for toxicological data using urine and blood samples.

The subjects were diagnosed by psychological autopsy which consisted of interviewing members of the family and a friend by means of clinical interviews using the Diagnostic and Statistical Manual of Mental Disorders (DSM-IV, 4th edition)/Structured Clinical Interview for DSM Disorders (SCID). The SCID was administered by a trained interviewer. The age range of both NC subjects and suicide victims was between 13 and 20 years, and included male and female subjects. Suicide was generally committed by gunshot, stabbing, hanging, or drowning, and some subjects had utilized violent methods. The mean age and postmortem interval of suicide subjects and NC subjects were similar and did not significantly differ. Although diagnoses varied overall, the majority of subjects had the diagnosis of MDD or adjustment or conduct disorders. Few (5 or 6) subjects had no history of psychopathology. The protein expression of TNF-α, IL-1β, and IL-6 in the PFC was determined using the ELISA method, and mRNA expression was determined using quantitative real-time PCR (qPCR).

Proinflammatory Cytokines in the Prefrontal Cortex of Teenage Suicide Victims and Normal Control Subjects

When we first compared the mRNA expression levels of IL-1β, IL-6, and TNF-α in Brodmann area (BA) 10 of teenage suicide subjects and NC subjects, we observed that mRNA levels of the proinflammatory cytokines IL-1β, IL-6, and TNF-α were significantly increased in the BA10 of teenage suicide subjects compared with NC subjects (Fig. 1) [53]. When we compared the protein levels of cytokines between teenage suicide victims and NC subjects, we found that the protein expression levels of IL-1β, IL-6, and TNF-α were significantly increased in the BA10 of PFC of suicide victims compared with NC subjects (Fig. 2).

Since this was probably the first study of proinflammatory cytokines in postmortem brain of suicide victims, we wanted to replicate and confirm this observation in another brain area. We, therefore, determined the protein expression of cytokines in BA8, an adjacent PFC area. We found that the protein expression of TNF-α and IL-1β in BA8 was also significantly increased in teenage suicide victims compared with NC subjects (Fig. 3).

These results thus suggested that the expression of the proinflammatory cytokines we studied was significantly increased in 2 adjacent PFC areas (BA10 and BA8) of teenage suicide victims compared with teenage NC subjects [53].

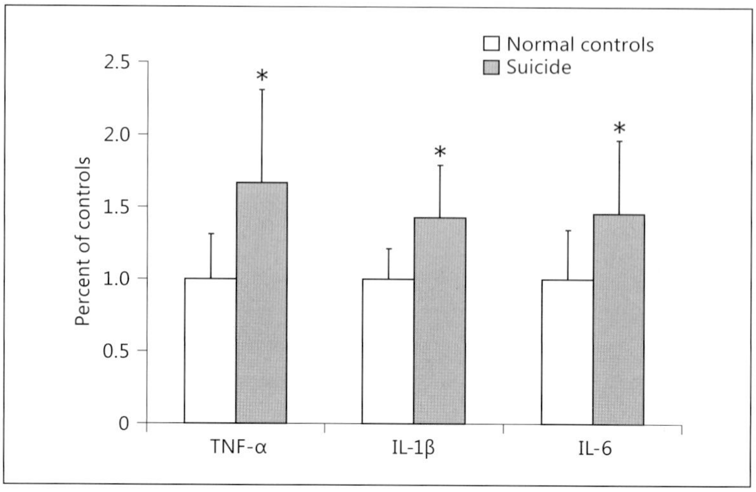

Fig. 1. mRNA levels of proinflammatory cytokines (mean ± SD) in the postmortem brain of teenage suicide victims and normal control subjects: tumor necrosis factor (TNF)-α, interleukin (IL)-1β, and IL-6 in the prefrontal cortex (Brodmann area 10) of normal control (NC) subjects ($n = 24$) and teenage suicide victims ($n = 24$). * $p < 0.01$.

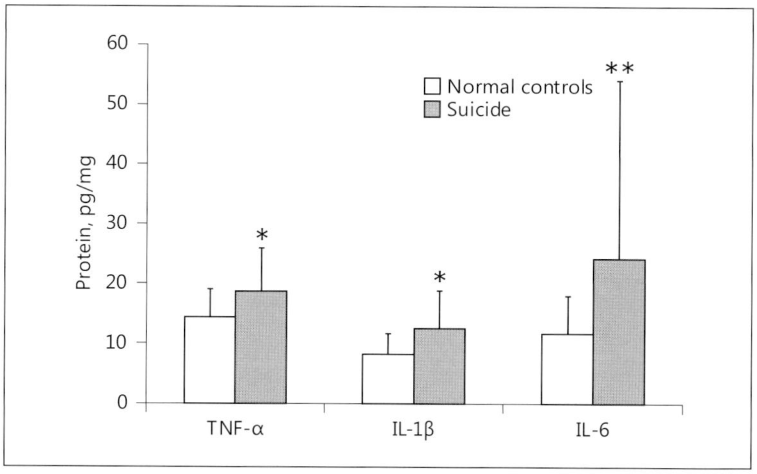

Fig. 2. Protein levels of proinflammatory cytokines (mean ± SD) in the postmortem brain of teenage suicide victims and normal control (NC) subjects: tumor necrosis factor (TNF)-α, interleukin (IL)-1β, and IL-6 in the prefrontal cortex (Brodmann area 10) of NC subjects ($n = 24$) and teenage suicide victims ($n = 24$). * $p < 0.01$, ** $p < 0.06$.

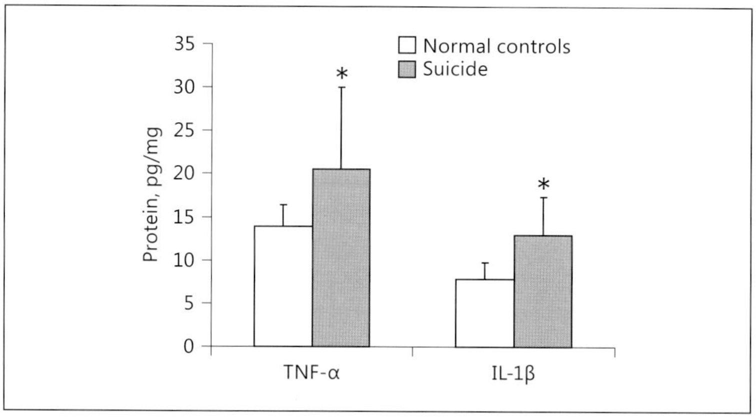

Fig. 3. Protein levels of proinflammatory cytokines (mean ± SD) in the postmortem brain of teenage suicide victims and normal control (NC) subjects: tumor necrosis factor (TNF)-α and interleukin (IL)-1β in the prefrontal cortex (Brodmann area 8) of NC subjects ($n = 24$) and teenage suicide victims ($n = 24$). * $p < 0.01$.

Cytokine Studies in the Brain of Adult Patients

There are only a few studies of cytokines in postmortem brain of adult subjects with mood disorders and suicide. Using qPCR, Tonelli et al. [49] determined the mRNA levels of TNF-α, IL-1β, and IL-6 in the BA11 area of postmortem brain samples obtained from suicide and control subjects. They did not find significant differences in the mRNA expression of IL-1β, IL-6, and TNF-α between suicide and control subjects in BA11. However, they found that there was a significant difference in the mRNA expression of IL-4 between suicide and NC subjects, and between women and men. When separated by gender, they found increased IL-4 expression in female suicide victims and increased IL-13 expression in male suicide victims compared to the respective controls. No significant difference in TNF-α was found in male and female suicide victims compared to NC subjects.

Dean et al. [54] determined the TNF-α concentration in the PFC (BA24 and BA46) obtained from MDD and NC subjects. In the brain, TNF-α exists in 2 forms, membrane-bound TNF (transmembrane) and soluble TNF. They found that transmembrane TNF was significantly increased in BA46 in MDD subjects but not in BA24. Soluble TNF was not significantly different in BA24 or BA46 between the MDD and control group.

Rao et al. [55] determined TNF-α, IL-1β, and IL receptor levels in postmortem PFC from patients with BPD disorders and NC subjects. They found significantly higher mRNA and protein levels of IL-1β and the IL-1receptor (IL1R) in the PFC of BPD patients compared with NC subjects.

In this respect, it is interesting to note that Shelton et al. [56] conducted a microarray analysis in postmortem brain (BA10) obtained from MDD and NC subjects. They found several genes that are related to apoptotic and inflammatory pathways to be overexpressed. In the inflammatory pathway, they found that the expression of cytokines IL-1α, IL-2, IL-3, IL-5, IL-8, IL-9, IL-10, IL-12α, IL-13, IL-15, IL-18, as well as lymphotoxin-α, a member of the TNF superfamily, was increased. This gene array study in general showed upregulation of both pro- and anti-inflammatory cytokines.

Toll-Like Receptors in Depression and Suicide

Besides cytokines, TLRs are other major mediators of immunity, particularly innate immunity [57]. TLRs are a family of innate immune system receptors (up to 11) that were first discovered in *Drosophila melanogaster* [57, 58]. Since then, many TLRs have been described. They play a developmental role during embryogenesis as well as a physiological and metabolic role in adults and are expressed in a variety of mammalian immune system-related cell types and the brain. TLRs recognize pathogen-associated molecular patterns and damage-associated molecular patterns [57–59]. When a pathogen or damage-associated molecular pattern ligand binds to a TLR, it initiates signaling circuits through the nuclear factor (NF)-κB pathway resulting in the formation of cytokines and chemokines that in turn initiate a specific adaptive immune response [60]. TLRs are widely expressed in the CNS and have been found in neurons, microglia, and astrocytes [61, 62]. However, among all the TLRs, TLR3 is the only TLR expressed in human neurons [62, 63]. TLRs are also mediators of CNS plasticity, generation and growth of neurons, development of neuronal circuits, and cognition [64, 65]. Among the TLRs, TLR3 and TLR4 exert broad effects on cognitive performance of hippocampal-dependent and -independent tasks in mice, and TLR3 deficiency confers superior performance in spatial working memory and contextual fear [63, 65, 66].

Because of the important role TLRs play in neuronal function, and because of their involvement in the production of cytokines and chemokines through MyD88 and MyD88-independent pathways, activating NF-κB and resulting in the

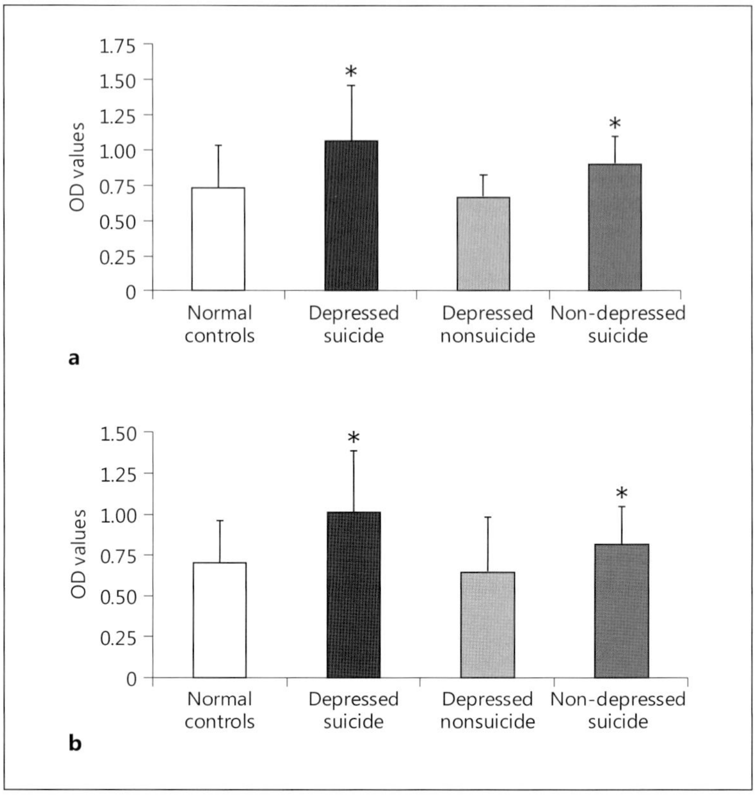

Fig. 4. Mean protein expression (mean ± SD) of Toll-like receptors TLR3 (**a**) and TLR4 (**b**) in the dorsolateral prefrontal cortex membrane fractions of normal control (NC) subjects ($n = 20$), depressed ($n = 22$) and nondepressed suicide victims ($n = 11$), and depressed nonsuicide subjects ($n = 12$). * $p < 0.05$. OD, optical density.

accumulation of chemokines and cytokines [60], we sought to determine whether abnormalities in TLR levels are associated with depression and/or suicide. For that purpose, we analyzed their protein and mRNA expression in PFC obtained from depressed and nondepressed suicide victims, depressed nonsuicide subjects, and NC subjects [67]. Of all the TLRs, we specifically determined the gene and protein expression of TLR3 and TLR4 in this study because (i) TLR3 is the only TLR present in human neurons [61]; (ii) it is associated with cognitive function in mice [63]; and (iii) both TLR3 and TLR4 are involved in neurite growth and other neuronal functions [68]. Both TLR3 and TLR4 are also unique compared to other TLRs because of their ability to activate IFN-regulatory factor through the MyD88-independent signaling pathway [68].

TLR3 and TLR4 in the Prefrontal Cortex of Depressed and Suicide Victims

We determined the protein and mRNA expression of TLR3 and TLR4 in 20 NC subjects, 22 depressed and 11 nondepressed suicide victims, and 12 depressed nonsuicide subjects [67]. We found that the protein expression of TLR3 and TLR4 was significantly increased in depressed suicide and nondepressed suicide victims compared with NC subjects (Fig. 4). However, there was no significant increase in protein expression in depressed nonsuicide compared with NC subjects. On the other hand, when we determined the mRNA expression of TLR3 and TLR4, we found that mRNA expression of TLR3 and TLR4 was significantly increased in depressed and nondepressed suicide victims, as well as in depressed nonsuicide subjects compared with NC subjects (Fig. 5). Thus, there was dissociation be-

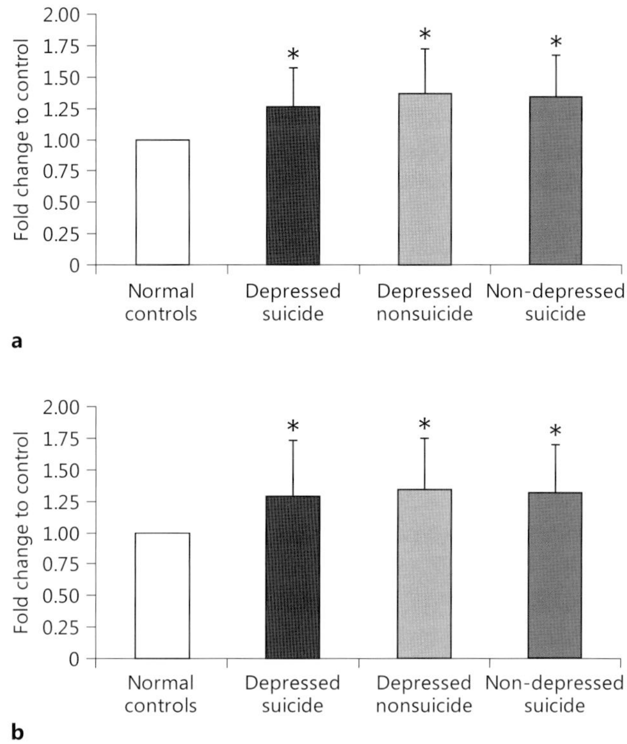

Fig. 5. mRNA expression (mean ± SEM) of Toll-like receptors TLR3 (**a**) and TLR4 (**b**) in the dorsolateral prefrontal cortex. * $p < 0.05$. **a** Normal control (NC) subjects 1.00 ± 0.09 ($n = 20$), depressed suicide victims 1.26 ± 0.06 ($n = 22$), depressed nonsuicide subjects 1.37 ± 0.10 ($n = 12$), and nondepressed suicide victims 1.35 ± 0.10 ($n = 11$). **b** NC 1.00 ± 0.09 ($n = 20$), depressed suicide victims 1.29 ± 0.09 ($n = 22$), depressed nonsuicide subjects 1.35 ± 0.12 ($n = 12$), and nondepressed suicide victims 1.32 ± 0.11 ($n = 11$).

tween protein and mRNA expression in depressed nonsuicide subjects such that the protein expression was not significantly different in depressed nonsuicide subjects compared with NC subjects, but the mRNA, similar to depressed and nondepressed suicide victims, was significantly higher in depressed nonsuicide subjects than NC subjects. TLR3 is the only TLR that does not signal through MyD88, but signals via TRIFF recruitment, causing the activation of the transcription factor IR3, leading to the expression of IFN-β and IFN-α, along with the activation of NF-κB, resulting in the production of cytokines and chemokines [60]. TLR3 is also the only TLR which is expressed in neurons and has significant effects on cognitive performance in hippocampal-dependent and -independent tasks in mice. TLR4 has also been implicated in the pathogenesis of cognitive impairment and alcoholism. Abnormal TLR3 and TLR4 expression thus suggests that innate immunity is impaired in depressed suicide subjects [67]. It is also possible that the observed increase in cytokines in the postmortem brain of suicide victims may be related to increased levels of some of the TLRs, such as TLR3 and TLR4.

These studies indicate abnormalities of innate immunity are found in the postmortem brain of depressed and suicide subjects and that cytokine abnormalities in suicide may be related to abnormalities of some specific TLRs.

Cytokines

As stated earlier, several studies indicate abnormalities of cytokines in mood disorders based on and confirming the original observation of sickness behavior induced by the administration of IFNs to either animals or cancer patients. Administration of IFNs to animals or patients induces a certain kind of behavior known as "sickness behavior" which includes cognitive changes. Although the symptoms of psychosis and mania have not been generally associated with IFN therapy, Du et al. [69] reported that some patients developed manic symptoms on IFN-α therapy which improved after the cessation of IFN-α therapy. Similarly, Greenberg et al. [70] have reported that treatment with IFNs caused the development of manic symptoms in melanoma patients. There are also several studies which suggest elevations of proinflammatory cytokine levels in BPD. For example, high levels of TNF-α and IL-6 have been reported in BPD patients, especially during mania, while IL-6 levels returned to baseline after treatment with mood stabilizers. These studies of cytokines in BPD have been reviewed in a meta-analysis by Munkholm et al. [71, 72].

Besides levels of proinflammatory cytokines, levels of soluble receptors have also been found to be increased in the serum of BPD patients. These soluble receptors play an important role in mediating the functions of proinflammatory cytokines either by competing with proinflammatory cytokines for their receptors or by acting directly on the membrane-bound cytokine receptors. Although the protein expression of cytokines has been studied in the serum of BPD patients, the gene expression of these inflammatory cytokines has generally not been studied in patients with mood disorders. Similarly, the levels of membrane-bound receptors have not been investigated in BPD patients either. The biological and behavioral effects of these cytokines on their target cells are mediated through membrane-bound receptors. Therefore, in order to appreciate and understand the functional consequences of altered cytokine receptors in BPD disorders, one really needs to study the levels of membrane-bound receptors.

We determined the mRNA levels of the membrane-bound receptors for IL-1, IL-6, and TNF-α in lymphocytes obtained from BPD and MDD patients and NC subjects [73, 74]. In addition, we also studied mRNA as well as protein expression levels of the proinflammatory cytokines IL-1β, IL-6, and TNF-α in lymphocytes and plasma of BPD and MDD patients and NC subjects [73, 74]. These studies were performed in blood obtained from MDD and BPD patients and NC subjects. The patients were hospitalized at the UIC (University of Illinois) Department of Psychiatry; NC subjects were not hospitalized. Blood was drawn from patients and NC subjects, and plasma and lymphocytes were isolated according to standard methods. Protein was determined in the plasma of these patients using an ELISA method, and mRNA was determined in lymphocytes using the qPCR method. Patients were diagnosed according to DSM-IV and SCID.

IL-1β, IL-6, TNF-α, and Their Membrane-Bound Receptors in Lymphocytes of Patients with Bipolar Disorder and Major Depression

We compared mRNA expression of the various receptors between NC and BPD subjects [73, 74]. Two types of IL-1 receptors have been shown and identified, known as IL1R1 and IL1R2. Of these two receptors, only IL1R1 initiates signal transduction when it interacts with IL-1β or IL-1α, although both of them use the same receptors. On the other hand, the IL1R2 receptors do not have signal transduction properties and normally function as a sink or decoy receptor. Thus, most of the signal transduction properties of the IL1Rs are mediated through IL1R1. A comparison of the mRNA expression of IL1R1 and IL1R2 revealed that the mRNA levels of IL1R1 were significantly increased in the lymphocytes of BPD (Fig. 6) and MDD (Fig. 7) patients compared to NC subjects, whereas we did not find any signif-

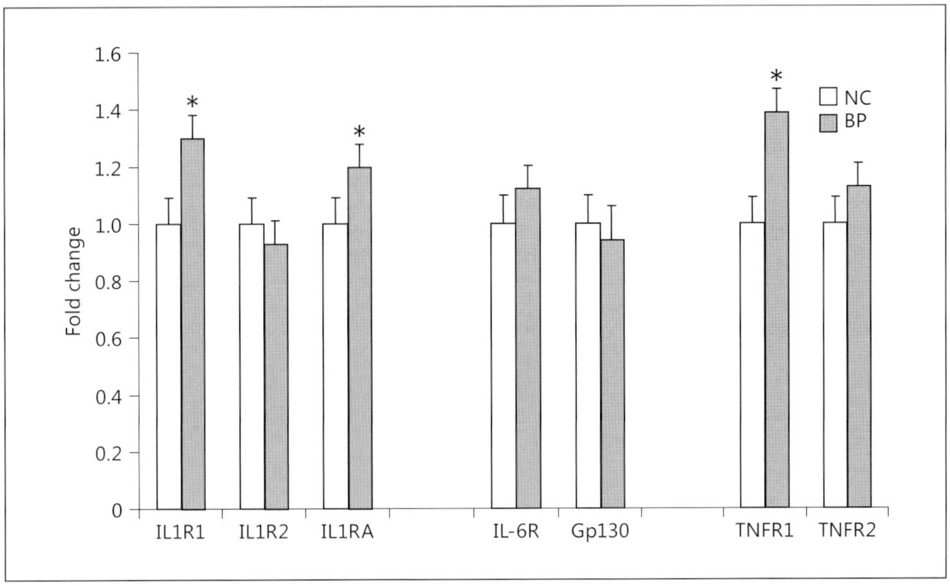

Fig. 6. mRNA expression (mean ± SEM) of receptors for proinflammatory cytokines, interleukin receptors IL1R1, IL1R2, IL1RA, and IL6R, Gp130, and tumor necrosis factor receptors TNFR1 and TNFR2, in the lymphocytes of bipolar disorder (BPD) patients and normal control (NC) subjects. * $p < 0.05$.

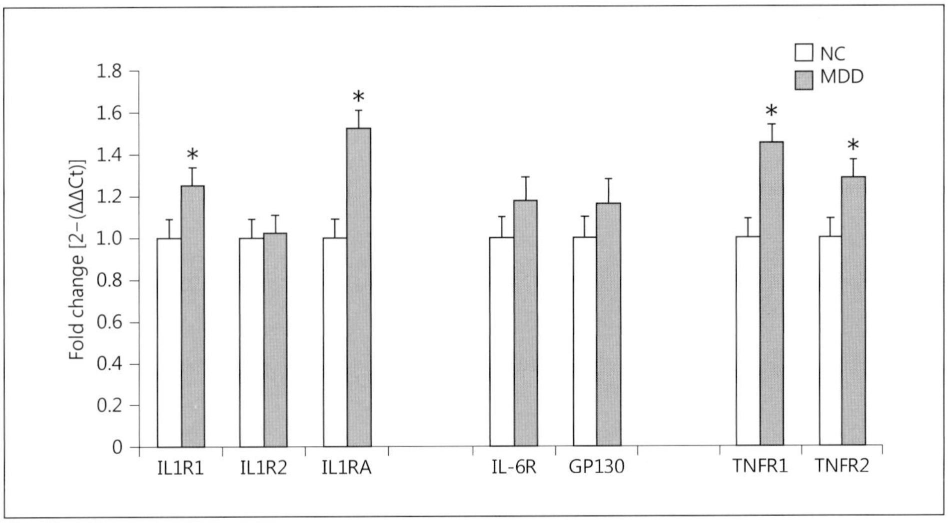

Fig. 7. mRNA expression (mean ± SEM) of receptors for proinflammatory cytokines, interleukin receptors IL1R1, IL1R2, IL1RA, and IL6R, Gp130, and tumor necrosis factor receptors TNFR1 and TNFR2, in the lymphocytes of patients with major depressive disorder (MDD) and normal control (NC) subjects. * $p < 0.05$.

icant difference in the mRNA expression of IL1R2 between BPD and MDD patients and NC subjects.

In addition to IL1Rs, the antagonist for IL receptors has also been identified. This antagonist competes with cytokines for binding to the same membrane-bound receptors, and the only example of this antagonist-type receptors identified is human IL1RA. IL1RA competes for binding and subsequent signaling of both forms of IL-1α and IL-1β. IL1RA levels have been found to be altered in the serum of BPD patients. When we compared the mRNA levels of IL1RA, we found that the gene expression of IL1RA was significantly increased in the lymphocytes of BPD (Fig. 6) and MDD (Fig. 7) patients compared with NC subjects.

Only one type of IL-6 receptor known as IL-6R has been identified. However, the biological effects of IL-6 are mediated through its receptor complex which consists of IL-6R and the signal-transducing protein Gp130. The signaling is initiated by the binding of IL-6 to IL-6R forming an IL-6R complex which then associates with a protein known as Gp130, thus initiating the activation of various signaling pathways. We determined the gene expression of both IL-6R and Gp130 in the lymphocytes of BPD patients and established that the gene expression levels of IL-6R or Gp130 were not significantly different in the lymphocytes of BPD (Fig. 6) and MDD (Fig. 7) patients compared with NC subjects.

Similar to the IL1R, there are two subtypes of receptors identified for TNF, known as TNFR1 and TNFR2. Both TNF-α and TNF-β act on these receptors to initiate their signaling. The signal transduction pathways which are activated by TNFR1 and TNFR2 are diverse and include G-protein-mediated activation of protein kinase A phospholipase C and phospholipase A_2. As opposed to the subtypes of receptors for IL1R, both TNFR1 and TNFR2 cause activation of transcription factors leading to the activation of the signaling pathway and effect the apoptotic process. We compared mRNA levels of TNFR1 and TNFR2 between BPD patients and NC subjects. mRNA expression of TNFR1, but not TNFR2, was significantly increased in the lymphocytes of BPD patients (Fig. 6) compared with NC subjects. On the other hand, comparing mRNA levels of TNFR1 and TNFR2 between MDD patients and NC subjects, mRNA levels of both TNFR1 and TNFR2 were significantly increased in MDD patients compared with NC subjects (Fig. 7).

These studies suggest both similarities and subtle differences in the expression of membrane-bound receptors in BPD and MDD patients. Both, MDD and BPD patients had increased mRNA levels for IL1R1 and TNFR1, but TNFR2 levels were increased in MDD patients but not in BPD patients compared with NC subjects.

mRNA and Protein Levels of Proinflammatory Cytokines

We then determined mRNA levels of the proinflammatory cytokines IL-1β, IL-6, and TNF-α, and found that the mRNA expression of IL-1β, IL-6, and TNF-α was significantly increased in lymphocytes of BPD (Fig. 8) and MDD (Fig. 9) patients compared with NC subjects. In order to examine if there was an association between mRNA and protein levels, we determined the protein levels of IL-1β, IL-6, and TNF-α, and found that the protein levels of IL-1β, IL-6, and TNF-α were significantly increased in lymphocytes of BPD (Fig. 10) and MDD (Fig. 11) patients compared with NC subjects. These results indicate that there is a good association in the alterations of both protein and mRNA expression of the studied proinflammatory cytokines in BPD and MDD patients. The observation of increased protein expression of IL-1β, IL-6, and TNF-α in the plasma of our BPD and MDD patients is consistent with many previous reports in BPD patients.

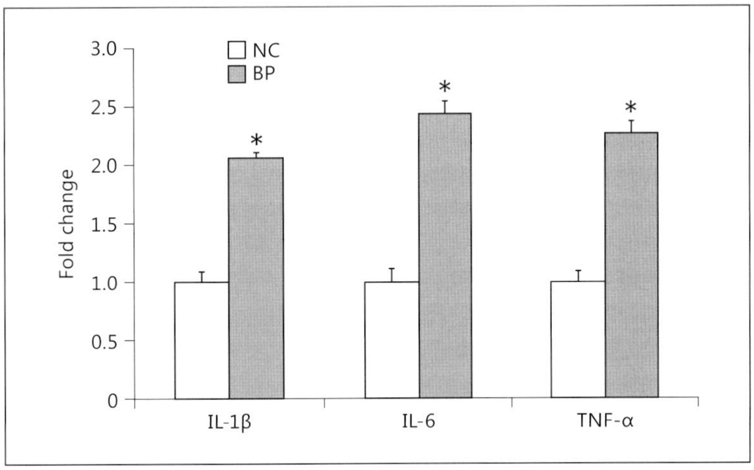

Fig. 8. mRNA expression (mean ± SEM) of proinflammatory cytokines, interleukin (IL)-1β, IL-6, and tumor necrosis factor (TNF)-α in the lymphocytes of bipolar disorder (BPD) patients and normal control (NC) subjects. * $p < 0.05$.

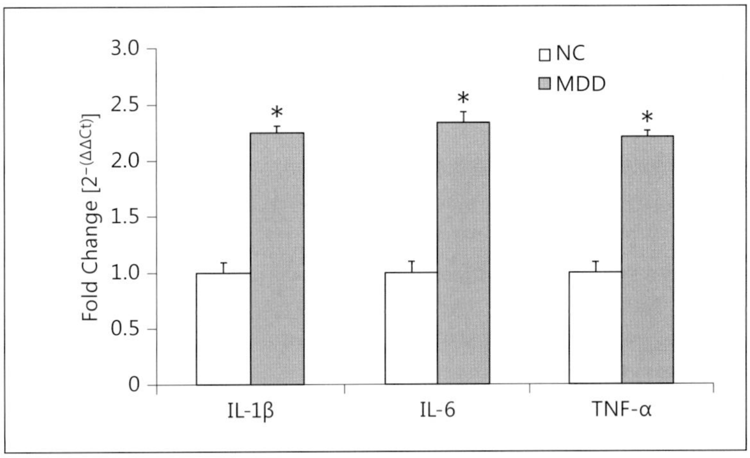

Fig. 9. mRNA expression (mean ± SEM) of proinflammatory cytokines, interleukin (IL)-1β, IL-6, and tumor necrosis factor (TNF)-α, in the lymphocytes of patients with major depressive disorder (MDD) and normal control (NC) subjects. * $p < 0.05$.

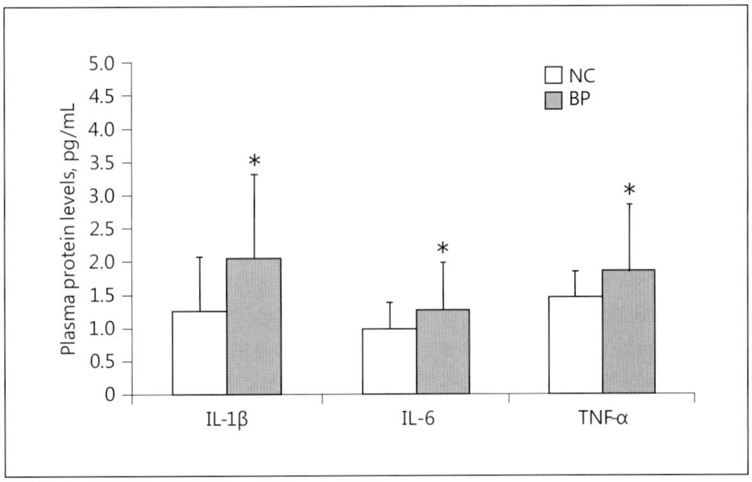

Fig. 10. Protein expression (mean ± SD) of interleukin (IL)-1β, IL-6, and tumor necrosis factor (TNF)-α in the plasma of bipolar disorder (BPD) patients and normal control (NC) subjects. * $p < 0.05$.

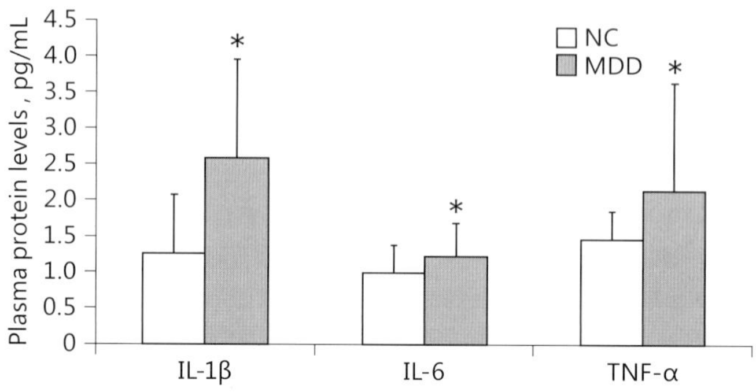

Fig. 11. Protein expression (mean ± SD) of interleukin (IL)-1β, IL-6, and tumor necrosis factor (TNF)-α in the plasma of patients with major depressive disorder (MDD) and normal control (NC) subjects. * $p < 0.05$.

Conclusion and Future Directions

In this study, we examined the role of cytokines and TLRs in the pathophysiology of mood disorders and suicide and examined the contributions of cytokines and TLRs in neuroprogression. Several previous studies suggest abnormalities of proinflammatory cytokines and their soluble receptors in the serum of MDD and BPD subjects. The biological effects of cytokines are mediated through their membrane-bound receptors. Although soluble receptors have been studied in MDD and BPD subjects, the membrane-bound receptors have not been included. Our study shows that the mRNA expression of the proinflammatory cytokines TNF-α, IL-1β, and IL-6 is significantly increased in the postmortem brain of teenage suicide victims compared with normal control subjects. Our studies also show abnormalities of two specific TLRs, TLR3 and TLR4, in the PFC of adult depressed suicide subjects. In addition, abnormalities of mRNA expression of membrane-bound receptors, such as IL1R1, TNFR1, and IL1RA, were observed in lymphocytes of MDD and BPD patients. In summary, these results demonstrate abnormalities of innate and adaptive immunity in postmortem brain of teenage and adult suicide victims, suggesting a contribution of proinflammatory cytokines and TLRs to neuroprogression. Although these are initial studies of cytokines and TLRs in postmortem brain, they do provide clear evidence of the involvement of cytokines and TLRs in neuroprogression. Future studies focusing on both adolescent and adult brain will be needed to further clarify the contribution of brain cytokines and TLRs not only in the pathophysiology of MDD, BPD, and suicide, but also to the neuroprogressive course of these disorders.

Acknowledgments

This research was supported by grants RO1-MH-56528 and RO1-MH-98554 from the National Institute of Mental Health, Rockville, MD, USA. The funding source had no involvement in the conduct of the research and/or preparation of the paper.

References

1 Anisman H, Merali Z: Cytokines, stress, and depressive illness. Brain Behav Immun 2002;16:513–524.

2 Hopkins SJ, Rothwell NJ: Cytokines and the nervous system. I: Expression and recognition. Trends Neurosci 1995;18:83–88.

3 Kronfol Z, Remick DG. Cytokines and the brain: implications for clinical psychiatry. Am J Psychiatry 2000;157:683–694.

4 Muller N, Ackenheil M: Psychoneuroimmunology and the cytokine action in the CNS: implications for psychiatric disorders. Prog Neuropsychopharmacol Biol Psychiatry 1998;22:1–33.

5 Bonaccorso S, Marino V, Biondi M, Grimaldi F, Ippoliti F, Maes M: Depression induced by treatment with interferon-alpha in patients affected by hepatitis C virus. J Affect Disord 2002;72:237–241.

6 Bonaccorso S, Marino V, Puzella A, Pasquini M, Biondi M, Artini M, et al: Increased depressive ratings in patients with hepatitis C receiving interferon-alpha-based immunotherapy are related to interferon-alpha-induced changes in the serotonergic system. J Clin Psychopharmacol 2002;22:86–90.

7 Bonaccorso S, Puzella A, Marino V, Pasquini M, Biondi M, Artini M, et al: Immunotherapy with interferon-alpha in patients affected by chronic hepatitis C induces an intercorrelated stimulation of the cytokine network and an increase in depressive and anxiety symptoms. Psychiatry Res 2001;105:45–55.

8 Collier J, Chapman R: Combination therapy with interferon-alpha and ribavirin for hepatitis C: practical treatment issues. BioDrugs 2001;15:225–238.

9 Hunt CM, Dominitz JA, Bute BP, Waters B, Blasi U, Williams DM: Effect of interferon-alpha treatment of chronic hepatitis C on health-related quality of life. Dig Dis Sci 1997;42:2482–2486.

10 Loftis JM, Hauser P: The phenomenology and treatment of interferon-induced depression. J Affect Disord 2004;82:175–190.

11 Malaguarnera M, Di Fazio I, Restuccia S, Pistone G, Ferlito L, Rampello L: Interferon alpha-induced depression in chronic hepatitis C patients: comparison between different types of interferon alpha. Neuropsychobiology 1998;37:93–97.

12 Pariante CM, Orru MG, Baita A, Farci MG, Carpiniello B: Treatment with interferon-alpha in patients with chronic hepatitis and mood or anxiety disorders. Lancet 1999;354:131–132.

13 Pavol MA, Meyers CA, Rexer JL, Valentine AD, Mattis PJ, Talpaz M: Pattern of neurobehavioral deficits associated with interferon alfa therapy for leukemia. Neurology 1995;45:947–950.

14 Renault PF, Hoofnagle JH, Park Y, Mullen KD, Peters M, Jones DB, et al: Psychiatric complications of long-term interferon alfa therapy. Arch Intern Med 1987;147:1577–1580.

15 Valentine AD, Meyers CA, Kling MA, Richelson E, Hauser P: Mood and cognitive side effects of interferon-alpha therapy. Semin Oncol 1998;25:39–47.

16 Yirmiya R, Weidenfeld J, Pollak Y, Morag M, Morag A, Avitsur R, et al: Cytokines, "depression due to a general medical condition," and antidepressant drugs. Adv Exp Med Biol 1999;461:283–316.

17 Anisman H, Merali Z: Anhedonic and anxiogenic effects of cytokine exposure. Adv Exp Med Biol 1999;461:199–233.

18 Berk M, Wadee AA, Kuschke RH, O'Neill-Kerr A: Acute phase proteins in major depression. J Psychosom Res 1997;43:529–534.

19 Kim YK, Suh IB, Kim H, Han CS, Lim CS, Choi SH, et al: The plasma levels of interleukin-12 in schizophrenia, major depression, and bipolar mania: effects of psychotropic drugs. Mol Psychiatry 2002;7:1107–1114.

20 Kubera M, Kenis G, Bosmans E, Zieba A, Dudek D, Nowak G, et al: Plasma levels of interleukin-6, interleukin-10, and interleukin-1 receptor antagonist in depression: comparison between the acute state and after remission. Pol J Pharmacol 2000;52:237–241.

21 Maes M: Evidence for an immune response in major depression: a review and hypothesis. Prog Neuropsychopharmacol Biol Psychiatry 1995;19:11–38.

22 Maes M, Bosmans E, De Jongh R, Kenis G, Vandoolaeghe E, Neels H: Increased serum IL-6 and IL-1 receptor antagonist concentrations in major depression and treatment resistant depression. Cytokine 1997;9:853–858.

23 Maes M, Meltzer HY, Bosmans E, Bergmans R, Vandoolaeghe E, Ranjan R, et al: Increased plasma concentrations of interleukin-6, soluble interleukin-6, soluble interleukin-2 and transferrin receptor in major depression. J Affect Disord 1995;34:301–309.

24 Musselman DL, Miller AH, Porter MR, Manatunga A, Gao F, Penna S, et al: Higher than normal plasma interleukin-6 concentrations in cancer patients with depression: preliminary findings. Am J Psychiatry 2001;158:1252–1257.

25 Nassberger L, Traskman-Bendz L: Increased soluble interleukin-2 receptor concentrations in suicide attempters. Acta Psychiatr Scand 1993;88:48–52.

26 O'Brien SM, Scott LV, Dinan TG: Cytokines: abnormalities in major depression and implications for pharmacological treatment. Hum Psychopharmacol 2004;19:397–403.

27 Sluzewska A, Rybakowski JK, Laciak M, Mackiewicz A, Sobieska M, Wiktorowicz K: Interleukin-6 serum levels in depressed patients before and after treatment with fluoxetine. Ann NY Acad Sci 1995;762:474–476.

28 Song C, Dinan T, Leonard BE: Changes in immunoglobulin, complement and acute phase protein levels in the depressed patients and normal controls. J Affect Disord 1994;30:283–288.

29 Thomas AJ, Davis S, Morris C, Jackson E, Harrison R, O'Brien JT: Increase in interleukin-1β in late-life depression. Am J Psychiatry 2005;162:175–177.

30 Tuglu C, Kara SH, Caliyurt O, Vardar E, Abay E: Increased serum tumor necrosis factor-alpha levels and treatment response in major depressive disorder. Psychopharmacology (Berl) 2003;170:429–433.

31 Weisse CS: Depression and immunocompetence: a review of the literature. Psychol Bull 1992;111:475–489.
32 Carpenter LL, Heninger GR, Malison RT, Tyrka AR, Price LH: Cerebrospinal fluid interleukin (IL)-6 in unipolar major depression. J Affect Disord 2004;79: 285–289.
33 Minami M, Kuraishi Y, Yamaguchi T, Nakai S, Hirai Y, Satoh M: Immobilization stress induces interleukin-1 beta mRNA in the rat hypothalamus. Neurosci Lett 1991;123:254–256.
34 Anisman H, Kokkinidis L, Merali Z: Further evidence for the depressive effects of cytokines: anhedonia and neurochemical changes. Brain Behav Immun 2002;16:544–556.
35 Connor TJ, Leonard BE: Depression, stress and immunological activation: the role of cytokines in depressive disorders. Life Sci 1998;62:583–606.
36 Leonard BE, Song C: Stress, depression, and the role of cytokines. Adv Exp Med Biol 1999;461:251–265.
37 Merrill JE, Murphy SP: Inflammatory events at the blood brain barrier: regulation of adhesion molecules, cytokines, and chemokines by reactive nitrogen and oxygen species. Brain Behav Immun 1997;11:245–263.
38 Tilders FJ, Schmidt ED: Cross-sensitization between immune and non-immune stressors. A role in the etiology of depression? Adv Exp Med Biol 1999;461: 179–197.
39 Kronfol Z, Silva J Jr, Greden J, Dembinski S, Gardner R, Carroll B: Impaired lymphocyte function in depressive illness. Life Sci 1983;33:241–247.
40 Calabrese JR, Skwerer RG, Barna B, Gulledge AD, Valenzuela R, Butkus A, et al: Depression, immunocompetence, and prostaglandins of the E series. Psychiatry Res 1986;17:41–47.
41 Herberman RB: Effect of alpha-interferons on immune function. Semin Oncol 1997;24(3 suppl 9):S9-S78–S9-S80.
42 Irwin M, Gillin JC: Impaired natural killer cell activity among depressed patients. Psychiatry Res 1987;20:181–182.
43 Maes M, Bosmans E, Suy E, Vandervorst C, DeJonckheere C, Raus J: Depression-related disturbances in mitogen-induced lymphocyte responses and interleukin-1 beta and soluble interleukin-2 receptor production. Acta Psychiatr Scand 1991; 84:379–386.
44 Herbert TB, Cohen S: Depression and immunity: a meta-analytic review. Psychol Bull 1993;113:472–486.
45 Irwin M: Immune correlates of depression. Adv Exp Med Biol 1999;461:1–24.
46 Steiner J, Bielau H, Brisch R, Danos P, Ullrich O, Mawrin C, et al: Immunological aspects in the neurobiology of suicide: elevated microglial density in schizophrenia and depression is associated with suicide. J Psychiatr Res 2008; 42:151–157.
47 Goodwin RD, Eaton WW: Asthma, suicidal ideation, and suicide attempts: findings from the Baltimore epidemiologic catchment area follow-up. Am J Public Health 2005;95:717–722.
48 Goodwin RD, Messineo K, Bregante A, Hoven CW, Kairam R: Prevalence of probable mental disorders among pediatric asthma patients in an inner-city clinic. J Asthma 2005;42: 643–647.
49 Tonelli LH, Stiller J, Rujescu D, Giegling I, Schneider B, Maurer K, et al: Elevated cytokine expression in the orbitofrontal cortex of victims of suicide. Acta Psychiatr Scand 2008;117: 198–206.
50 Lindqvist D, Janelidze S, Hagell P, Erhardt S, Samuelsson M, Minthon L, et al: Interleukin-6 is elevated in the cerebrospinal fluid of suicide attempters and related to symptom severity. Biol Psychiatry 2009;66:287–292.
51 Black C, Miller BJ: Meta-analysis of cytokines and chemokines in suicidality: distinguishing suicidal versus nonsuicidal patients. Biol Psychiatry 2015;78: 28–37.
52 Schobitz B, De Kloet ER, Holsboer F: Gene expression and function of interleukin 1, interleukin 6 and tumor necrosis factor in the brain. Prog Neurobiol 1994;44:397–432.
53 Pandey GN, Rizavi HS, Ren X, Fareed J, Hoppensteadt DA, Roberts RC, et al: Proinflammatory cytokines in the prefrontal cortex of teenage suicide victims. J Psychiatr Res 2012;46:57–63.
54 Dean B, Tawadros N, Scarr E, Gibbons AS: Regionally-specific changes in levels of tumour necrosis factor in the dorsolateral prefrontal cortex obtained postmortem from subjects with major depressive disorder. J Affect Disord 2010; 120:245–248.
55 Rao JS, Harry GJ, Rapoport SI, Kim HW: Increased excitotoxicity and neuroinflammatory markers in postmortem frontal cortex from bipolar disorder patients. Mol Psychiatry 2010;15:384–392.
56 Shelton RC, Claiborne J, Sidoryk-Wegrzynowicz M, Reddy R, Aschner M, Lewis DA, et al: Altered expression of genes involved in inflammation and apoptosis in frontal cortex in major depression. Mol Psychiatry 2011;16:751–762.
57 Kawai T, Akira S: TLR signaling. Semin Immunol 2007;19:24–32.
58 Piccinini AM, Midwood KS: DAMPening inflammation by modulating TLR signalling. Mediators Inflamm 2010; 2010:672395.
59 Klune JR, Dhupar R, Cardinal J, Billiar TR, Tsung A: HMGB1: endogenous danger signaling. Mol Med 2008;14:476–484.
60 O'Neill LA: How Toll-like receptors signal: what we know and what we don't know. Curr Opin Immunol 2006;18:3–9.
61 Lafon M, Megret F, Lafage M, Prehaud C: The innate immune facet of brain: human neurons express TLR-3 and sense viral dsRNA. J Mol Neurosci 2006; 29:185–194.
62 Bsibsi M, Persoon-Deen C, Verwer RW, Meeuwsen S, Ravid R, Van Noort JM: Toll-like receptor 3 on adult human astrocytes triggers production of neuroprotective mediators. Glia 2006;53:688–695.
63 Okun E, Griffioen K, Barak B, Roberts NJ, Castro K, Pita MA, et al: Toll-like receptor 3 inhibits memory retention and constrains adult hippocampal neurogenesis. Proc Natl Acad Sci USA 2010; 107:15625–15630.
64 Ma Y, Li J, Chiu I, Wang Y, Sloane JA, Lu J, et al: Toll-like receptor 8 functions as a negative regulator of neurite outgrowth and inducer of neuronal apoptosis. J Cell Biol 2006;175:209–215.
65 Okun E, Griffioen KJ, Mattson MP: Toll-like receptor signaling in neural plasticity and disease. Trends Neurosci 2011; 34:269–281.
66 Crews FT, Qin L, Sheedy D, Vetreno RP, Zou J: High mobility group box 1/Toll-like receptor danger signaling increases brain neuroimmune activation in alcohol dependence. Biol Psychiatry 2013; 73:602–612.

67 Pandey GN, Rizavi HS, Ren X, Bhaumik R, Dwivedi Y: Toll-like receptors in the depressed and suicide brain. J Psychiatr Res 2014;53:62–68.
68 Okun E, Griffioen KJ, Lathia JD, Tang SC, Mattson MP, Arumugam TV: Toll-like receptors in neurodegeneration. Brain Res Rev 2009;59:278–292.
69 Du J, Gould TD, Manji HK: Neurotrophic signaling in mood disorders; in Finkel T, Gutkind JS (eds): Signal Transduction and Human Disease. Hoboken, Wiley, 2003.
70 Greenberg DB, Jonasch E, Gadd MA, Ryan BF, Everett JR, Sober AJ, et al: Adjuvant therapy of melanoma with interferon-alpha-2b is associated with mania and bipolar syndromes. Cancer 2000;89:356–362.
71 Munkholm K, Brauner JV, Kessing LV, Vinberg M: Cytokines in bipolar disorder vs. healthy control subjects: a systematic review and meta-analysis. J Psychiatr Res 2013;47:1119–1133.
72 Munkholm K, Vinberg M, Vedel Kessing L: Cytokines in bipolar disorder: a systematic review and meta-analysis. J Affect Disord 2013;144:16–27.
73 Pandey GN, Ren X, Rizavi HS, Zhang H: Abnormal gene expression of proinflammatory cytokines and their receptors in the lymphocytes of patients with bipolar disorder. Bipolar Disord 2015;17:636–644.
74 Rizavi HS, Ren X, Zhang H, Bhaumik R, Pandey GN: Abnormal gene expression of proinflammatory cytokines and their membrane-bound receptors in the lymphocytes of depressed patients. Psychiatry Res 2016;240:314–320.

Ghanshyam N. Pandey, PhD
Department of Psychiatry
University of Illinois at Chicago
1601 West Taylor Street
Chicago, IL 60612 (USA)
E-Mail Gpandey@psych.uic.edu

Halaris A, Leonard BE (eds): Neuroprogression in Psychiatric Disorders.
Mod Trends Pharmacopsychiatry. Basel, Karger, 2017, vol 31, pp 96–106 (DOI: 10.1159/000470810)

Towards an Integrated View of Early Molecular Changes Underlying Vulnerability to Social Stress in Psychosis

Henry Barron[a] · Sina Hafizi[a] · Romina Mizrahi[a–d]

[a] Research Imaging Centre and [b] Campbell Family Mental Health Research Institute, Centre for Addiction and Mental Health, and [c] Department of Psychiatry and [d] Institute of Medical Science, University of Toronto, Toronto, ON, Canada

Abstract

Psychotic disorders are heterogeneous and complex, involving many putative causal factors interacting along the course of disease development. Many of the factors implicated in the pathogenesis of psychosis also appear to be involved in disease onset and subsequent neuroprogression. Herein, we highlight the pertinent literature implicating inflammation and oxidative stress in the pathogenesis of psychosis, and the potential contribution of N-methyl-D-aspartate receptors (NMDARs). We also emphasize the role of peripubertal social stress in psychosis, and the ways in which hippocampal dysfunction can mediate dysregulation of the hypothalamic-pituitary-adrenal axis and cortisol release. Finally, we propose a model wherein inflammation and oxidative stress act as a first hit, producing altered parvalbumin interneuron development, NMDAR hypofunction, microglial priming, and sensitivity to a second hit of peripubertal social stress. With a greater understanding of how these factors interact, it may be possible to detect, prevent, and treat psychosis more effectively.

Introduction

Schizophrenia is a debilitating mental disorder characterized by positive symptoms (e.g., abnormal perceptions and beliefs), negative symptoms (e.g., anhedonia and social withdrawal), and cognitive deficits. It is a multifactorial illness usually diagnosed in late adolescence or early adulthood that has a complex developmental course beginning long before the first episode of psychosis and progressing long thereafter [1, 2]. Herein, we will focus on the molecular basis of psychosis development and neuroprogression, first describing the N-methyl-D-aspartate receptor (NMDAR) model of psychosis, and then outlining the literature on the important roles of inflammation, oxidative stress, and social stress. We have chosen these factors because they appear to have important interactions in the development, onset, and progression of psychosis (Fig. 1), although there are other neurochemical/functional/genetic alterations also involved [2–4].

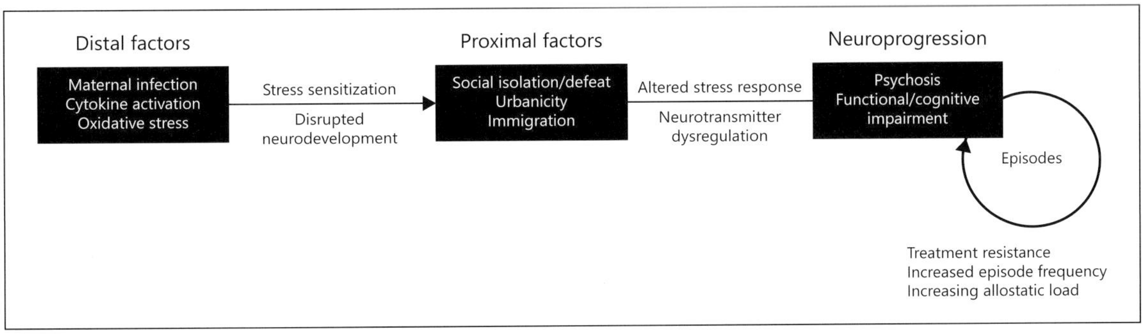

Fig. 1. Factors suggested to contribute to the neurodevelopment and neuroprogression of psychosis. Distal factors, which generally include prenatal insults, play an important role in predisposing the child to later social stress. Social stress further impacts the development of the brain, leading to an altered stress response and dysregulation of neurotransmitter systems such as glutamate and dopamine. Onset may begin with mild symptoms, as in the clinical high-risk population, but can progress to psychosis as episodes compound existing biological and psychosocial impairments. Eventually, neuroprogression may lead to treatment resistance and further deterioration unless psychosocial or pharmacological interventions divert the trajectory away from psychosis towards resilience and health.

NMDAR Hypofunction and Psychosis

There is a great deal of interest in the function of NMDARs expressed on cortical fast-spiking GABAergic interneurons, also known as parvalbumin interneurons (PVIs). PVI NMDAR hypofunction provides a potential common target for the inflammatory, oxidative, and social stress pathways that will be discussed in this chapter. In addition to being affected by metabolic stressors, NMDAR hypofunction may also instigate further damage after the onset of psychosis. These receptors appear to help regulate the same stressors that caused their dysfunction in the first place, suggesting they may, in part, account for neuroprogression at the molecular level [5]. These factors are briefly outlined here to provide context for the remainder of the chapter.

NMDARs are ionotropic channels that are gated by the neurotransmitter glutamate. They play a key role in mediating long-term potentiation and other synaptic modifications that are dependent on the level of neuronal activity [6]. NMDAR alterations were first implicated in psychosis through studies of NMDAR antagonists, such as phencyclidine and ketamine, showing that their effects on humans were similar to symptoms seen in schizophrenia [7]. Following these findings, other lines of evidence emerged linking NMDAR hypofunction to schizophrenia [8, 9]. Disruptions in NMDARs are especially relevant in PVIs because these neurons regulate pyramidal neurons in relevant brain areas. Interestingly, even mild antagonism of NMDARs can lead to lasting alterations in the PVI number [10]. Moreover, PVIs are the last subset of interneurons to develop, and they undergo significant maturation during puberty, ostensibly an important period in the onset of psychosis [11, 12].

Inflammation and Psychosis

Inflammation has been implicated in several psychiatric (e.g., depression) and neurodegenerative disorders, and its role in psychosis is still under intense study [13–15]. Multiple lines of evidence from epidemiological, preclinical, and clinical studies have repeatedly shown signs of immune alterations in the pathogenesis of schizophrenia.

The effect of inflammation on psychosis may be mediated by its impact on PVI NMDARs over multiple stages of development [16].

Epidemiological Studies

Epidemiological studies have shown an association between maternal infection during pregnancy and the subsequent development of psychosis in the child. The first to describe this connection were studies of the 1957 influenza pandemic which showed increases in psychiatric hospital admissions for psychosis among those who were in their second trimester of fetal development during the outbreak [17, 18]. Later studies found that the relative risk of developing schizophrenia following a first-trimester influenza infection can be as high as 7-fold [19]. These findings have since been replicated in several other infectious diseases, and studies suggest that it is the mother's immune reaction, rather than the infectious agent itself, that increases the risk of schizophrenia [13, 20]. It is estimated that if maternal influenza infections were eliminated, as many as 14–21% of schizophrenia cases would be prevented, suggesting that alterations in the immune system are likely involved in the pathogenesis of psychosis [21].

Evidence from Preclinical Models

Rodent models have also provided supporting evidence for the role of immune activation in the development of psychosis, including the disruption of PVI development and NMDAR function. There are several maternal immune activation (MIA) models used to study this process, including simulated viral infection with polyriboinosini c:polyribocytidylic acid, poly(I:C), simulated bacterial infection with lipopolysaccharide (LPS), and frank infection with live pathogens [13]. Noninfectious MIA models such as poly(I:C) or LPS have reliably produced a psychosis-like phenotype in rodent models, including characteristic anatomical and behavioral alterations [22, 23]. In addition, multiple changes in key neurotransmitter systems have been noted in the MIA model, including hippocampal NMDAR hypofunction and enhanced sensitivity to acute dopaminergic stimulation, both of which have been noted in schizophrenia [24].

Early immune activation not only appears to affect normal brain maturation in the developing fetus, it also may have a role in predisposing the brain to future inflammation. Studies have shown that microglia, the most numerous brain immune cells, may be "primed" if activated during fetal development, causing them to be more easily activated later in life [25, 26]. Prenatal microglial activation can also alter glutamatergic synaptic function later in life, probably through their direct interaction with NMDAR-expressing neurons [27, 28].

Adding another layer of complexity, proinflammatory cytokines can alter the activity of enzymes, such as tryptophan 2,3-dioxygenase and indoleamine 2,3-dioxygenase (IDO), leading to altered tryptophan metabolism to favor kynurenine and its subsequent conversion to quinolinic acid (QUIN), an NMDAR agonist, and kynurenic acid (KYNA), an NMDAR antagonist. In psychosis, multiple studies have demonstrated a shift towards the KYNA side of the catabolic pathway, as evidenced by elevated KYNA levels [29–32] and a decreased QUIN/KYNA ratio in CSF [33]. KYNA elevations in psychosis are contrary to findings in depression, where induction of IDO appears to favor the QUIN catabolic pathway, leading to elevations in the QUIN/KYNA ratio [34]. The apparent disparity between the two illnesses may be explained by alterations in IDO and its downstream enzymes, such as kynurenine 3-monooxygenase (KMO), which shunts kynurenine into the QUIN synthesis pathway. Consistent with this hypothesis, KMO enzyme activity was reduced in the prefrontal cortex in postmortem samples of patients with psychosis, which may explain relative elevations in KYNA compared to depression [35]. However, more research is required to fully understand the differences in tryptophan catabolism between psychiatric illnesses.

In rodent models of psychosis, the effects of KYNA administration appear to mimic those of NMDAR antagonists, such as ketamine and phencyclidine [36]. Furthermore, KYNA may have a permanent impact on the developing brain – rodents exposed to KYNA during early development continue to display higher levels as adults and present psychosis-like symptoms [37]. Taken together, these findings suggest that inflammation may prime the brain for NMDAR hypofunction by increasing the level of KYNA in psychosis.

In vivo Human Studies

In studies of patients with schizophrenia, there is evidence of a chronic inflammatory state with an increase in proinflammatory cytokines and reductions in anti-inflammatory cytokines [38, 39]. Polymorphisms in some proinflammatory cytokine gene complexes, e.g., interleukin (IL)-1, may be linked to increased likelihood of developing schizophrenia [40]. Furthermore, although imaging studies of neuroinflammation using positron emission tomography have yielded inconclusive results in participants with psychosis [41, 42], increases in peripheral inflammatory markers have been observed in first-episode psychosis and are linked to the severity of psychopathology [43, 44]. Finally, treatment with antipsychotic medication may act, at least in part, through an anti-inflammatory mechanism by modifying cytokine levels [39, 45]. These findings suggest that inflammation may be linked to neuroprogression in psychosis.

Oxidative Stress and Psychosis

Oxidative stress from reactive oxygen species (ROS) has also been implicated in the pathogenesis of psychosis and may be an important factor in the onset and later progression of the illness. ROS are an unavoidable by-product of aerobic metabolism, produced primarily in the mitochondria of cells throughout the human body. These chemical species are strongly implicated in the process of aging due to their ability to damage cellular components such as lipids and proteins [46]. Under normal circumstances, endogenous antioxidants, such as glutathione (GSH), neutralize these factors and protect human tissues from excess damage [47]. Without sufficient GSH to balance ROS, neurotoxicity can occur, which may explain some of the structural and functional changes occurring in psychosis [48].

Preclinical Studies

A number of animal studies have begun to tease apart the complex time-dependent relationship between oxidative balance and NMDAR hypofunction in the development of psychosis. Without protective antioxidants in development, damage can ensue; studies have shown that GSH depletion in mice during postnatal days 5–16 leads to permanent psychosis-like deficits [49]. Interestingly, PVIs appear to be particularly vulnerable to oxidative stress. This may be due to the especially high metabolic demands accompanying their fast firing rate and the fact that they rely on ensheathing perineuronal nets for protection during development, which are themselves vulnerable to oxidative stress [5, 50]. Furthermore, GSH deficits can reduce NMDAR function, which may in turn lead to decreases in the cortical PVI number [10, 51]. Conversely, increasing GSH levels has a beneficial effect – treatment with the GSH-replenishing precursor N-acetylcysteine has been shown to prevent PVI abnormalities in various rodent models of psychosis [52, 53]. The level of GSH may normally be regulated by a feedback loop between NMDARs and the glutamate cysteine ligase (GCL) catalytic subunit, a component of the enzyme responsible for GSH synthesis. Deletion of the GCL catalytic subunit triggers oxidative stress leading to impairments in cortical plasticity that can cause long-lasting failure of PVIs [54]. The expression of the GCL catalytic subunit is regulated by NMDAR activity in response to higher synaptic activity, an important protective function whose absence can cause apoptosis in cortical neurons [55, 56]. In turn, the activity of NMDAR is regulated by the oxidative status at its GRIN2A subunit [57]. All of this points to a close relationship between GSH and NMDARs in mitigating oxidative damage in the rodent brain, particularly in vulnerable fast-spiking PVIs.

Clinical Studies

In patients with psychosis, our understanding of the effects of oxidative stress is limited. Some studies have found evidence of excess protein and lipid oxidation in blood, CSF, and postmortem tissue as well as alterations in central and peripheral GSH levels [47, 58–60]. Furthermore, there appears to be a correlation between GSH levels and the severity of psychotic symptoms in schizophrenia. Recent clinical trials have shown that N-acetylcysteine administration in schizophrenia can reduce negative and cognitive symptoms, including mismatch negativity, an indicator of NMDAR function [61–63]. Gene and protein alterations related to oxidative stress have also been noted. For instance, familial variations in the GCL modifier, a subunit of the GSH synthesis enzyme, may increase the risk of schizophrenia up to 4-fold [64]. On the other hand, levels of oxidative markers appear to vary significantly based on clinical status in cross-sectional studies [47]. Furthermore, a recent meta-analysis found no effect of clinical status on GSH synthesis [65]. Thus, although there seem to be perturbations in oxidative balance in clinical studies, replication of findings and large-scale longitudinal studies are needed in order to understand changes in the oxidative status over the course of psychosis.

Connections between Inflammation and Oxidative Stress

Oxidative stress and inflammation interact at multiple levels, and it is likely that many of the deleterious effects of inflammation are mediated by oxidative stress and vice versa. Oxidative stress induces inflammation via activation of nuclear factor (NF)-κB [66, 67]. The immune system is also a source of oxidative stress. For example, microglia generate ROS to destroy pathogens, which can also be toxic for neurons and can sensitize the brain to more damage by activating signaling molecules such as NF-κB, consistent with the neuroprogression of psychosis [68]. Proinflammatory cytokines, such as IL-1β, IL-6, and tumor necrosis factor (TNF)-α, induce production of free radicals through NF-κB [69]. Studies also show that the development of psychosis in immune activation models may, in fact, be mediated by oxidative stress due to low levels of GSH [70, 71]. Specifically, one study using an MIA model showed elevations in multiple markers of oxidative stress in the hippocampus of male offspring, including a decreased ratio of GSH to its reduced form, GSSG. In this same study, replenishing GSH by administering N-acetylcysteine reduced oxidative markers back to baseline and delayed detrimental effects in male offspring [71]. Taken together, these findings suggest that immune activation and oxidative stress may have a close reciprocal relationship in psychosis development and neuroprogression.

Social Stress and Psychosis

Stress, particularly social stress, is another important etiologic factor implicated in multiple developmental stages of psychosis [72–76]. Studies suggest that this relationship is mostly mediated via the dysregulation of the hypothalamic-pituitary-adrenal (HPA) axis, which dictates the release of the stress hormone cortisol. In youth at risk for psychosis, high baseline cortisol levels predict clinically significant symptoms after 2 years [77]. Moreover, in those with schizophrenia, the level of cortisol has been positively associated with the severity of psychotic symptoms [78]. It is worth noting, however, that some studies show that cortisol is not consistently elevated in psychosis but rather appears to display disturbed HPA axis regulation; patients with psychosis, for example, display a blunted response to dexamethasone suppression, lower cortisol awakening response, and flattening of diurnal cortisol variations [79–82].

Social Stress, Hypothalamic-Pituitary-Adrenal Axis Dysregulation, Inflammation, and NMDAR Hypofunction

Despite the reported links between cortisol, HPA axis dysregulation, and the increased risk of psychosis, the exact mechanisms are poorly understood. However, one major theory implicates hippocampal dysfunction as the driving factor. This comes from several lines of evidence reviewed extensively elsewhere [83]. Key findings are as follows:

(a) evidence from primates suggests that the hippocampus plays a key role in regulating the HPA axis through negative feedback by decreasing the corticotropin-releasing hormone level; (b) elevated cortisol levels are known to damage the hippocampus; (c) there is an inverse relationship between cortisol secretion and hippocampal volume in humans; (d) in schizophrenia, hippocampal volume is reduced more than in any other brain region; and (e) the magnitude of hippocampal volume reduction is correlated with the severity of cognitive symptoms in schizophrenia [83].

Inflammation may play an important role in mediating these alterations in the hippocampus. It has long been known that social stress is capable of increasing inflammatory markers, both in acute and chronic stress [84]. Acute psychosocial stress challenges increase the levels of both cortisol and ILs in healthy volunteers, participants with negative affect, and individuals exposed to early life stress [85–87]. In terms of chronic stress, those raised in a harsh family have been shown to have higher levels of peripheral markers of inflammation, as do those who self-report as being lonely or having low social status [88–90]. It is possible that the inflammation related to stress can lead to microglial activation, which may further damage PVIs in the hippocampus [91]. Once an inflammatory state is induced, it may give feedback to cause further dysregulation of the stress response by disrupting the normal anti-inflammatory properties of glucocorticoids [92–94]. Interestingly, reducing stress through the pubertal administration of benzodiazepines was able to prevent the onset of psychotic symptoms and elevations in dopamine in the methylazoxymethanol rodent model of psychosis [95].

Towards a Common Molecular Pathway

Studies combining measures of oxidative stress, inflammation, and social stress with PVI dysfunction provide further insight into the complex relationships described so far and provide the fodder for an integrated model. In general, these studies have suggested that neonatal oxidative stress or immune activation make the developing brain more susceptible to the effects of social stress during puberty. For instance, one animal study combined oxidative stress, NMDAR hypofunction, and social stress along development, showing that selective deletion of an NMDAR subunit increased oxidative stress in a way that was exacerbated by later social isolation [96]. Another study testing a similar hypothesis found that pharmacologically inhibiting the formation of ROS prevented PVI deficits induced by social isolation [97]. In terms of neuroinflammation, one study demonstrated that maternal treatment with poly(I:C) followed by a peripubertal stress task led to a synergistic increase in offspring levels of inflammatory cytokines, psychosis-like symptoms, and microglial activation in response to stress [98].

All of these preclinical studies appear to support a disease pathway similar to the "two-hit"

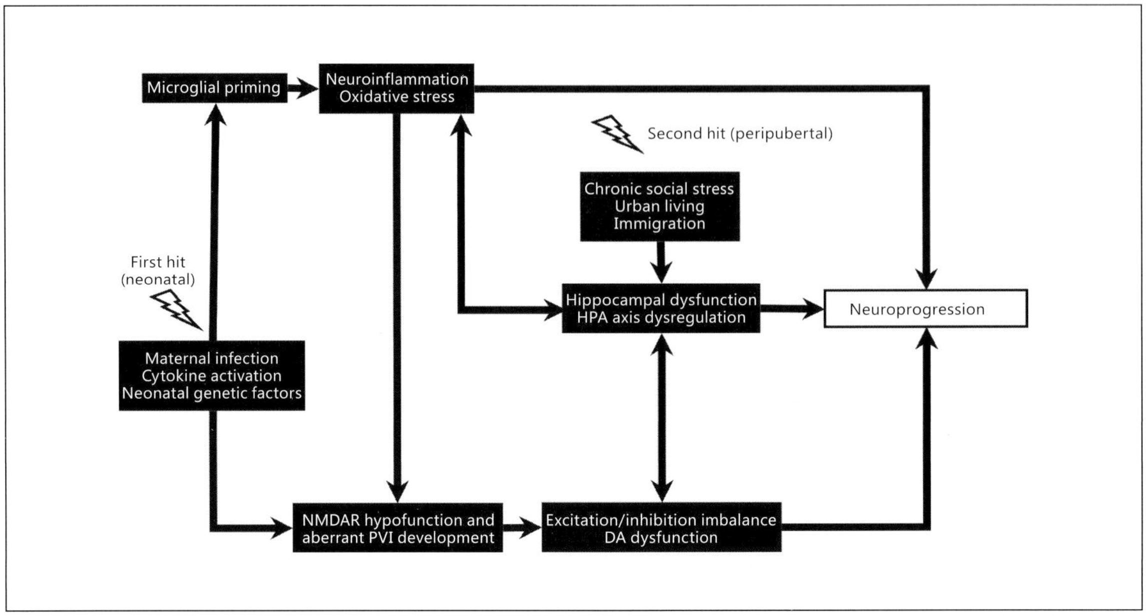

Fig. 2. Proposed model linking the molecular and psychosocial mechanisms involved in the development and neuroprogression of psychosis. In the "first hit," prenatal insults due to maternal cytokine activation and oxidative stress may alter the development of parvalbumin interneurons (PVIs) and predispose the developing brain to further inflammation by priming microglia for subsequent overactivation. Chronic social stress during puberty provides a "second hit" that triggers hippocampal dysfunction and hypothalamic-pituitary-adrenal (HPA) axis dysregulation, triggering the primed microglia, and leading to inflammation, oxidative stress, and further PVI dysfunction. All this culminates in excitation/inhibition imbalances and altered control of dopamine (DA) pathways, leading to the first psychotic symptoms and later neuroprogression through the accumulation of allostatic load.

model of psychosis (Fig. 2). In summary, a maternal inflammatory state may cause microglial activation and priming, redox imbalance, and vulnerabilities in highly sensitive hippocampal and cortical PVIs. These factors increase vulnerability in the affected individual such that during crucial periods of neuronal growth and development, stressors can cause overactivation of the HPA axis, microglial activation, and further PVI damage. The hippocampus might thus be negatively affected by both PVI dysfunction and cortisol levels, and is no longer able to provide negative feedback to the HPA axis [99]. The activation of oxidative and inflammatory mechanisms throughout the onset and course of the disease may contribute to disease progression.

Conclusion

Herein, we discussed the molecular mechanisms underlying the pathogenesis and neuroprogression of psychosis. There are still many unanswered questions, especially considering that many studies are in preclinical models of the disease. However, advances in imaging techniques, longitudinal studies, and clinical trials may help to refine our understanding of the illness in living humans. Indeed, there are already a number of clinical trials that take advantage of the mechanisms described here, including NMDAR agonists, antioxidant therapies, anti-inflammatory treatments, and psychosocial interventions to prevent stress/

oxidative stress and help prevent the worsening of symptoms over time [2, 5, 13, 100]. In the not-too-distant future, clinicians may be able to reliably identify early psychosis in vulnerable individuals before it has fully emerged, and initiate preventative therapies or provide social support. In doing so, it may be possible to reduce the considerable societal and personal burden of schizophrenia.

References

1 Fusar-Poli P, Borgwardt S, Bechdolf A, Addington J, Riecher-Rossler A, Schultze-Lutter F, Keshavan M, Wood S, Ruhrmann S, Seidman LJ, Valmaggia L, Cannon T, Velthorst E, De Haan L, Cornblatt B, Bonoldi I, Birchwood M, McGlashan T, Carpenter W, McGorry P, Klosterkotter J, McGuire P, Yung A: The psychosis high-risk state: a comprehensive state-of-the-art review. JAMA Psychiatry 2013;70:107–120.

2 Davis J, Moylan S, Harvey BH, Maes M, Berk M: Neuroprogression in schizophrenia: pathways underpinning clinical staging and therapeutic corollaries. Aust NZ J Psychiatry 2014;48: 512–529.

3 Fusar-Poli P, Meyer-Lindenberg A: Striatal presynaptic dopamine in schizophrenia, Part II: meta-analysis of [^{18}F/^{11}C]-DOPA PET studies. Schizophr Bull 2013;39:33–42.

4 Howes OD, Kapur S: The dopamine hypothesis of schizophrenia: version III – the final common pathway. Schizophr Bull 2009;35:549–562.

5 Hardingham GE, Do KQ: Linking early-life NMDAR hypofunction and oxidative stress in schizophrenia pathogenesis. Nat Rev Neurosci 2016;17:125–134.

6 Malenka RC, Bear MF: LTP and LTD: an embarrassment of riches. Neuron 2004; 44:5–21.

7 Krystal JH, Karper LP, Seibyl JP, Freeman GK, Delaney R, Bremner JD, Heninger GR, Bowers MB Jr, Charney DS: Subanesthetic effects of the noncompetitive NMDA antagonist, ketamine, in humans. Psychotomimetic, perceptual, cognitive, and neuroendocrine responses. Arch Gen Psychiatry 1994; 51:199–214.

8 Schwarcz R, Bruno JP, Muchowski PJ, Wu HQ: Kynurenines in the mammalian brain: when physiology meets pathology. Nat Rev Neurosci 2012;13: 465–477.

9 Dalmau J, Lancaster E, Martinez-Hernandez E, Rosenfeld MR, Balice-Gordon R: Clinical experience and laboratory investigations in patients with anti-NMDAR encephalitis. Lancet Neurol 2011; 10:63–74.

10 Wang X, Pinto-Duarte A, Sejnowski TJ, Behrens MM: How Nox2-containing NADPH oxidase affects cortical circuits in the NMDA receptor antagonist model of schizophrenia. Antioxid Redox Signal 2013;18:1444–1462.

11 Gulyas AI, Megias M, Emri Z, Freund TF: Total number and ratio of excitatory and inhibitory synapses converging onto single interneurons of different types in the CA1 area of the rat hippocampus. J Neurosci 1999;19:10082–10097.

12 Sullivan EM, O'Donnell P: Inhibitory interneurons, oxidative stress, and schizophrenia. Schizophr Bull 2012;38: 373–376.

13 Feigenson KA, Kusnecov AW, Silverstein SM: Inflammation and the two-hit hypothesis of schizophrenia. Neurosci Biobehav Rev 2014;38:72–93.

14 Miller AH, Maletic V, Raison CL: Inflammation and its discontents: the role of cytokines in the pathophysiology of major depression. Biol Psychiatry 2009; 65:732–741.

15 Khansari PS, Sperlagh B: Inflammation in neurological and psychiatric diseases. Inflammopharmacology 2012;20:103–107.

16 Giovanoli S, Weber L, Meyer U: Single and combined effects of prenatal immune activation and peripubertal stress on parvalbumin and reelin expression in the hippocampal formation. Brain Behav Immun 2014;40:48–54.

17 Barr CE, Mednick SA, Munk-Jorgensen P: Exposure to influenza epidemics during gestation and adult schizophrenia. A 40-year study. Arch Gen Psychiatry 1990;47:869–874.

18 Mednick SA, Machon RA, Huttunen MO, Bonett D: Adult schizophrenia following prenatal exposure to an influenza epidemic. Arch Gen Psychiatry 1988; 45:189–192.

19 Brown AS, Begg MD, Gravenstein S, Schaefer CA, Wyatt RJ, Bresnahan M, Babulas VP, Susser ES: Serologic evidence of prenatal influenza in the etiology of schizophrenia. Arch Gen Psychiatry 2004;61:774–780.

20 Brown AS, Patterson PH: Maternal infection and schizophrenia: implications for prevention. Schizophr Bull 2011;37: 284–290.

21 Brown AS: Prenatal infection as a risk factor for schizophrenia. Schizophr Bull 2006;32:200–202.

22 Piontkewitz Y, Arad M, Weiner I: Abnormal trajectories of neurodevelopment and behavior following in utero insult in the rat. Biol Psychiatry 2011;70: 842–851.

23 Meyer U, Feldon J: To poly(I: C) or not to poly(I: C): advancing preclinical schizophrenia research through the use of prenatal immune activation models. Neuropharmacology 2012;62:1308–1321.

24 Meyer U, Feldon J, Schedlowski M, Yee BK: Towards an immuno-precipitated neurodevelopmental animal model of schizophrenia. Neurosci Biobehav Rev 2005;29:913–947.

25 Bland ST, Beckley JT, Young S, Tsang V, Watkins LR, Maier SF, Bilbo SD: Enduring consequences of early-life infection on glial and neural cell genesis within cognitive regions of the brain. Brain Behav Immun 2010;24: 329–338.

26 Juckel G, Manitz MP, Brune M, Friebe A, Heneka MT, Wolf RJ: Microglial activation in a neuroinflammational animal model of schizophrenia – a pilot study. Schizophr Res 2011;131:96–100.

27 Hayashi Y, Ishibashi H, Hashimoto K, Nakanishi H: Potentiation of the NMDA receptor-mediated responses through the activation of the glycine site by microglia secreting soluble factors. Glia 2006;53:660–668.
28 Takaki J, Fujimori K, Miura M, Suzuki T, Sekino Y, Sato K: L-Glutamate released from activated microglia downregulates astrocytic L-glutamate transporter expression in neuroinflammation: the "collusion" hypothesis for increased extracellular L-glutamate concentration in neuroinflammation. J Neuroinflammation 2012;9:1.
29 Nilsson LK, Linderholm KR, Engberg G, Paulson L, Blennow K, Lindstrom LH, Nordin C, Karanti A, Persson P, Erhardt S: Elevated levels of kynurenic acid in the cerebrospinal fluid of male patients with schizophrenia. Schizophr Res 2005;80:315–322.
30 Linderholm KR, Skogh E, Olsson SK, Dahl ML, Holtze M, Engberg G, Samuelsson M, Erhardt S: Increased levels of kynurenine and kynurenic acid in the CSF of patients with schizophrenia. Schizophr Bull 2012;38:426–432.
31 Erhardt S, Blennow K, Nordin C, Skogh E, Lindstrom LH, Engberg G: Kynurenic acid levels are elevated in the cerebrospinal fluid of patients with schizophrenia. Neurosci Lett 2001;313:96–98.
32 Schwarcz R, Bruno JP, Muchowski PJ, Wu H-Q: Kynurenines in the mammalian brain: when physiology meets pathology. Nat Rev Neurosci 2012;13:465–477.
33 Kegel ME, Bhat M, Skogh E, Samuelsson M, Lundberg K, Dahl ML, Sellgren C, Schwieler L, Engberg G, Schuppe-Koistinen I, Erhardt S: Imbalanced kynurenine pathway in schizophrenia. Int J Tryptophan Res 2014;7:15–22.
34 Erhardt S, Lim CK, Linderholm KR, Janelidze S, Lindqvist D, Samuelsson M, Lundberg K, Postolache TT, Traskman-Bendz L, Guillemin GJ, Brundin L: Connecting inflammation with glutamate agonism in suicidality. Neuropsychopharmacology 2013;38:743–752.
35 Sathyasaikumar KV, Stachowski EK, Wonodi I, Roberts RC, Rassoulpour A, McMahon RP, Schwarcz R: Impaired kynurenine pathway metabolism in the prefrontal cortex of individuals with schizophrenia. Schizophr Bull 2011;37:1147–1156.
36 Erhardt S, Schwieler L, Emanuelsson C, Geyer M: Endogenous kynurenic acid disrupts prepulse inhibition. Biol Psychiatry 2004;56:255–260.
37 Pocivavsek A, Wu HQ, Elmer GI, Bruno JP, Schwarcz R: Pre- and postnatal exposure to kynurenine causes cognitive deficits in adulthood. Eur J Neurosci 2012;35:1605–1612.
38 Miller BJ, Buckley P, Seabolt W, Mellor A, Kirkpatrick B: Meta-analysis of cytokine alterations in schizophrenia: clinical status and antipsychotic effects. Biol Psychiatry 2011;70:663–671.
39 Maes M, Bosmans E, Ranjan R, Vandoolaeghe E, Meltzer HY, De Ley M, Berghmans R, Stans G, Desnyder R: Lower plasma CC16, a natural anti-inflammatory protein, and increased plasma interleukin-1 receptor antagonist in schizophrenia: effects of antipsychotic drugs. Schizophr Res 1996;21:39–50.
40 Ellman LM, Deicken RF, Vinogradov S, Kremen WS, Poole JH, Kern DM, Tsai WY, Schaefer CA, Brown AS: Structural brain alterations in schizophrenia following fetal exposure to the inflammatory cytokine interleukin-8. Schizophr Res 2010;121:46–54.
41 Carter CS, Bullmore ET, Harrison P: Is there a flame in the brain in psychosis? Biol Psychiatry 2014;75:258–259.
42 Kenk M, Selvanathan T, Rao N, Suridjan I, Rusjan P, Remington G, Meyer JH, Wilson AA, Houle S, Mizrahi R: Imaging neuroinflammation in gray and white matter in schizophrenia: an in vivo PET study with [^{18}F]-FEPPA. Schizophr Bull 2015;41:85–93.
43 Fan X, Pristach C, Liu EY, Freudenreich O, Henderson DC, Goff DC: Elevated serum levels of C-reactive protein are associated with more severe psychopathology in a subgroup of patients with schizophrenia. Psychiatry Res 2007;149:267–271.
44 Fernandes B, Steiner J, Bernstein H, Dodd S, Pasco J, Dean O, Nardin P, Goncalves C, Berk M: C-reactive protein is increased in schizophrenia but is not altered by antipsychotics: meta-analysis and implications. Mol Psychiatry 2016;21:554–564.
45 Maes M, Bosmans E, Kenis G, De Jong R, Smith RS, Meltzer HY: In vivo immunomodulatory effects of clozapine in schizophrenia. Schizophr Res 1997;26:221–225.
46 Turrens JF: Mitochondrial formation of reactive oxygen species. J Physiol 2003;552:335–344.
47 Flatow J, Buckley P, Miller BJ: Meta-analysis of oxidative stress in schizophrenia. Biol Psychiatry 2013;74:400–409.
48 Prabakaran S, Swatton JE, Ryan MM, Huffaker SJ, Huang JT, Griffin JL, Wayland M, Freeman T, Dudbridge F, Lilley KS, Karp NA, Hester S, Tkachev D, Mimmack ML, Yolken RH, Webster MJ, Torrey EF, Bahn S: Mitochondrial dysfunction in schizophrenia: evidence for compromised brain metabolism and oxidative stress. Mol Psychiatry 2004;9:684–697, 643.
49 Jacobsen JP, Rodriguiz RM, Mork A, Wetsel WC: Monoaminergic dysregulation in glutathione-deficient mice: possible relevance to schizophrenia? Neuroscience 2005;132:1055–1072.
50 Cabungcal JH, Steullet P, Morishita H, Kraftsik R, Cuenod M, Hensch TK, Do KQ: Perineuronal nets protect fast-spiking interneurons against oxidative stress. Proc Natl Acad Sci USA 2013;110:9130–9135.
51 Steullet P, Neijt HC, Cuenod M, Do KQ: Synaptic plasticity impairment and hypofunction of NMDA receptors induced by glutathione deficit: relevance to schizophrenia. Neuroscience 2006;137:807–819.
52 Cabungcal JH, Counotte DS, Lewis EM, Tejeda HA, Piantadosi P, Pollock C, Calhoon GG, Sullivan EM, Presgraves E, Kil J, Hong LE, Cuenod M, Do KQ, O'Donnell P: Juvenile antioxidant treatment prevents adult deficits in a developmental model of schizophrenia. Neuron 2014;83:1073–1084.
53 Otte DM, Sommersberg B, Kudin A, Guerrero C, Albayram O, Filiou MD, Frisch P, Yilmaz O, Drews E, Turck CW, Bilkei-Gorzo A, Kunz WS, Beck H, Zimmer A: N-acetyl cysteine treatment rescues cognitive deficits induced by mitochondrial dysfunction in G72/G30 transgenic mice. Neuropsychopharmacology 2011;36:2233–2243.
54 Morishita H, Cabungcal JH, Chen Y, Do KQ, Hensch TK: Prolonged period of cortical plasticity upon redox dysregulation in fast-spiking interneurons. Biol Psychiatry 2015;78:396–402.

55 Papadia S, Soriano FX, Leveille F, Martel MA, Dakin KA, Hansen HH, Kaindl A, Sifringer M, Fowler J, Stefovska V, McKenzie G, Craigon M, Corriveau R, Ghazal P, Horsburgh K, Yankner BA, Wyllie DJ, Ikonomidou C, Hardingham GE: Synaptic NMDA receptor activity boosts intrinsic antioxidant defenses. Nat Neurosci 2008;11:476–487.
56 Baxter PS, Bell KF, Hasel P, Kaindl AM, Fricker M, Thomson D, Cregan SP, Gillingwater TH, Hardingham GE: Synaptic NMDA receptor activity is coupled to the transcriptional control of the glutathione system. Nat Commun 2015;6:6761.
57 Lipton SA, Choi YB, Takahashi H, Zhang D, Li W, Godzik A, Bankston LA: Cysteine regulation of protein function – as exemplified by NMDA-receptor modulation. Trends Neurosci 2002;25:474–480.
58 Gawryluk JW, Wang JF, Andreazza AC, Shao L, Young LT: Decreased levels of glutathione, the major brain antioxidant, in post-mortem prefrontal cortex from patients with psychiatric disorders. Int J Neuropsychopharmacol 2011;14:123–130.
59 Do KQ, Trabesinger AH, Kirsten-Kruger M, Lauer CJ, Dydak U, Hell D, Holsboer F, Boesiger P, Cuenod M: Schizophrenia: glutathione deficit in cerebrospinal fluid and prefrontal cortex in vivo. Eur J Neurosci 2000;12:3721–3728.
60 Wu JQ, Kosten TR, Zhang XY: Free radicals, antioxidant defense systems, and schizophrenia. Prog Neuropsychopharmacol Biol Psychiatry 2013;46:200–206.
61 Lavoie S, Murray MM, Deppen P, Knyazeva MG, Berk M, Boulat O, Bovet P, Bush AI, Conus P, Copolov D, Fornari E, Meuli R, Solida A, Vianin P, Cuenod M, Buclin T, Do KQ: Glutathione precursor, N-acetyl-cysteine, improves mismatch negativity in schizophrenia patients. Neuropsychopharmacology 2008;33:2187–2199.
62 Turetsky BI, Calkins ME, Light GA, Olincy A, Radant AD, Swerdlow NR: Neurophysiological endophenotypes of schizophrenia: the viability of selected candidate measures. Schizophr Bull 2007;33:69–94.
63 Matsuzawa D, Hashimoto K: Magnetic resonance spectroscopy study of the antioxidant defense system in schizophrenia. Antioxid Redox Signal 2011;15:2057–2065.
64 Tosic M, Ott J, Barral S, Bovet P, Deppen P, Gheorghita F, Matthey ML, Parnas J, Preisig M, Saraga M, Solida A, Timm S, Wang AG, Werge T, Cuenod M, Do KQ: Schizophrenia and oxidative stress: glutamate cysteine ligase modifier as a susceptibility gene. Am J Hum Genet 2006;79:586–592.
65 Zhang M, Zhao Z, He L, Wan C: A meta-analysis of oxidative stress markers in schizophrenia. Sci China Life Sci 2010;53:112–124.
66 Hayden MS, Ghosh S: Signaling to NF-κB. Genes Dev 2004;18:2195–2224.
67 Bitanihirwe BK, Woo TU: Oxidative stress in schizophrenia: an integrated approach. Neurosci Biobehav Rev 2011;35:878–893.
68 Block ML, Zecca L, Hong JS: Microglia-mediated neurotoxicity: uncovering the molecular mechanisms. Nat Rev Neurosci 2007;8:57–69.
69 Liu G-H, Qu J, Shen X: NF-κB/p65 antagonizes Nrf2-ARE pathway by depriving CBP from Nrf2 and facilitating recruitment of HDAC3 to MafK. Biochim Biophys Acta 2008;1783:713–727.
70 Lante F, Meunier J, Guiramand J, De Jesus Ferreira MC, Cambonie G, Aimar R, Cohen-Solal C, Maurice T, Vignes M, Barbanel G: Late N-acetylcysteine treatment prevents the deficits induced in the offspring of dams exposed to an immune stress during gestation. Hippocampus 2008;18:602–609.
71 Lante F, Meunier J, Guiramand J, Maurice T, Cavalier M, de Jesus Ferreira MC, Aimar R, Cohen-Solal C, Vignes M, Barbanel G: Neurodevelopmental damage after prenatal infection: role of oxidative stress in the fetal brain. Free Radic Biol Med 2007;42:1231–1245.
72 Mizrahi R: Social stress and psychosis risk: common neurochemical substrates? Neuropsychopharmacology 2016;41:666–674.
73 Akdeniz C, Tost H, Streit F, Haddad L, Wust S, Schafer A, Schneider M, Rietschel M, Kirsch P, Meyer-Lindenberg A: Neuroimaging evidence for a role of neural social stress processing in ethnic minority-associated environmental risk. JAMA Psychiatry 2014;71:672–680.
74 Boydell J, van Os J, McKenzie K, Allardyce J, Goel R, McCreadie RG, Murray RM: Incidence of schizophrenia in ethnic minorities in London: ecological study into interactions with environment. Br Med J 2001;323:1336–1338.
75 Morgan C, Fearon P: Social experience and psychosis insights from studies of migrant and ethnic minority groups. Epidemiol Psichiatr Soc 2007;16:118–123.
76 van Os J, Hanssen M, Bijl RV, Vollebergh W: Prevalence of psychotic disorder and community level of psychotic symptoms: an urban-rural comparison. Arch Gen Psychiatry 2001;58:663–668.
77 Walker EF, Trotman HD, Pearce BD, Addington J, Cadenhead KS, Cornblatt BA, Heinssen R, Mathalon DH, Perkins DO, Seidman LJ, Tsuang MT, Cannon TD, McGlashan TH, Woods SW: Cortisol levels and risk for psychosis: initial findings from the North American prodrome longitudinal study. Biol Psychiatry 2013;74:410–417.
78 Walder DJ, Walker EF, Lewine RJ: Cognitive functioning, cortisol release, and symptom severity in patients with schizophrenia. Biol Psychiatry 2000;48:1121–1132.
79 Carroll BJ: Dexamethasone suppression test: a review of contemporary confusion. J Clin Psychiatry 1985;46:13–24.
80 Thompson KN, Phillips LJ, Komesaroff P, Yuen HP, Wood SJ, Pantelis C, Velakoulis D, Yung AR, McGorry PD: Stress and HPA-axis functioning in young people at ultra high risk for psychosis. J Psychiatr Res 2007;41:561–569.
81 Mondelli V, Dazzan P, Hepgul N, Di Forti M, Aas M, D'Albenzio A, Di Nicola M, Fisher H, Handley R, Marques TR, Morgan C, Navari S, Taylor H, Papadopoulos A, Aitchison KJ, Murray RM, Pariante CM: Abnormal cortisol levels during the day and cortisol awakening response in first-episode psychosis: the role of stress and of antipsychotic treatment. Schizophr Res 2010;116:234–242.
82 Berger M, Kraeuter AK, Romanik D, Malouf P, Amminger GP, Sarnyai Z: Cortisol awakening response in patients with psychosis: systematic review and meta-analysis. Neurosci Biobehav Rev 2016;68:157–166.
83 Walker E, Mittal V, Tessner K: Stress and the hypothalamic pituitary adrenal axis in the developmental course of schizophrenia. Annu Rev Clin Psychol 2008;4:189–216.
84 Calcia MA, Bonsall DR, Bloomfield PS, Selvaraj S, Barichello T, Howes OD: Stress and neuroinflammation: a systematic review of the effects of stress on microglia and the implications for mental illness. Psychopharmacology (Berl) 2016;233:1637–1650.

85 Yamakawa K, Matsunaga M, Isowa T, Kimura K, Kasugai K, Yoneda M, Kaneko H, Ohira H: Transient responses of inflammatory cytokines in acute stress. Biol Psychol 2009;82:25–32.
86 Carroll JE, Low CA, Prather AA, Cohen S, Fury JM, Ross DC, Marsland AL: Negative affective responses to a speech task predict changes in interleukin (IL)-6. Brain Behav Immun 2011;25:232–238.
87 Carpenter LL, Gawuga CE, Tyrka AR, Lee JK, Anderson GM, Price LH: Association between plasma IL-6 response to acute stress and early-life adversity in healthy adults. Neuropsychopharmacology 2010;35:2617–2623.
88 Miller GE, Chen E: Harsh family climate in early life presages the emergence of a proinflammatory phenotype in adolescence. Psychol Sci 2010;21:848–856.
89 Coelho R, Viola TW, Walss-Bass C, Brietzke E, Grassi-Oliveira R: Childhood maltreatment and inflammatory markers: a systematic review. Acta Psychiatr Scand 2014;129:180–192.
90 Jaremka LM, Fagundes CP, Peng J, Bennett JM, Glaser R, Malarkey WB, Kiecolt-Glaser JK: Loneliness promotes inflammation during acute stress. Psychol Sci 2013;24:1089–1097.
91 Nair A, Bonneau RH: Stress-induced elevation of glucocorticoids increases microglia proliferation through NMDA receptor activation. J Neuroimmunol 2006;171:72–85.
92 Rhen T, Cidlowski JA: Antiinflammatory action of glucocorticoids – new mechanisms for old drugs. N Engl J Med 2005;353:1711–1723.
93 Frank MG, Miguel ZD, Watkins LR, Maier SF: Prior exposure to glucocorticoids sensitizes the neuroinflammatory and peripheral inflammatory responses to *E. coli* lipopolysaccharide. Brain Behav Immun 2010;24:19–30.
94 Pace TW, Miller AH: Cytokines and glucocorticoid receptor signaling. Relevance to major depression. Ann NY Acad Sci 2009;1179:86–105.
95 Du Y, Grace AA: Peripubertal diazepam administration prevents the emergence of dopamine system hyperresponsivity in the MAM developmental disruption model of schizophrenia. Neuropsychopharmacology 2013;38:1881–1888.
96 Belforte JE, Zsiros V, Sklar ER, Jiang Z, Yu G, Li Y, Quinlan EM, Nakazawa K: Postnatal NMDA receptor ablation in corticolimbic interneurons confers schizophrenia-like phenotypes. Nat Neurosci 2010;13:76–83.
97 Schiavone S, Sorce S, Dubois-Dauphin M, Jaquet V, Colaianna M, Zotti M, Cuomo V, Trabace L, Krause KH: Involvement of NOX2 in the development of behavioral and pathologic alterations in isolated rats. Biol Psychiatry 2009;66:384–392.
98 Giovanoli S, Engler H, Engler A, Richetto J, Voget M, Willi R, Winter C, Riva MA, Mortensen PB, Feldon J: Stress in puberty unmasks latent neuropathological consequences of prenatal immune activation in mice. Science 2013;339:1095–1099.
99 Mizrahi R, Addington J, Rusjan PM, Suridjan I, Ng A, Boileau I, Pruessner JC, Remington G, Houle S, Wilson AA: Increased stress-induced dopamine release in psychosis. Biol Psychiatry 2012;71:561–567.
100 Stafford MR, Jackson H, Mayo-Wilson E, Morrison AP, Kendall T: Early interventions to prevent psychosis: systematic review and meta-analysis. BMJ 2013;346:f185.

Romina Mizrahi, MD, PhD
PET Centre, Research Imaging Centre
Centre for Addiction and Mental Health
250 College Street
Toronto, ON M5T 1R8 (Canada)
E-Mail romina.mizrahi@camhpet.ca

Halaris A, Leonard BE (eds): Neuroprogression in Psychiatric Disorders.
Mod Trends Pharmacopsychiatry. Basel, Karger, 2017, vol 31, pp 107–123 (DOI: 10.1159/000470811)

Neurodegeneration, Neuroregeneration, and Neuroprotection in Psychiatric Disorders

Siu W. Tang[a, b] · Daiga M. Helmeste[a] · Brian E. Leonard[a, c]

[a]Institute of Brain Medicine, Hong Kong, SAR, China; [b]University of California, Irvine, CA, USA; [c]National University of Ireland, Galway, Ireland

Abstract

Prevention of deterioration of brain function over time is important in the long-term management of chronic brain disorders such as dementia, schizophrenia, and mood disorders. Although the possibility of neurogenesis in the adult human brain is attractive, and there are psychiatric drugs proven to be effective inducers of neurogenesis in animals, we have yet to see their utility in clinical practice. The terms neurodegeneration and neuroregeneration are often used in a nonspecific manner. Neuroregeneration may mean neurogenesis, dendritogenesis, spinogenesis, or axonogenesis. The term "neuroprotection" is attractive clinically and may involve different mechanisms. Many causative and protective factors of neurodegeneration and neuroregeneration have been proposed. However, the specificity of these factors and agents and differential neuronal vulnerability factors have generally been ignored in past studies. It is also hard to separate disease-modifying from "neuroprotective" effects of a drug. The application of stringent long-term neuroanatomical, neurochemical, neurophysiological, and therapeutic efficacy criteria should improve future research in this important area.

Introduction

The brain is a dynamic organ where degeneration and regeneration are in a continuum. Clinicians are well aware of the progressive nature of the neurodegenerative processes in patients suffering from Alzheimer (AD) and Parkinson disease (PD). While treatment may slow down the degenerative process and shift the degeneration curve to the right for a few months or 1–2 years at best [1], restoration of original brain function is generally accepted to be impossible. In other brain disorders, such as psychosis and mood disorders, the subject of neurodegeneration, regeneration, and neuroprotection has been of great interest to clinicians for years. The term neuroprogression has been coined to encompass these concepts. This chapter is our attempt to evaluate the rapidly accumulating literature in this area. Since a comprehensive review would far exceed the space limitations of this chapter, we have chosen to focus only on information which has clinical relevance and to identify some important areas for further research.

Neurodegeneration and Neuroregeneration

Neurodegeneration is a nonspecific term. Degeneration could range from extensive neuronal death and shrinkage of brain matter, as seen in late stage AD, to various degrees of degeneration in fine neuronal structures, such as dendrites, spines, and axons without neuronal death in other situations.

By contrast, adult brain neurogenesis has been shown to occur in areas of the brain that are essential for the behavioral effects of antidepressant drugs [2] and that play a role in memory function and the animal's emotional state [3, 4] in a strain-dependent manner [4]. There is some evidence for neurogenesis in adult human brain [5, 6], but Rakic [7, 8] has argued that these human studies have suffered from methodological problems and that many studies did not meet basic criteria and few met strict criteria for neurogenesis.

Chemotherapy which inhibits neurogenesis has been reported to be associated with cognitive defects and depression which might be prevented by fluoxetine in humans [9, 10]. However, chemotherapy treatment also has nonneurogenesis-related effects on the brain, and reversal of depression by fluoxetine could be independent of or unrelated to neurogenesis. Also, if the neurogenesis-promoting action of fluoxetine is robust, then it should be able to counter the cognitive and cellular effects of cytotoxic agents such as 5-fluorouracil, which is not the case [10].

Traditional antidepressants, powerful in inducing neurogenesis in rats, are also unable to reverse memory loss and cognitive defects in patients suffering from neurodegenerative disorders, such as AD, Huntington disease (HD), or PD [11]. Memory and cognitive impairments often remain clinical challenges after mood improvement in patients treated with traditional antidepressant drugs, and this deficit is being addressed with new antidepressant drugs such as vortioxetine. In summary, drugs clearly shown to have a clear neurogenesis action in animal studies, including antidepressants, atypical antipsychotics, and electroconvulsive treatment (ECT), are not effective in reversing neurodegeneration and do not have a rescuing effect for memory and cognitive defects. Antidegenerative action of these drugs and antidepressants, such as fluoxetine and paroxetine, may not be related to neurogenesis but to other actions of the antidepressant, such as glial cell activation, increase in serotonin (5-HT) concentration, or direct 5-HT receptor effects [12–15]. Future research needs to address this important discrepancy.

In contrast to neurogenesis, growth of spines (spinogenesis), dendrites (dendritogenesis), and axons (axonogenesis) may be more clinically relevant. Like leaves, branches, and roots of trees [16, 17], they change in response to environmental challenges [18].

In a series of experiments, we have found that hypercortisolemia induced by cortisol injection in rats caused loss of spines and shortening of dendrites and, when hypercortisolemia is sustained, neuronal death. Treatment with antidepressants, such as paroxetine, induced neurogenesis as labeled by bromodeoxyuridine and restoration of the dendritic arbor and spines [19]. Exercise like wheel running appeared to antagonize the hypercortisolemia effect and restored neurogenesis [20]. Reorganization of dendrites after antidepressant treatment has been reported in many other similar studies [21–24].

In a transgenic mouse model of early AD, Roy et al. [25] have shown that amnesia is correlated with progressive reduction in spine density of hippocampal dentate gyrus engram cells before amyloid plaque deposition. Optogenetic induction of long-term potentiation at these perforant path synapses restored spine density and long-term memory. By contrast, ablation of dentate engram cells containing restored spine density prevented improvement of long-term memory. Rescue of spine density in hippocampal engram cells may thus be an effective treatment for memory loss in the early stages. Again, antidepressants and atypical antipsychotic drugs – while powerful in inducing hippocampal neurogenesis, spinogenesis, and dendritogenesis – are ineffective.

Neurodegeneration and Gross Brain Volume Changes

The phenomenon of "hypofrontality" was reported decades ago in schizophrenia, depression, and attention deficit hyperactivity and addictive disorders [26, 27]. However, hypofrontality itself may not have localizing value as the primary cause of hypofrontality may or may not be in the frontal lobe.

Numerous studies over the past few decades have shown a reduction in brain volume in patients suffering from brain disorders such as dementia, schizophrenia, bipolar mood disorders, and major depression. While one may argue that brain volume changes reflect significant reduction in dendritic arbors and/or neuronal death, reversible brain volume changes have also been reported which are independent of tissue atrophy [28]. Furthermore, reduced brain volume in schizophrenia and mood disorders resulting in enlarged ventricular volume (ventricle/brain ratio) has been reported only for some but not all patients [29]. The suggestion that a late stage or severe form of the illness with significant neuronal death contributes to a measureable reduction in brain substance has been questioned [30–33]. Any deviation from controls might also be too small to be reliably revealed by common diagnostic imaging techniques [34]. Furthermore, confounding factors are many, in particular the mathematical method of comparing brain images between controls and patients, using measures such as the ventricle/brain ratio and the force fitting of patients' brains into a standard template for statistical analysis.

Accepting that brain substance reduction from neurodegeneration does occur in some patients suffering from functional psychiatric disorders, it would be useful to identify the possible causes. It would also be important to find out if such patients are more resistant to treatment [35, 36], have poorer prognosis, or whether early treatment would have a neuroprotective value thereby arresting the neuroprogressive course of the illness. However, it may be hard to distinguish between a primary "neuroprotective" drug treatment effect and their secondary psychosocial effects. Psychosocial effects include secondary consequences (poor school performance from poor concentration or marital disharmony from unstable mood and social withdrawal) and tertiary complications (such as school dropout resulting in lower social standing). Early treatment of psychiatric disorders obviously would help to minimize these psychosocial effects, thereby at least delaying the neuroprogressive course of the illness.

Inflammation, Depression, and Neurodegeneration

Abnormal glial activities have been proposed to be the cause of AD dementia, and glial cells are also targets for new drug development [37]. Antidepressant drugs, including fluoxetine, paroxetine, and some others, have been shown to be neuroprotective, perhaps through activation of microglia or their anti-inflammatory properties [12–15, 38–40]. A direct or indirect effect of fluoxetine and paroxetine on microglia, independent of their actions on 5-HT reuptake and 5-HT receptors, has been suspected to be the foundation of their neuroprotective effect. It is important to point out that antidepressant drugs, including lithium, have not been shown unequivocally to be neuroprotective in any neurodegenerative disorders despite much enthusiasm [41].

On the other hand, the frequency of AD and other neurodegenerative disorders may increase

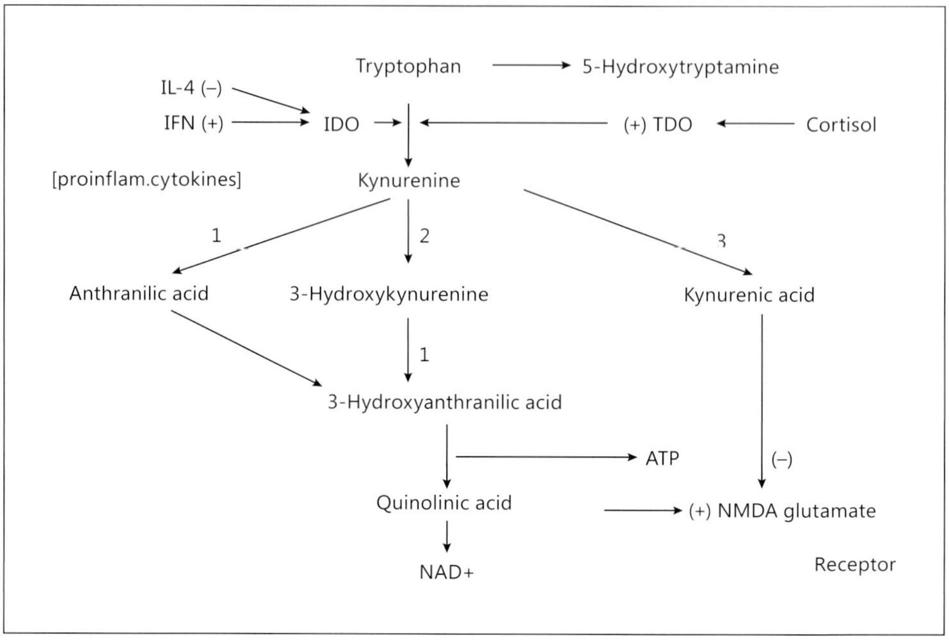

Fig. 1. The tryptophan-kynurenine pathway that links depression with chronic low-grade inflammation. IDO, indoleamine 2,3-dioxygenase in spleen, blood, kidney, and brain; TDO, tryptophan 2,3-dioxygenase in liver; 1 =, kynureninase; 2, kynurenine oxygenase; 3, kynurenine aminotransferase; 4, tryptophan decarboxylase. Adapted from Leonard [49].

in patients with depression [42], and a history of depression is also an important risk factor for dementia [43]. Amyloid plaques and neurofibrillary tangles are more frequent in patients with AD comorbid with major depression [44, 45], and neurodegenerative changes in the hippocampus and prefrontal cortex have been reported in patients with major depression [46]. Thus, neurodegenerative changes and severe cognitive deficits may be an outcome of major depression in some patients. How chronic inflammation may contribute to such changes has been reviewed by Leonard [47–49] and Leonard and Myint [50] (Fig. 1).

The implication that the tryptophan-kynurenine pathway contributes to neuroprogression has been studied for over 40 years [51, 52]. Briefly, under nonstressful conditions, tryptophan is primarily metabolized to kynurenine by tryptophan 2,3-dioxygenase (TDO) under the influence of circulating glucocorticoids [53]. However, under stressful and inflammatory conditions, tryptophan is metabolized by indoleamine 2,3-dioxygenase (IDO), which is widely distributed in immune cells, lung, kidney, and brain [54, 55]. Anti-inflammatory cytokines reduce the activity of IDO.

Approximately 60% of brain kynurenine arises from the blood [56], where it is further metabolized to either the excitotoxic metabolite quinolinic acid, which acts as an agonist at N-methyl D-aspartate (NMDA)-glutamate receptors, or to the neuroprotective metabolite kynurenic acid, which acts as an antagonist at the NMDA receptor (NMDAR) [57]. Under noninflammatory/nonstress conditions, there is a balance between the end products of the neurodegenerative and neuroprotective pathways [58].

The distribution of the tryptophan-kynurenine pathway in the brain is concentrated mainly in astrocytes and microglia [59, 60]. In astrocytes, kynurenine is mainly synthesized to kynurenic acid by kynurenine amino transferase whereas in microglia it is primarily converted to 3-hydroxykynurenine and quinolinic acid [61, 62]. Under noninflammatory conditions, the end product of kynurenine metabolism is kynurenic acid, but following the activation of microglia by stress or inflammation, the neurodegenerative pathway predominates [63]. This is further enhanced under inflammatory conditions by the apoptosis of astrocytes induced by quinolinic acid [63]. This results in the exposure of neurons to the excitotoxic kynurenine metabolites and may, thereby, contribute to neurodegeneration in some patients with chronic schizophrenia or depression.

Quinolinic acid is also an important substrate for the formation of nicotinamide adenine dinucleotide (NAD+). As NAD+ is a key component of the respiratory chain, chronic pathological changes that reduce its formation are liable to have adverse consequences for neurons that are critically dependent on high energy sources.

In major depressive disorder, Steiner et al. [64] have shown that the quinolinic acid concentration in microglia from subregions of the anterior cingulate gyrus is increased. There was also an increase in plasma 3-hydroxykynurenine [65].

Approximately 99% of tryptophan that is not used for protein and serotonin synthesis is metabolized to NAD+ via the tryptophan-kynurenine pathway [56, 66]. This situation would be compounded by a reduction in the availability of insulin, a key factor in the transport of glucose into neurons [67]. There is evidence that insulin receptor resistance is a frequent feature of depression and other major psychiatric disorders and age-related pathology associated with dementia [68, 69]. Possible mitochondrial dysfunction in depression may also result in decreased synthesis of adenosine triphosphate and related high-energy molecules with an associated increase in oxidative damage [70] and contribution to the inflammatory damages.

In summary, neurodegenerative changes in depression may be caused by the activation of the neurotoxic portion of the tryptophan-kynurenine pathway, which is activated by proinflammatory cytokines and glucocorticoids. These effects are compounded by the metabolic changes associated with the changes in energy metabolism caused by mitochondrial dysfunction. The reported neuroprotective action of antidepressant drugs may be through their anti-inflammatory property or action on microglia.

Spine and Dendritic Changes in Neurodegeneration and Neuroprotection

Both in vivo and in vitro imaging of spines and dendrites is of value clinically [71–74]. Pyramidal neurons, abundant in the frontal cortex, hippocampus, and amygdala, are important in neurodegeneration and neuroregeneration research. Pyramidal neurons receive inputs from acetylcholine (ACh), GABA, 5-HT, norepinephrine (NE), and dopamine (DA) neurons (Fig. 2).

Pyramidal neurons may be important targets for the cognitive action of a drug. The antidepressant drug vortioxetine shows both antidepressant and cognitive-enhancing actions, and the cognitive effects are postulated to be through its action on pyramidal neurons [75–77]. This leads to an interesting possibility, in that degenerating pyramidal neurons and their ACh inputs may be responsible for the memory and cognitive defects reported in mood disorders, while impaired catecholamine (DA and NE) and indoleamine neurons (5-HT) and or disturbances in their inputs into pyramidal neurons are responsible for low mood, loss of interest and enjoyment, and withdrawal symptoms. This may be an important and interesting area of future research (Fig. 2).

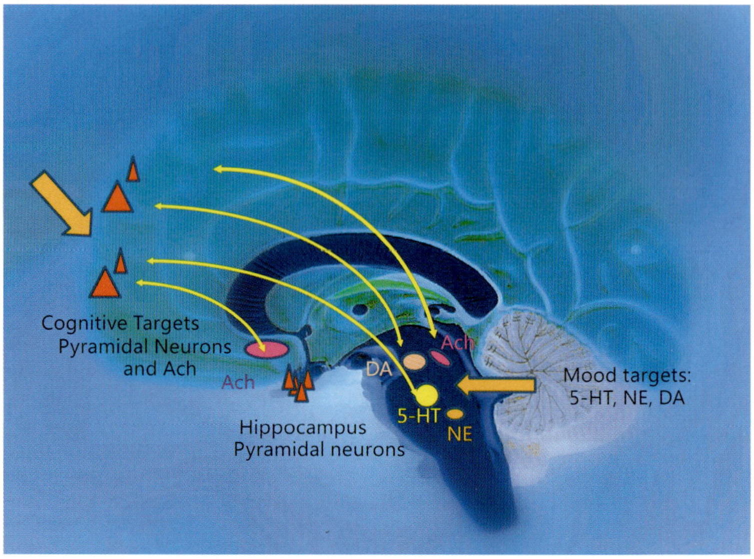

Fig. 2. Serotonin (5-HT), dopamine (DA), norepinephrine (NE), and acetylcholine (ACh) inputs to pyramidal neurons.

Drugs, Life Style, and Exercise in Neurodegeneration and Neuroregeneration

Most drugs and chemicals do not *generally* cause neurodegeneration or neuroregeneration in brain substance at least as far as can be determined by standard clinical imaging techniques.

The effects of alcohol on brain function range from transient memory loss to serious Wernicke-Korsakoff syndrome in severe cases. The statement 'the more alcohol consumed, the smaller the total brain volume' implies its possible potent neurodegenerative action [78].

In addition, brain damage associated with volume change has been reported for multiple substance abuse [79, 80], methamphetamine [81], and cocaine abuse [82]. However, there is no consistency or specificity in brain volume changes.

Lithium treatment has been shown to increase brain volume [83–87], and this may be related to its reported effect on dendritic arbor enhancement or to other unrelated effects such as fluid retention.

Endurance exercise slows hippocampal volume decline and is associated with cognitive improvement in humans [88]. Acquisition or training in specific skills (taxi driver navigator skills, juggling, and acrobatic training) has been associated with brain volume increases in relevant areas as well [89, 90].

Other studies on lifestyle factors such as reading or other forms of brain exercise such as tai chi, games, various types of health foods, vitamins, or antioxidants have been reported to change neuronal spine numbers or enhance neuronal health. Their effects, if any, seem to be general rather than specific to a particular brain neurodegenerative disorder. These studies are too numerous, and many times anecdotal, and, therefore, they will not be reviewed.

The Hypercortisolemia-Neurotoxic Hypothesis of Depression

As discussed, many antidepressant drugs have been shown to induce dendritic extensions and increases in spine number as well as neurogenesis in animals [91–95]. An elevation in brain-derived neurotropic factor (BDNF) levels has been

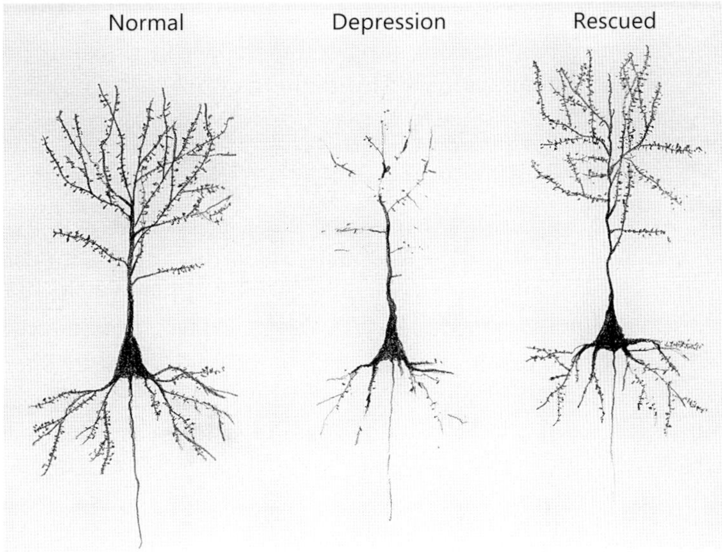

Fig. 3. The hypercortisolemia neurotoxic hypothesis. Hypercortisolemia induced degeneration of dendrites and spines and rescued by antidepressant drugs.

proposed to be the responsible mechanism, but these effects may be separate and independent [96, 97]. The important role of mTOR (mammalian target of rapamycin) in neuroprotection and as a future drug target has been raised in recent studies [98–100].

The foundation of the hypercortisolemia-neurotoxic hypothesis of depression [22] is based on stress-induced elevation of cortisol levels resulting in a neurotoxic environment for the brain. Spines and dendrites begin to degenerate, with subsequent apoptosis. Antidepressant drugs reverse the process by inducing an elevation in BDNF levels, mitigating the neurotoxic environment created by the high cortisol concentrations, and gradually repairing the damaged neurons (Fig. 3). Data supporting this attractive hypothesis include many animal studies in the early 2000s, demonstrating rats suffering from induced neurodegeneration by cortisol administration were "rescued" by antidepressant drugs, such as paroxetine [19, 91, 101]. This hypothesis has led to the suggestion that antidepressant drugs are neuroprotective in their actions, both in antagonizing the neurotoxic effect of cortisol and also in repairing the damage once it had occurred.

Many drugs other than antidepressants and activities such as exercise also have been shown to elevate BDNF [102]. Exercise induces neurogenesis and improves learning and depressive behavior in stressed rats [103] and thus should be neuroprotective as well.

However, there are many problems with this hypothesis. First of all, neither exercise nor many of these other "neurogenic drugs" are effective antidepressants by themselves. Secondly, these drugs, including antidepressants, while they may be capable of inducing neurogenesis, are not effective in correcting memory and cognitive defects in dementia. Thirdly, most studies have focused on dendrites and spines of pyramidal neurons and not on the axonal terminals of serotonergic or catecholaminergic neurons. Finally, antidepressant drugs are as effective in the elderly as in the young patients, though neurogenesis is supposed to be minimal in the elderly. Neurogenesis (in the sense of new neurons produced) as the basis for "antidepressant" action is thus questionable [91].

The neuroprotective effects of antidepressant drugs, if any, may be general and independent of their antidepressant effect. The neurotoxic effect from hypercortisolemia in some patients may also be nonspecific or even unrelated to depression. Not all patients suffering from mood disorders have hypercortisolemia. The antidepressant fluoxetine has been shown to be neuroprotective in situations other than depression. It attenuated the hypercortisolemia seen in alcoholism in an animal model [104]. In posttraumatic stress disorders, cortisol levels were generally low or inconsistent among patients [105]. A recent study has reported the relationship between blood cortisol levels and brain volume in the elderly [106], but, as discussed, antidepressant drugs are just as effective in the elderly.

Other Drugs in Neurodegeneration

Some other drugs should be mentioned. Ketamine was reported to have a fast antidepressant action, but its mechanism of action is unclear. Various hypotheses attempting to explain its fast antidepressant action have all suffered from specificity issues. For example, it is claimed to work as an NMDAR antagonist and to potentiate glutamate transmission at AMPA receptors resulting in antidepressant action. However, ketamine had similar affinity for NMDAR and the high-affinity state of DA D_2 receptors [107], plus a relatively high affinity for the high-affinity state of 5-HT_2 receptors. Thus, ketamine is not selective for NMDAR. Ketamine increases DA release following a single dose administration in drug-naive animals, and 5-hydroxyindole acetic acid levels in both drug-naive and ketamine-pretreated animals [108]. Since DA and 5-HT receptors are implicated in antidepressant action, it is difficult to argue that the NMDA selectivity is solely responsible for the antidepressant action of ketamine.

Another argument related to neuroregeneration is that ketamine induces rapid enhancement of synaptic structure and function in the cortex, concomitantly with antidepressant behavioral effects in rodents. It was found to rapidly activate the mTOR pathway, leading to increased synaptic signaling proteins. Blockade of mTOR signaling prevented both ketamine-induced synaptogenesis and antidepressant behavioral responses [109]. However, induction of synaptogenesis is also not specific to ketamine though it may act faster through a direct effect on mTOR. Many antidepressant drugs, atypical antipsychotics, and lithium induce fast synaptogenesis in rats but do not exert a fast antidepressant effect. Other data suggest that ketamine may be neurotoxic, and in developing animals it has been reported to cause interference in brain GABA neuronal networks. Ketamine reduces amyloid β-protein degradation by suppressing neprilysin expression in astrocytes. All these actions would counter any potential benefits of its use in dementia [110].

While the existing literature has focused on spine and dendritic changes, it should be noted that axonal degeneration is as important, but there are few studies in this area [111–115]. Two drugs have been reported to cause extensive, long-term, or even irreversible 5-HT neuronal degeneration. These include the banned weight reduction agent, fenfluramine, and its alternate form, dexfenfluramine, and the drug of abuse, ecstasy (MDMA). Their damaging effects on 5-HT neurons and axon terminals [116] are well established and are supported by a substantial body of data [117–119]. Both drugs release axonal 5-HT and block the reuptake mechanism. As a result, synaptic 5-HT concentrations increase and give a "high" feeling from the elevated 5-HT concentrations. The one-way release or depletion of 5-HT and possible influx of DA into the 5-HT terminals is neurodegenerative. The neurodegeneration is specific for 5-HT axon terminals. A self-repairing neuroregenerative process seems absent, and the neurodegeneration has been shown to be long term or irreversible [120–123]. Some have argued against the neurodegenerative effects of MDMA based on questions of confounding factors in earlier research, such as potential interactions from multidrug use [124, 125].

As both fenfluramine and MDMA have been shown to damage 5-HT neurons by depleting the axonal 5-HT content, one may consider the enhancement of functional 5-HT through antidepressants to effect a rescue. Yet, considering that the neurotoxic effects of both fenfluramine and MDMA stem from their 5-HT-releasing action, and that serotonin reuptake inhibitors further increase synaptic 5-HT concentrations, compounded by the fact that all are CYP 2D6 substrates, serious adverse pharmacodynamic as well as pharmacokinetic interactions may result [126, 127].

In summary, the neurodegenerative actions and neuroprotective action of both psychiatric drugs and drugs of abuse is complex. Many of these compounds have multiple actions. On the other hand, different psychiatric disorders have different pathology while neurodegeneration and neuroregeneration are general downstream phenomena with differential upstream etiology. Thus, any hypothesis to explain the mechanism of a specific psychiatric disorder, such as depression, psychosis, or dementia, or their response to treatment with antidepressants, for example, will need to address the specificity issue. For example, if hypercortisolemia and/or inflammation are/is the cause of depression, and neurodegeneration is the basis of symptoms, why did it produce mood disorder and not psychosis or dementia when the dendrites and spines were damaged? Also, why are there no data showing specificity in the damaged neuronal types such as pyramidal, 5-HT, NE, and DA neurons? All these important questions cannot be answered without further research.

Electroconvulsive Treatment and Brain Volume

Both electrical and magnetic stimulation [128–130] have been shown to affect the activities of neurons. The effect of ECT on brain volume was opposite in different areas: a decrease in the dendritic arbor in the basolateral amygdala as the basis of recovery [131] but an increase in hippocampal volume through neurogenesis or synaptogenesis [132–138]. This supports the notion that neurodegeneration and neuroregeneration cannot be attributed to dendritogenesis or spinogenesis in a simple cause-effect manner. It also raises the interesting question of whether ECT is neuroprotective or, as many have worried, neurodegenerative, as ECT may induce memory loss.

Neuronal Vulnerability and Specificity Issues in Neurodegeneration and Regeneration

Specific neuronal degeneration seems to occur in certain brain disorders, and certain neurons seem to be more vulnerable to neurodegeneration than others, for example, pigmented DA neurons in the basal ganglia in PD. This could be a function of their iron or metal content [139; Youdim MB, pers. commun.], while work by Marey-Semper et al. [140] has suggested the existence of a constitutive metabolic deficiency in DA neurons from the substantia nigra but not those in the ventral tegmental area or hypothalamus. Pissadaki and Bolam [141] have proposed that the energy cost of action potential propagation in substantia nigra DA neurons is very high, enhancing susceptibility for PD. With this in mind, endurance exercise (which is also known to increase brain oxygenation levels) has been found to be neuroprotective in animal models of DA deficiency in PD [142].

However, pigmented neurons are not more susceptible to neurodegeneration in other brain disorders such as depression, AD, or under general neurotoxic, inflammatory, or hypercortisolemic situations. Therefore, pigmentation itself, or metal content, could only be one of the many possible factors for neurodegeneration in a particular neurodegenerative disorder and not others. The following is a small sample of the complex considerations involved. Mitochondrial dysfunction has been implicated in affective spectrum disorders [143] implying a brain energy metabolism issue. APOE4, a strong risk factor for AD, is associated with altered energy metabolism in specific brain regions [144], which would be disadvantageous to neurons with high energy requirements in those regions. In addition to the AD risk, APOE4 is one of a number of genetic factors affecting outcome after traumatic brain injury [145], and its effects likely go hand in hand with lifestyle management issues. There is cell-selective neurotoxicity of the Aβ peptide in AD [146] which requires membrane binding for toxicity. Calcium channel blocker effects may differ, depending on the etiology of specific dementias, i.e., vascular dementia versus AD [147]. Elevated oxidative stress levels and impaired energy metabolism have been proposed to be a common shared neurotoxicity in several neurodegenerative disorders, including AD, amyotrophic lateral sclerosis, and PD [148]. While oxidative stress is a general rather than specific factor, the vulnerability of the neuron may still be determined by other innate factors, as not all neurons begin to degenerate simultaneously, and some neurons are more resistant than others even in the late disease stage [149, 150]. This might have important clinical implications, as a neuroprotective approach may also be specific to a specific neurodegenerative disorder.

Other examples reflecting specific susceptibility include GABA degeneration in AD and HD. GABAergic neuronal degeneration has been reported in both disorders and early in the HD case [151, 152]; again, the mechanism is unknown and may not occur in other brain disorders.

The size of neurons seems to matter in certain neurodegenerative disorders. For example, in HD, striatal medium-sized spiny neurons degenerated while large cholinergic interneurons were relatively spared though their ability to release ACh was impaired [153]. Again, size alone is not a general neurodegenerative factor.

It could be argued that 5-HT neurons in mood disorders or DA neurons in schizophrenia are not in and of themselves the cause of the disorder but rather neurons that become "deregulated" or "released" from other or higher controls. Antidepressant or antipsychotic drugs only treat the disorder targeting lower end targets. Therefore, neurodegeneration or neuroregeneration issues in catecholamine and indoleamine neurons may be irrelevant. Again, literature in this area is missing. In the fenfluramine and MDMA axonal degeneration situation, the neurodegeneration process is specific to the 5-HT terminals while other catecholamine neurons are spared [120–123]. This specificity of neurodegeneration contrasts with the spreading type of neurodegeneration in AD and PD where degeneration seems widespread and may involve multiple neuronal types [154].

Many drugs which target a specific brain disorder, such as antidepressants and lithium in mood disorders, or ACh esterase inhibitors in AD, have been claimed to be "neuroprotective" for that disorder [155, 156]. However, it is difficult to separate the disease-modifying therapeutic effect of a drug from its true neuroprotective effect. Rasagiline, for example, was claimed to have a neuroprotective action in PD [157], but as stated by Ahlskog and Uitti [158]: "nearly all the dopamine-active drugs used to treat PD have similarly been reported to demonstrate in vivo and in vitro evidence of neuroprotective action, including all the dopamine agonists and even levodopa. In fact, it has been proposed that early treatment with any dopaminergic drug may have a long-term favorable effect in PD" [159–161].

The term "neuroprotective" obviously needs to be qualified as to its "protecting effect." Many brain disorders, such as mood disorders, schizophrenia, and dementia, are chronic. Prevention of apoptosis in vitro may not be the same as "protecting" a neuron from degeneration in vivo over time. What the drug protects against may not be clear in vivo. Does the drug protect neurons from the degenerating process over time, or does the drug protect neurons from neurotoxins like the toxic oligomers in AD or the proposed toxic hypercortisolemia in mood disorders?

Clinically, we sometimes encounter the loss of efficacy of drug treatment over time. At times, the disorder progresses, and the clinical manifestation and/or symptoms change(s), and the clinical diagnosis is changed accordingly. Epigenetic mechanisms could be in play in some cases [Kennedy, 2015, pers. commun.], while gradual spreading of neuronal pathway damage or malfunctioning could be a highly possible cause. How we could possibly neuroprotect the brain under these circumstances, and how we could translate laboratory animal data into clinical research and application is definitely an interesting area to follow in future research.

Conclusion

Signs and symptoms of neurodegeneration signal that a brain disorder is progressing. The cause of neurodegeneration could be a general one or specific to the brain disorder itself or both. When cognitive and psychiatric symptoms appear in a neurodegenerative disorder, they could be from an upset in the balance of neurotransmitter pathways from a single pathology in one or multiple neurotransmitter pathways, or spread of a general degenerative process from single or multiple general causative factors.

Genetic and additional etiologic factors of the disorder and neuronal susceptibility may determine the specificity of neurodegeneration. Other factors, such as the cortisol level and inflammation, provide the stage for the etiologic factors to act, and combined give rise to symptoms specific to the disorder. Primary treatment should ideally be aimed at the etiology of the disorder while supplementary and complementary neuroprotection measures may be important in all brain disorders. Neuroprotective agents, on the other hand, may be specific to a degenerative brain disorder as specific neuronal types seem to be damaged in a specific disorder. In addition to drugs, lifestyle management, such as regular exercise, is important also, from the early developmental stage, including the prenatal stage, in building up a healthy reserve to sustain later neurodegenerative disorders when they happen. Lastly, stress management should exert additional neuroprotective effects thereby creating a less hostile internal environment for the brain.

References

1 Cortes F, Portet F, Touchon J, Vellas B: Six and 18-month changes in mild to moderate Alzheimer's patients treated with acetylcholinesterase inhibitors: what can we learn for clinical outcomes of therapeutic trials? J Nutr Health Aging 2007;11:330–337.
2 Santarelli L, Saxe M, Gross C, Surget A, Battaglia F, Dulawa S, Weisstaub N, Lee J, Duman R, Arancio O, Belzung C, Hen R: Requirement of hippocampal neurogenesis for the behavioral effects of antidepressants. Science 2003;301:805–809.
3 Kheirbek MA, Hen R: Add neurons, subtract anxiety. Sci Am 2014;311:62–67.
4 Miller BR, Hen R: The current state of the neurogenic theory of depression and anxiety. Curr Opin Neurobiol 2015;30:51–58.
5 Spalding KL, Bergmann O, Alkass K, Bernard S, Salehpour M, Huttner HB, Boström E, Westerlund I, Vial C, Buchholz BA, Possnert G, Mash DC, Druid H, Frisén J: Dynamics of hippocampal neurogenesis in adult humans. Cell 2013;156:1219–1227.
6 Bergmann O, Spalding KL, Frisén J: Adult neurogenesis in humans. Cold Spring Harb Perspect Biol 2015;7:a018994.
7 Rakic P: Adult neurogenesis in mammals: an identity crisis. J Neurosci 2002;22:614–618.
8 Rakic P: Neurogenesis in adult primate neocortex: an evaluation of the evidence. Nat Rev Neurosci 2002;3:65–71.
9 Lyons L, ElBeltagy M, Umka J, Markwick R, Startin C, Bennett G, Wigmore P: Fluoxetine reverses the memory impairment and reduction in proliferation and survival of hippocampal cells caused by methotrexate chemotherapy. Psychopharmacology (Berl) 2011;215:105–115.

10 Lyons L, ElBeltagy M, Bennett G, Wigmore P: Fluoxetine counteracts the cognitive and cellular effects of 5-fluorouracil in the rat hippocampus by a mechanism of prevention rather than recovery. PLoS One 2012;7:e30010.
11 Ernst A, Frisén J: Adult neurogenesis in humans – common and unique traits in mammals. PLoS Biol 2015;13:e1002045.
12 Li Y, Hu X, Liu Y, Bao Y, An L: Nimodipine protects dopaminergic neurons against inflammation-mediated degeneration through inhibition of microglial activation. Neuropharmacology 2009;56:580–589.
13 Chung YC, Kim SR, Jin BK: Paroxetine prevents loss of nigrostriatal dopaminergic neurons by inhibiting brain inflammation and oxidative stress in an experimental model of PD disease. J Immunol 2010;185:1230–1237.
14 Ubhi K, Inglis C, Mante M, Patrick C, Adame A, Spencer B, Rockenstein E, May V, Winkler J, Masliah E: Fluoxetine ameliorates behavioral and neuropathological deficits in a transgenic model mouse of α-synucleinopathy. Exp Neurol 2012;234:405–416.
15 Handa J, Sekiguchi M, Krupkova O, Konno S: The effect of serotonin-noradrenaline reuptake inhibitor duloxetine on the intervertebral disk-related radiculopathy in rats. Eur Spine J 2016;25: 877–887.
16 Jan YN, Jan LY: Branching out: mechanisms of dendritic arborization. Nat Rev Neurosci 2010;11:316–328.
17 Heiman MG, Shaham S: Twigs into branches: how a filopodium becomes a dendrite. Curr Opin Neurobiol 2010;20: 86–91.
18 McEwen BS: Stress, sex, and neural adaptation to a changing environment: mechanisms of neuronal remodeling. Ann NY Acad Sci 2010; 1204(suppl):E38–E59.
19 Qiu G, Helmeste DM, Samaranayake AN, Lau WM, Lee TM, Tang SW, So KF: Modulation of the suppressive effect of corticosterone on adult rat hippocampal cell proliferation by paroxetine. Neurosci Bull 2007;23:131–136.
20 Yau SY, Li A, Zhang ED, Christie BR, Xu A, Lee TM, So KF: Sustained running in rats administered corticosterone prevents the development of depressive behaviors and enhances hippocampal neurogenesis and synaptic plasticity without increasing neurotrophic factor levels. Cell Transplant 2014;23:481–492.
21 Kitahara Y, Ohta K, Hasuo H, Shuto T, Kuroiwa M, Sotogaku N, Togo A, Nakamura K, Nishi A: Chronic fluoxetine induces the enlargement of perforant path-granule cell synapses in the mouse dentate gyrus. PLoS One 2016; 11:e0147307.
22 Duman CH, Duman RS: Spine synapse remodeling in the pathophysiology and treatment of depression. Neurosci Lett 2015;601:20–29.
23 Noorafshan A, Abdollahifar MA, Karbalay-Doust S, Asadi-Golshan R, Rashidian-Rashidabadi A: Sertraline and curcumin prevent stress-induced morphological changes of dendrites and neurons in the medial prefrontal cortex of rats. Folia Neuropathol 2015;53:69–79.
24 Chen F, du Jardin KG, Waller JA, Sanchez C, Nyengaard JR, Wegener G: Vortioxetine promotes early changes in dendritic morphology compared to fluoxetine in rat hippocampus. Eur Neuropsychopharmacol 2016;26:234–245.
25 Roy DS, Arons A, Mitchell TI, Pignatelli M, Ryan TJ, Tonegawa S: Memory retrieval by activating engram cells in mouse models of early Alzheimer's disease. Nature 2016;531:508–512.
26 Williamson P: Hypofrontality in schizophrenia: a review of the evidence. Can J Psychiatry 1987;32:399–404.
27 Galynker II, Cai J, Ongseng F, Finestone H, Dutta E, Serseni D: Hypofrontality and negative symptoms in major depressive disorder. J Nucl Med 1998;39: 608–612.
28 Zahr NM, Mayer D, Rohlfing T, Orduna J, Luong R, Sullivan EV, Pfefferbaum A: A mechanism of rapidly reversible cerebral ventricular enlargement independent of tissue atrophy. Neuropsychopharmacology 2013;38:1121–1129.
29 Frodl T, Jäger M, Smajstrlova I, Born C, Bottlender R, Palladino T, Reiser M, Möller HJ, Meisenzahl EM: Effect of hippocampal and amygdala volumes on clinical outcomes in major depression: a 3-year prospective magnetic resonance imaging study. J Psychiatry Neurosci 2008;33:423–430.
30 Velakoulis D, Pantelis C, McGorry PD, Dudgeon P, Brewer W, Cook M, Desmond P, Bridle N, Tierney P, Murrie V, Singh B, Copolov D: Hippocampal volume in first-episode psychoses and chronic schizophrenia: a high-resolution magnetic resonance imaging study. Arch Gen Psychiatry 1999;56:133–141.
31 Velakoulis D, Wood SJ, Wong MT, McGorry PD, Yung A, Phillips L, Smith D, Brewer W, Proffitt T, Desmond P, Pantelis C: Hippocampal and amygdala volumes according to psychosis stage and diagnosis: a magnetic resonance imaging study of chronic schizophrenia, first-episode psychosis, and ultra-high-risk individuals. Arch Gen Psychiatry 2006; 63:139–149.
32 Kasai K, Shenton ME, Salisbury DF, Hirayasu Y, Lee CU, Ciszewski AA, Yurgelun-Todd D, Kikinis R, Jolesz FA, McCarley RW: Progressive decrease of left superior temporal gyrus gray matter volume in patients with first-episode schizophrenia. Am J Psychiatry 2003; 160:156–164.
33 Takahashi T, Wood SJ, Soulsby B, McGorry PD, Tanino R, Suzuki M, Velakoulis D, Pantelis C: Follow-up MRI study of the insular cortex in first-episode psychosis and chronic schizophrenia. Schizophr Res 2009;108:49–56.
34 Van Horn JD, McManus IC: Ventricular enlargement in schizophrenia. a meta-analysis of studies of the ventricle:brain ratio (VBR). Br J Psychiatry 1992;160: 687–697.
35 Phillips LJ, Velakoulis D, Pantelis C, Wood S, Yuen HP, Yung AR, Desmond P, Brewer W, McGorry PD: Non-reduction in hippocampal volume is associated with higher risk of psychosis. Schizophr Res 2002;58:145–158.
36 Puri BK: Progressive structural brain changes in schizophrenia. Expert Rev Neurother 2010;10:33–42.
37 Lauterbach EC: Repurposing psychiatric medicines to target activated microglia in anxious mild cognitive impairment and early Parkinson's disease. Am J Neurodegener Dis 2016;5:29–51.
38 Hannestad J: Comment on "Paroxetine prevents loss of nigrostriatal dopaminergic neurons by inhibiting brain inflammation and oxidative stress in an experimental model of Parkinson's disease." J Immunol 2010;185: 4966.
39 Miguelez C, Grandoso L, Ugedo L: Locus coeruleus and dorsal raphe neuron activity and response to acute antidepressant administration in a rat model of Parkinson's disease. Int J Neuropsychopharmacol 2011;14:187–200.

40 Politis M, Wu K, Loane C, Quinn NP, Brooks DJ, Oertel WH, Björklund A, Lindvall O, Piccini P: Serotonin neuron loss and nonmotor symptoms continue in Parkinson's patients treated with dopamine grafts. Sci Transl Med 2012;4: 128ra41.
41 Leeds PR, Yu F, Wang Z, Chiu CT, Zhang Y, Leng Y, Linares GR, Chuang DM: A new avenue for lithium: intervention in traumatic brain injury. ACS Chem Neurosci 2014;5:422–433.
42 Geerlings MI, Schoevers RA, Beekman AT, Jonker C, Deeg DJ, Schmand B, Ader HJ, Bouter LM, Van Tilburg W: Depression and risk of cognitive decline and Alzheimer's disease. Results of two prospective community-based studies in the Netherlands. Br J Psychiatry 2000; 176:568–575.
43 Jorm AF: History of depression as a risk factor for dementia: an update. Aust NZ J Psychiatry 2001;35:776–781.
44 Rapp MA, Schnaider-Beeri M, Grossman HT, Sano M, Perl DP, Purohit DP, Gorman JM, Haroutunian V: Increased hippocampal plaques and tangles in patients with Alzheimer disease with a lifetime history of major depression. Arch Gen Psychiatry 2006;63:161–167.
45 Sun X, Steffens DC, Au R, Folstein M, Summergrad P, Yee J, Rosenberg I, Mwamburi DM, Qiu WQ: Amyloid-associated depression: a prodromal depression of Alzheimer disease? Arch Gen Psychiatry 2008;65:542–550.
46 Sheline YI: Neuroimaging studies of mood disorder effects on the brain. Biol Psychiatry 2003;54:338–352.
47 Leonard BE: Changes in the immune system in depression and dementia. Int J Dev Neurosci 2001;19:305–312.
48 Leonard BE: Inflammation, depression and dementia: are they connected? Neurochem Res 2007;32:1749–1756.
49 Leonard BE: Pain, depression and inflammation: are interconnected causative factors involved; in Finn DP, Leonard BE (eds): Pain in Psychiatric Disorders. Mod Trends Pharmacopsychiatry. Basel, Karger, 2015, vol 30, pp 22–35.
50 Leonard BE, Myint AM: Changes in the immune system in depression and dementia: causal or coincidental effects? Dialogues Clin Neurosci 2006;8:163–174.
51 Lapin IP, Oxenkrug GF: Intensification of the central serotonergic processes as possible determinants of the thymoleptic effect. Lancet 1969;1:32–39.
52 Lapin IP: Kynurenines as probable participants in depression. Pharmakopsychiatr Neuropsychopharmakol 1973;6: 273–279.
53 Satyanarana U, Rao BS: Dietary tryptophan levels and the enzymes of tryptophan NAD pathway. Br J Nutr 1980;43: 107–113.
54 Carlin JM, Borden EC, Sondel PM, Byrne GI: Interferon-induced indoleamine 2,3-dioxygenase activity in human mononuclear phagocytes. J Leukoc Biol 1989;45:29–34.
55 Taylor MW, Feng GS: Relationship between interferon gamma, indoleamine dioxygenase, and tryptophan catabolism. FASEB J 1991;5:2516–2522.
56 Gal EM, Sherman AD: L-Kyurenine: its synthesis and possible regulatory function in brain. Neurochem Res 1980;5: 223–239.
57 Myint AM, Kim YK: Cytokine-serotonin interaction through IDO: a neurodegeneration hypothesis of depression. Med Hypotheses 2003;61:519–525.
58 Perkins MN, Stone TW: An iontophoretic investigation of the action of convulsant kynurenines and their interaction with the endogenous excitant quinolinic acid. Brain Res 1982;247: 184–187.
59 Grant RS, Kapoor V: Murine glial cells regenerate NAD, after peroxide-induced depletion, using either nicotinic acid, nicotinamide, or quinolinic acid as substrates. J Neurochem 1998;70:1759–1763.
60 Grant RS, Naif H, Espinosa M, Kapoor V: IDO induction in IFN-gamma activated astrocytes: a role in improving cell viability during oxidative stress. Redox Rep 2005;5:101–104.
61 Guillemin GT, Lim CK, Smythe GA, Garner B, Kapoor V, Takikawa O, Brew BJ: Characterization of the kynurenine pathway in human neurons. J Neurosci 2007;27:12884–12892.
62 Guillemin GT, Smythe G, Takikawa O, Brew BJ: Expression of indoleamine 2,3-dioxygenase and production of quinolinic acid by human microglia, astrocytes, and neurons. Glia 2005;49: 15–23.
63 Guillemin GT, Wang L, Brew BJ: Quinolinic acid selectively induces apoptosis of human astrocytes: potential role in AIDS dementia complex. J Neuroinflammation 2005;2:16.
64 Steiner J, Walter M, Gos T, Guillemin GJ, Bernstein HG, Sarnyai Z, Mawrin C, Brisch R, Bielau H, Meyer zu Schwabedissen L, Bogerts B, Myint AM: Severe depression is associated with increased microglial quinolinic acid in subregions of the anterior cingulate gyrus: evidence for an immune-modulated glutamatergic neurotransmission? J Neuroinflammation 2011;8:94.
65 Myint AM: Inflammation, neurotoxins and psychiatric disorders; in Halaris A, Leonard BE (eds): Inflammation in Psychiatry. Mod Trends Pharmacopsychiatry. Basel, Karger, 2013, vol 28, pp 61–74.
66 Han Q, Tao DA, Li J: Structure, expression and function of kynurenine aminotransferase in human and rodent brain. Cell Mol Life Sci 2010;67:353–368.
67 Oxenkrug GF: Insulin resistance and dysregulation of tryptophan-kynurenine-NAD pathways. Mol Neurobiol 2013;48:294–301.
68 Lee YJ, Han SB, Nam SY, Oh KW, Hong JT: Inflammation and Alzheimer's disease. Arch Pharm Res 2010;33:1539–1556.
69 Lee YJ, Kim JE, Hwang IS, Kwak MH, Lee JH, Jung YJ, An BS, Kwon HS, Kim BC, Kim SJ, Kim JM, Hwang DY: Alzheimer's phenotypes induced by overexpression of human presenilin 2 mutant proteins stimulate significant changes in key factors of glucose metabolism. Mol Med Rep 2013;7:1571–1578.
70 Sas K, Robotka H, Toldie J, Vecsei L: Mitochondria, metabolic disturbance, oxidative stress and the kynurenine system with a focus on neurodegenerative disorders. J Neurol Sci 2007;257:221–239.
71 Stosiek C, Garaschuk O, Holthoff K, Konnerth A: In vivo two-photon calcium imaging of neuronal networks. Proc Natl Acad Sci USA 2003;100:7319–7324.
72 Helmchen F: Two-photon functional imaging of neuronal activity; in Frostig RD (ed): In vivo Optical Imaging of Brain Function, ed 2. Boca Raton, CRC Press/Taylor & Francis, 2009, chapt 2.
73 Kawakami R, Sawada K, Sato A, Hibi T, Kozawa Y, Sato S, Yokoyama H, Nemoto T: Visualizing hippocampal neurons with in vivo two-photon microscopy using a 1030 nm picosecond pulse laser. Sci Rep 2013;3:1014.

74 Kawakami R, Sawada K, Kusama Y, Fang YC, Kanazawa S, Kozawa Y, Sato S, Yokoyama H, Nemoto T: In vivo two-photon imaging of mouse hippocampal neurons in dentate gyrus using a light source based on a high-peak power gain-switched laser diode. Biomed Opt Express 2015;6:891–901.
75 Chen F, du Jardin KG, Waller JA, Sanchez C, Nyengaard JR, Wegener G: Vortioxetine promotes early changes in dendritic morphology compared to fluoxetine in rat hippocampus. Eur Neuropsychopharmacol 2016;26: 234–245.
76 Dale E, Pehrson AL, Jeyarajah T, Li Y, Leiser SC, Smagin G, Olsen CK, Sanchez C: Effects of serotonin in the hippocampus: how SSRIs and multimodal antidepressants might regulate pyramidal cell function. CNS Spectr 2016;21:143–161.
77 Waller JA, Chen F, Sanchez C: Vortioxetine promotes maturation of dendritic spines in vitro: a comparative study in hippocampal cultures. Neuropharmacology 2016;103:143–154.
78 Paul CA, Au R, Fredman L, Massaro JM, Seshadri S, Decarli C, Wolf PA: Association of alcohol consumption with brain volume in the Framingham study. Arch Neurol 2008;65:1363–1367.
79 Liu X, Matochik JA, Cadet JL, London ED: Smaller volume of prefrontal lobe in polysubstance abusers: a magnetic resonance imaging study. Neuropsychopharmacology 1998;18:243–252.
80 Fowler JS, Kroll C, Ferrieri R, Alexoff D, Logan J, Dewey SL, Schiffer W, Schlyer D, Carter P, King P, Shea C, Xu Y, Muench L, Benveniste H, Vaska P, Volkow ND: PET studies of *d*-methamphetamine pharmacokinetics in primates: comparison with *l*-methamphetamine and (–)-cocaine. J Nucl Med 2007;48:1724–1732.
81 Kim SJ, Lyoo IK, Hwang J, Sung YH, Lee HY, Lee DS, Jeong DU, Renshaw PF: Frontal glucose hypometabolism in abstinent methamphetamine users. Neuropsychopharmacology 2005;30:1383–1391.
82 Bolla KI, Eldreth DA, London ED, Kiehl KA, Mouratidis M, Contoreggi C, Matochik JA, Kurian V, Cadet JL, Kimes AS, Funderburk FR, Ernst M: Orbitofrontal cortex dysfunction in abstinent cocaine abusers performing a decision-making task. Neuroimage 2003;19: 1085–1094.
83 Sassi RB, Brambilla P, Hatch JP, Nicoletti MA, Mallinger AG, Frank E, Kupfer DJ, Keshavan MS, Soares JC: Reduced left anterior cingulate volumes in untreated bipolar patients. Biol Psychiatry 2004;56:467–475.
84 Caetano SC, Kaur S, Brambilla P, Nicoletti M, Hatch JP, Sassi RB, Mallinger AG, Keshavan MS, Kupfer DJ, Frank E, Soares JC: Smaller cingulate volumes in unipolar depressed patients. Biol Psychiatry 2006;59:702–706.
85 Friedman SD, Dager SR, Parow A, Hirashima F, Demopulos C, Stoll AL, Lyoo IK, Dunner DL, Renshaw PF: Lithium and valproic acid treatment effects on brain chemistry in bipolar disorder. Biol Psychiatry 2004;56:340–348.
86 Lyoo IK, Dager SR, Kim JE, Yoon SJ, Friedman SD, Dunner DL, Renshaw PF: Lithium-induced gray matter volume increase as a neural correlate of treatment response in bipolar disorder: a longitudinal brain imaging study. Neuropsychopharmacology 2010;35:1743–1750.
87 Hafeman DM, Chang KD, Garrett AS, Sanders EM, Phillips ML: Effects of medication on neuroimaging findings in bipolar disorder: an updated review. Bipolar Disord 2012;14:375–410.
88 Erickson KI, Voss MW, Prakash RS, Basak C, Szabo A, Chaddock L, Kim JS, Heo S, Alves H, White SM, Wojcicki TR, Mailey E, Vieira VJ, Martin SA, Pence BD, Woods JA, McAuley E, Kramer AF: Exercise training increases size of hippocampus and improves memory. Proc Natl Acad Sci USA 2011;108:3017–3022.
89 Maguire EA, Gadian DG, Johnsrude IS, Good CD, Ashburner J, Frackowiak RS, Frith CD: Navigation-related structural change in the hippocampi of taxi drivers. Proc Natl Acad Sci USA 2000;97: 4398–4403.
90 Sampaio-Baptista C, Scholz J, Jenkinson M, Thomas AG, Filippini N, Smit G, Douaud G, Johansen-Berg H: Gray matter volume is associated with rate of subsequent skill learning after a long term training intervention. Neuroimage 2014;96:158–166.
91 Tang SW, Helmeste D, Leonard B: Is neurogenesis relevant in depression and in the mechanism of antidepressant drug action? A critical review. World J Biol Psychiatry 2012;13:402–412.
92 Kitahara Y, Ohta K, Hasuo H, Shuto T, Kuroiwa M, Sotogaku N, Togo A, Nakamura K, Nishi A: Chronic fluoxetine induces the enlargement of perforant path-granule cell synapses in the mouse dentate gyrus. PLoS One 2016;11: e0147307.
93 Noorafshan A, Abdollahifar M, Karbalay-Doust S, Asadi-Golshan R, Rashidian-Rashidabadi A: Sertraline and curcumin prevent stress-induced morphological changes of dendrites and neurons in the medial prefrontal cortex of rats. Folia Neuropathol 2015;53:69–79.
94 Chen F, du Jardin KG, Waller JA, Sanchez C, Nyengaard JR, Wegener G: Vortioxetine promotes early changes in dendritic morphology compared to fluoxetine in rat hippocampus. Eur Neuropsychopharmacol 2016;26:234–245.
95 Nava N, Treccani G, Alabsi A, Kaastrup Mueller H, Elfving B, Popoli M, Wegener G, Nyengaard JR: Temporal dynamics of acute stress-induced dendritic remodeling in medial prefrontal cortex and the protective effect of desipramine. Cereb Cortex 2017;27:694–705.
96 Sairanen M, Lucas G, Ernfors P, Castrén M, Castrén E: Brain-derived neurotrophic factor and antidepressant drugs have different but coordinated effects on neuronal turnover, proliferation, and survival in the adult dentate gyrus. J Neurosci 2005;25:1089–1094.
97 Sakata K, Mastin JR, Duke SM, Vail MG, Overacre AE, Dong BE, Jha S: Effects of antidepressant treatment on mice lacking brain-derived neurotrophic factor expression through promoter IV. Eur J Neurosci 2013;37:1863–1874.
98 Park SW, Lee JG, Seo MK, Lee CH, Cho HY, Lee BJ, Seol W, Kim YH: Differential effects of antidepressant drugs on mTOR signalling in rat hippocampal neurons. Int J Neuropsychopharmacol 2014;17:1831–1846.
99 Chiu CT, Scheuing L, Liu G, Liao HM, Linares GR, Lin D, Chuang DM: The mood stabilizer lithium potentiates the antidepressant-like effects and ameliorates oxidative stress induced by acute ketamine in a mouse model of stress. Int J Neuropsychopharmacol 2014;18: pyu102.

100 Goldschmit Y, Kanner S, Zacs M, Frisca F, Pinto AR, Currie PD, Pinkas-Kramarski R: Rapamycin increases neuronal survival, reduces inflammation and astrocyte proliferation after spinal cord injury. Mol Cell Neurosci 2015;68:82–91.
101 Lau WM, Qiu G, Helmeste DM, Lee TM, Tang SW, So KF, Tang SW: Corticosteroid decreases subventricular zone cell proliferation, which could be reversed by paroxetine. Restor Neurol Neurosci 2007;25:17–23.
102 Tang SW, Chu E, Hui T, Helmeste D, Law C: Influence of exercise on serum brain-derived neurotrophic factor concentrations in healthy human subjects. Neurosci Lett 2008;431:62–65.
103 Yau SY, Lau BW, Tong JB, Wong R, Ching YP, Qiu G, Tang SW, Lee TM, So KF: Hippocampal neurogenesis and dendritic plasticity support running-improved spatial learning and depression-like behaviour in stressed rats. PLoS One 2011;6:e24263.
104 Hu J, Xia Y, Wu Z, Liu L, Tang C: Fluoxetine might alleviate brain damage and hypercortisolemia related to chronic alcohol in rats. J Stud Alcohol Drugs 2010;71:290–294.
105 Meewisse ML, Reitsma JB, de Vries GJ, Gersons BP, Olff M: Cortisol and post-traumatic stress disorder in adults: systematic review and meta-analysis. Br J Psychiatry 2007;191:387–392.
106 Geerlings MI, Sigurdsson S, Eiriksdottir G, Garcia ME, Harris TB, Gudnason V, Launer LJ: Salivary cortisol, brain volumes, and cognition in community-dwelling elderly without dementia. Neurology 2015;85:976–983.
107 Kapur S, Seeman P: NMDA receptor antagonists ketamine and PCP have direct effects on the dopamine D_2 and serotonin 5-HT_2 receptors – implications for models of schizophrenia. Mol Psychiatry 2002;7:837–844.
108 Lindefors N, Barati S, O'Connor WT: Differential effects of single and repeated ketamine administration on dopamine, serotonin and GABA transmission in rat medial prefrontal cortex. Brain Res 1997;759:205–212.
109 Li N, Lee B, Liu RJ, Banasr M, Dwyer JM, Iwata M, Li XY, Aghajanian G, Duman RS: mTOR-dependent synapse formation underlies the rapid antidepressant effects of NMDA antagonists. Science 2010;329:959–964.
110 Yamamoto N, Arima H, Naruse K, Kasahara R, Taniura H, Hirate H, Sugiura T, Suzuki K, Sobue K: Ketamine reduces amyloid β-protein degradation by suppressing neprilysin expression in primary cultured astrocytes. Neurosci Lett 2013;545:54–58.
111 Stegeman S, Jolly LA, Premarathne S, Gecz J, Richards LJ, Mackay-Sim A, Wood SA: Loss of Usp9x disrupts cortical architecture, hippocampal development and TGFβ-mediated axonogenesis. PLoS One 2013;8:e68287.
112 Graciarena M, Dambly-Chaudière C, Ghysena A: Dynamics of axonal regeneration in adult and aging zebrafish reveal the promoting effect of a first lesion. Proc Natl Acad Sci USA 2014; 111:1610–1615.
113 Wu J, Chen H, Li H, Tang Y, Yang L, Cao S, Qin D: Antidepressant potential of chlorogenic acid-enriched extract from *Eucommia ulmoides* oliver bark with neuron protection and promotion of serotonin release through enhancing synapsin I expression. Molecules 2016; 21:E260.
114 Kharebava G, Rashid MA, Lee JW, Sarkar S, Kevala K, Kim HY: N-docosahexaenoylethanolamine regulates hedgehog signaling and promotes growth of cortical axons. Biol Open 2015;4:1660–1670.
115 Liu D, Wei N, Man H-Y, Lu Y, Zhu L-Q, Wang J-Z: The MT2 receptor stimulates axonogenesis and enhances synaptic transmission by activating Akt signaling. Cell Death Differ 2015; 22:583–596.
116 Booij L, Soucy JP, Young SN, Regoli M, Gravel P, Diksic M, Leyton M, Pihl RO, Benkelfat C: Brain serotonin synthesis in MDMA (ectasy) polydrug users: an alpha-[^{11}C]methyl-1-tryptophan study. J Neurochem 2014;131:634–644.
117 McCann UD, Szabo Z, Scheffel U, Dannals RF, Ricaurte GA: Positron emission tomographic evidence of toxic effect of MDMA ("ecstasy") on brain serotonin neurons in human beings. Lancet 1998;352:1433–1437.
118 McCann UD, Szabo Z, Vranesic M, Seckin E, Wand G, Duval A, Dannals RF, Ricaurte GA: Quantitative positron emission tomography studies of the serotonin transporter in humans previously treated with the appetite suppressants fenfluramine or dexfenfluramine. Mol Imaging Biol 2007;9: 151–157.
119 Thomasius R, Zapletalova P, Petersen K, Buchert R, Andresen B, Wartberg L, Nebeling B, Schmoldt A: Mood, cognition and serotonin transporter availability in current and former ecstasy (MDMA) users: the longitudinal perspective. J Psychopharmacol 2006;20: 211–225.
120 Insel TR, Battaglia G, Johannessen JN, Marra S, De Souza EB: 3,4-Methylenedioxymethamphetamine ("ecstasy") selectively destroys brain serotonin terminals in rhesus monkeys. J Pharmacol Exp Ther 1989;249:713–720.
121 Battaglia G, Sharkey J, Kuhar MJ, de Souza EB: Neuroanatomic specificity and time course of alterations in rat brain serotonergic pathways induced by MDMA (3,4-methylenedioxymethamphetamine): assessment using quantitative autoradiography. Synapse 1991;8:249–260.
122 De Souza EB, Battaglia G, Insel TR: Neurotoxic effect of MDMA on brain serotonin neurons: evidence from neurochemical and radioligand binding studies. Ann NY Acad Sci 1990;600: 682–697.
123 Armstrong BD, Noguchi KK: The neurotoxic effects of 3,4-methylenedioxymethamphetamine (MDMA) and methamphetamine on serotonin, dopamine, and GABA-ergic terminals: an in vitro autoradiographic study in rats. Neurotoxicology 2004;25:905–914.
124 Gouzoulis-Mayfrank E, Daumann J: Neurotoxicity of methylenedioxyamphetamines (MDMA; ecstasy) in humans: how strong is the evidence for persistent brain damage? Addiction 2006;101:348–361.
125 Mueller F, Lenz C, Steiner M, Dolder PC, Walter M, Lang UE, Liechti ME, Borgwardt S: Neuroimaging in moderate MDMA use: a systematic review. Neurosci Biobehav Rev 2016;62:21–34.
126 Oesterheld JR, Armstrong SC, Cozza KL: Ecstasy: pharmacodynamic and pharmacokinetic interactions. Psychosomatics 2004;45:84–87.
127 Oesterheld JR, Armstrong SC, Cozza KL: Ecstasy: pharmacodynamic and pharmacokinetic interactions. Psychosomatics 2004;45:84–87.

128 Vlachos A, Müller-Dahlhaus F, Rosskopp J, Lenz M, Ziemann U, Deller T: Repetitive magnetic stimulation induces functional and structural plasticity of excitatory postsynapses in mouse organotypic hippocampal slice cultures. J Neurosci 2012;32:17514–17523.
129 Lenz M, Platschek S, Priesemann V, Becker D, Willems LM, Ziemann U, Deller T, Müller-Dahlhaus F, Jedlicka P, Vlachos A: Repetitive magnetic stimulation induces plasticity of excitatory postsynapses on proximal dendrites of cultured mouse CA1 pyramidal neurons. Brain Struct Funct 2015; 220:3323–3337.
130 Murphy SC, Palmer LM, Nyffeler T, Müri RM, Larkum ME: Transcranial magnetic stimulation (TMS) inhibits cortical dendrites. Elife 2016; 5:e13598.
131 Khaleel N, Roopa R, Smitha JS, Andrade C: Electroconvulsive therapy attenuates dendritic arborization in the basolateral amygdala. J ECT 2013; 29:156–157.
132 Nordanskog P, Dahlstrand U, Larsson MR, Larsson EM, Knutsson L, Johanson A: Increase in hippocampal volume after electroconvulsive therapy in patients with depression: a volumetric magnetic resonance imaging study. J ECT 2010;26:62–67.
133 Nordanskog P, Larsson MR, Larsson EM, Johanson A: Hippocampal volume in relation to clinical and cognitive outcome after electroconvulsive therapy in depression. Acta Psychiatr Scand 2014;129:303–311.
134 Tendolkar I, van Beek M, van Oostrom I, Mulder M, Janzing J, Voshaar RO, van Eijndhoven P: Electroconvulsive therapy increases hippocampal and amygdala volume in therapy refractory depression: a longitudinal pilot study. Psychiatry Res 2013;214:197–203.
135 Bouckaert F, Sienaert P, Obbels J, Dols A, Vandenbulcke M, Stek M, Bolwig T: ECT: its brain enabling effects: a review of electroconvulsive therapy-induced structural brain plasticity. J ECT 2014;30:143–151.
136 Abbott CC, Jones T, Lemke NT, Gallegos P, McClintock SM, Mayer AR, Bustillo J, Calhoun VD: Hippocampal structural and functional changes associated with electroconvulsive therapy response. Transl Psychiatry 2014;4:e483.
137 Jorgensen A, Magnusson P, Hanson LG, Kirkegaard T, Benveniste H, Lee H, Svarer C, Mikkelsen JD, Fink-Jensen A, Knudsen GM, Paulson OB, Bolwig TG, Jorgensen MB: Regional brain volumes, diffusivity, and metabolite changes after electroconvulsive therapy for severe depression. Acta Psychiatr Scand 2016;133:154–164.
138 Joshi SH, Espinoza RT, Pirnia T, Shi J, Wang Y, Ayers B, Leaver A, Woods RP, Narr KL: Structural plasticity of the hippocampus and amygdala induced by electroconvulsive therapy in major depression. Biol Psychiatry 2016;79: 282–292.
139 Gerlach M, Double KL, Youdim MB, Riederer P: Potential sources of increased iron in the substantia nigra of parkinsonian patients. J Neural Transm Suppl 2006;70:133–142.
140 Marey-Semper I, Gelman M, Lévi-Strauss M: The high sensitivity to rotenone of striatal dopamine uptake suggests the existence of a constitutive metabolic deficiency in dopaminergic neurons from the substantia nigra. Eur J Neurosci 1993;5:1029–1034.
141 Pissadaki EK, Bolam JP: The energy cost of action potential propagation in dopamine neurons: clues to susceptibility in Parkinson's disease. Front Comput Neurosci 2013;7:13.
142 Zigmond MJ, Cameron JL, Leak RK, Mirnics K, Russell VA, Smeyne RJ, Smith AD: Triggering endogenous neuroprotective processes through exercise in models of dopamine deficiency. Parkinsonism Relat Disord 2009;15(suppl 3):S42–S45.
143 Gardner A, Boles RG: Beyond the serotonin hypothesis: mitochondria, inflammation and neurodegeneration in major depression and affective spectrum disorders. Prog Neuropsychopharmacol Biol Psychiatry 2011;35: 730–743.
144 Perkins M, Wolf AB, Chavira B, Shonebarger D, Meckel JP, Leung L, Ballina L, Ly S, Saini A, Bucky Jones T, Vallejo J, Jentarra G, Valla J: Altered energy metabolism pathways in the posterior cingulate in young adult apolipoprotein E ε4 carriers. J Alzheimers Dis 2016;53:95–106.
145 Diaz-Arrastia R, Baxter VK: Genetic factors in outcome after traumatic brain injury: what the human genome project can teach us about brain trauma. J Head Trauma Rehabil 2006;21: 361–374.
146 Simakova O, Arispe NJ: The cell-selective neurotoxicity of the Alzheimer's Aβ peptide is determined by surface phosphatidylserine and cytosolic ATP levels. Membrane binding is required for Aβ toxicity. J Neurosci 2007;27: 13719–13729.
147 Nimmrich V, Eckert A: Calcium channel blockers and dementia. Br J Pharmacol 2013;169:1203–1210.
148 Finsterwald C, Magistretti PJ, Lengacher S: Astrocytes: new targets for the treatment of neurodegenerative diseases. Curr Pharm Des 2015;21: 3570–3581.
149 Wang X, Michaelis EK: Selective neuronal vulnerability to oxidative stress in the brain. Front Aging Neurosci 2010;2:12.
150 Gao H-M, Jiang J, Wilson B, Zhang W, Hong J-S, Liu B: Microglial activation-mediated delayed and progressive degeneration of rat nigral dopaminergic neurons: relevance to Parkinson's disease. J Neurochem 2002;81:1285–1297.
151 Hardy J, Cowburn R, Barton A, Reynolds G, Dodd P, Wester P, O'Carroll AM, Lofdahl E, Winblad B: A disorder of cortical GABAergic innervation in Alzheimer's disease. Neurosci Lett 1987;73:192–196.
152 Reynolds GP, Pearson SJ: Brain GABA levels in asymptomatic Huntington's disease. N Engl J Med 1990;323:682.
153 Holley SM, Joshi PR, Parievsky A, Galvan L, Chen JY, Fisher YE, Huynh MN, Cepeda C, Levine MS: Enhanced GABAergic inputs contribute to functional alterations of cholinergic interneurons in the R6/2 mouse model of Huntington's disease. eNeuro 2015;2: e0008.
154 Zweig RM, Disbrow EA, Javalkar V: Cognitive and psychiatric disturbances in parkinsonian syndromes. Neurol Clin 2016;34:235–246.
155 Guzior N, Wieckowska A, Panek D, Malawska B: Recent development of multifunctional agents as potential drug candidates for the treatment of Alzheimer's disease. Curr Med Chem 2015;22:373–404.

156 Zha GF, Zhang CP, Qin HL, Jantan I, Sher M, Amjad MW, Hussain MA, Hussain Z, Bukhari SN: Biological evaluation of synthetic α,β-unsaturated carbonyl based cyclohexanone derivatives as neuroprotective novel inhibitors of acetylcholinesterase, butyrylcholinesterase and amyloid-β aggregation. Bioorg Med Chem 2016; 24:2352–2359.
157 Mandel S, Weinreb O, Amit T, Youdim MB: Mechanism of neuroprotective action of the anti-Parkinson drug rasagiline and its derivatives. Brain Res Brain Res Rev 2005;48:379–387.
158 Ahlskog JE, Uitti RJ: Rasagiline, Parkinson neuroprotection, and delayed-start trials: still no satisfaction? Neurology 2010;74:1143–1148.
159 Cotzias GC, Miller ST, Tang LC, Papavasiliou PS: Levodopa, fertility, and longevity. Science 1977;196:549–551.
160 Schapira AH, Obeso J: Timing of treatment initiation in Parkinson's disease: a need for reappraisal? Ann Neurol 2006;59:559–562.
161 Ahlskog JE: Challenging conventional wisdom: the etiologic role of dopamine oxidative stress in Parkinson's disease. Mov Disord 2005;20:271–282.

Professor Siu W. Tang
Psychiatry North Campus Zot 1681
University of California, Irvine
Irvine, CA 92697-1681 (USA)
E-Mail swtang@uci.edu

Halaris A, Leonard BE (eds): Neuroprogression in Psychiatric Disorders.
Mod Trends Pharmacopsychiatry. Basel, Karger, 2017, vol 31, pp 124–151 (DOI: 10.1159/000470812)

The Contribution of Adult Hippocampal Neurogenesis to the Progression of Psychiatric Disorders

Rachel A. Kohman[a] • Justin S. Rhodes[b]

[a]Department of Psychology, University of North Carolina Wilmington, Wilmington, NC, and [b]Department of Psychology, The Beckman Institute, University of Illinois, Urbana, IL, USA

Abstract

New neurons are continuously formed in the adult hippocampus of the human, nonhuman primate, and rodent throughout life though rates of neurogenesis precipitously decline with age to near zero levels at the end of the natural life span. Since its discovery in the 1960s, a large number of studies have documented numerous environmental and genetic factors which regulate adult neurogenesis. Chief among the positive regulators of neurogenesis are exercise and antidepressant drugs. Chief among the negative regulators of neurogenesis besides age are stress and inflammation. To the extent that many psychiatric disorders are comorbid with or causally related to stress and inflammation, decreased neurogenesis could be a partial contributor to the pathophysiology of the disorders. However, the functional significance of new neurons in behavior has yet to be established and is currently a hotly debated topic. Therefore, it is not clear whether changes in neurogenesis that occur alongside psychiatric illnesses are a cause or a consequence of the mediating factors such as stress, drug abuse, and inflammation, which are complexly involved in the disorders. It will be important moving forward to use modern technologies capable of instantaneously inactivating cohorts of new neurons to test their functional significance in behavior and the etiology of mental illnesses.

Introduction

In this chapter, we consider the recent literature on the role of adult hippocampal neurogenesis (and lack thereof) in psychiatric illnesses. We first carefully define adult hippocampal neurogenesis and briefly review the current understanding of the functional significance of adult neurogenesis in behavior, a topic that is still under considerable debate. Followed by a brief review of the positive regulators of neurogenesis, the majority of the chapter emphasizes known risk factors of psychiatric illnesses that are confirmed to negatively regulate neurogenesis, including the effects of stress and inflammation. Lastly, we discuss the evidence in support of a role of adult hippocampal neurogenesis in schizophrenia and major depressive disorder as well as currently employed pharmacological treatments.

Adult Hippocampal Neurogenesis

Prior to the 1960s, it was generally thought that new neurons were not formed in the adult human brain [1]. Instead, it was thought that a person was born with all the neurons they would ever have and that all one could do as an adult was to lose neurons through environmental exposures and experiences [2]. However, it was not until the late 1980s to early 1990s when immunohistochemical technologies became available that convinced the entire scientific community that neurogenesis continues to occur in certain regions of the adult human brain throughout life (Fig. 1) [3]. Remarkably, in two regions of the brain, levels of neurogenesis can be quite high, such as in the hippocampus and the olfactory bulb [4]. Because of the crucial importance of the hippocampus in learning and memory, since the 1990s, much attention has been devoted to understanding the regulatory factors and functional significance of adult hippocampal neurogenesis. Therefore, in the remainder of this chapter when we use the term adult neurogenesis, we are referring to adult hippocampal neurogenesis unless otherwise noted.

The mammalian hippocampus is crucially important for associative learning [5]. This has been demonstrated extensively in human and animal lesion and functional imaging studies [6, 7]. The hippocampus is well positioned for a function in forming associations between different types of stimuli because it is the first place in the brain that receives information from all the sensory modalities [5]. The information first converges on a region of the hippocampus known as the dentate gyrus. The granule neurons of the dentate gyrus receive the information through their dendrites which contact terminals of afferent entorhinal cortical neurons (Fig. 2). The entorhinal cortical neurons carry processed sensory information that is segregated by modality.

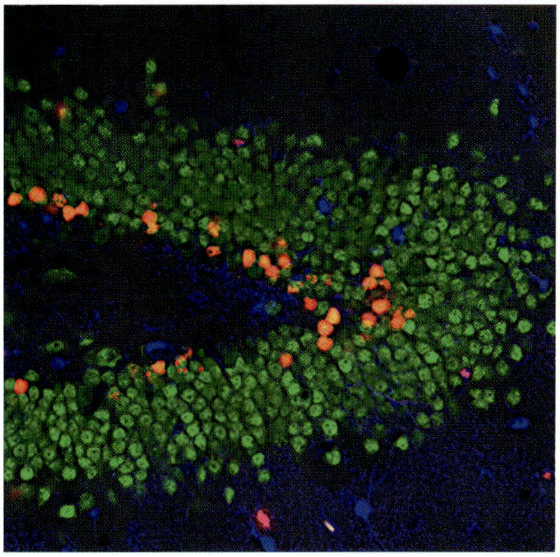

Fig. 1. Measuring adult hippocampal neurogenesis. Photograph of a section through the granule cell layer of the dentate gyrus of the hippocampus of a mouse, triple fluorescent stained to visualize adult neurogenesis. The mouse was given an intraperitoneal injection of bromodeoxyuridine (BrdU) to label dividing cells approximately 1 month before the mouse was euthanized. The brain was then sectioned and stained using immunohistochemistry. The red fluorescent label was conjugated to an antibody against BrdU and, therefore, detects newly formed cells approximately 1 month old. The green label was conjugated to an antibody against neuronal nuclear protein, to detect mature neurons. The blue label was conjugated to an antibody against S100β to detect mature astrocytes. The cells in the inside layer adjacent to the hilus that are yellow/orange have both the BrdU indicating newly formed cell that subsequently differentiated into a mature neuron, an episode of neurogenesis. A few cells outside the granule layer labeled purple indicate newly formed astrocytes.

One of the features of the mapping from the entorhinal cortex to the dentate that is thought to be critical for its function as an integrator of information is that the granule neurons are far more numerous than the relatively sparse inputs

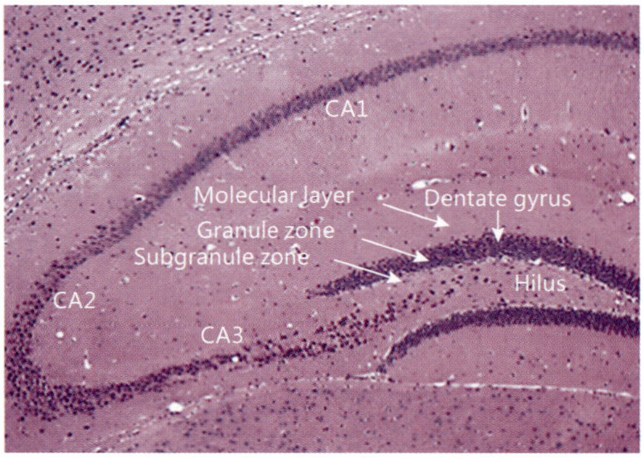

Fig. 2. Anatomy of the hippocampus. The hippocampus is composed of the dentate gyrus, CA1–CA3 fields. Information flowing into the hippocampus makes first contact on dendrites of granule neurons of the dentate gyrus which project into the molecular layer. Axons of granule neurons project to CA3 through the hilus.

from the entorhinal cortex [8]. This feature of few to many mapping is thought to enable unique representations even for very similar contexts or pairings of multiple related stimuli. This ability to distinguish subtle differences between similar contexts is referred to as pattern separation. The idea that the granule neurons of the dentate gyrus play a crucial role in pattern separation has recently received a great deal of attention [9–12]. The granule neurons, primarily excitatory and glutamatergic, project to the CA3 region of the hippocampus, a region of the hippocampus that has been widely studied for its role in long-term potentiation and integration of the sensory information into unique representations [5].

For the purposes of this review, an episode of adult hippocampal neurogenesis will be defined as an event in which a cell is born in the subgranule zone of an adult organism AND subsequently survives AND differentiates into a granule neuron AND functionally integrates into the hippocampal circuitry. The entire process from proliferation to integration is estimated to take approximately 4 weeks or longer [13]. The inside layer of the dentate gyrus adjacent to the hilus is referred to as the subgranule zone; it contains the stem cells and progenitor cells that asymmetrically divide to produce one daughter cell that has the potential to differentiate into a neuron or glial cell while the other cell retains its regenerative capacity for additional cycles of asymmetric division (Fig. 2). Eventually, the progenitor cell differentiates into an astrocyte [14]. A majority of the newly formed daughter cells do not survive [15]. Of the remaining cells that survive, a majority differentiate into granule neurons (approximately 80–90%) [16]. As they mature, they move to the outer edges of the granule cell layer. Approximately, 10–15% of the surviving cells differentiate into astrocytes, and remaining cells differentiate into oligodendrocytes or remain undifferentiated for an extended period [17]. Using methods which label the cells in vivo (e.g., by expressing green fluorescent protein), it has been estimated that it takes approximately 3–4 weeks for a newly divided cell to differentiate into a functional neuron [13]. Because the process of adult hippocampal neurogenesis involves multiple steps, a treatment or factor can affect adult hippocampal neurogenesis by affecting any or multiple stages in the process. For example, a treatment might increase proliferation of dividing cells but have no influence on survival or vice versa. Alternatively, a treatment might have no influence on proliferation or survival but might increase differentiation and integration. In all cases, net neurogenesis would increase.

Neurogenesis across the Life Span

Perhaps the most crucial factor to consider when evaluating levels of adult hippocampal neurogenesis is the age of the animal at the time when the measurement was taken. Age is crucially important to consider when interpreting group differences in neurogenesis because even slight differences in age distribution will confound the response unless measures are taken to account

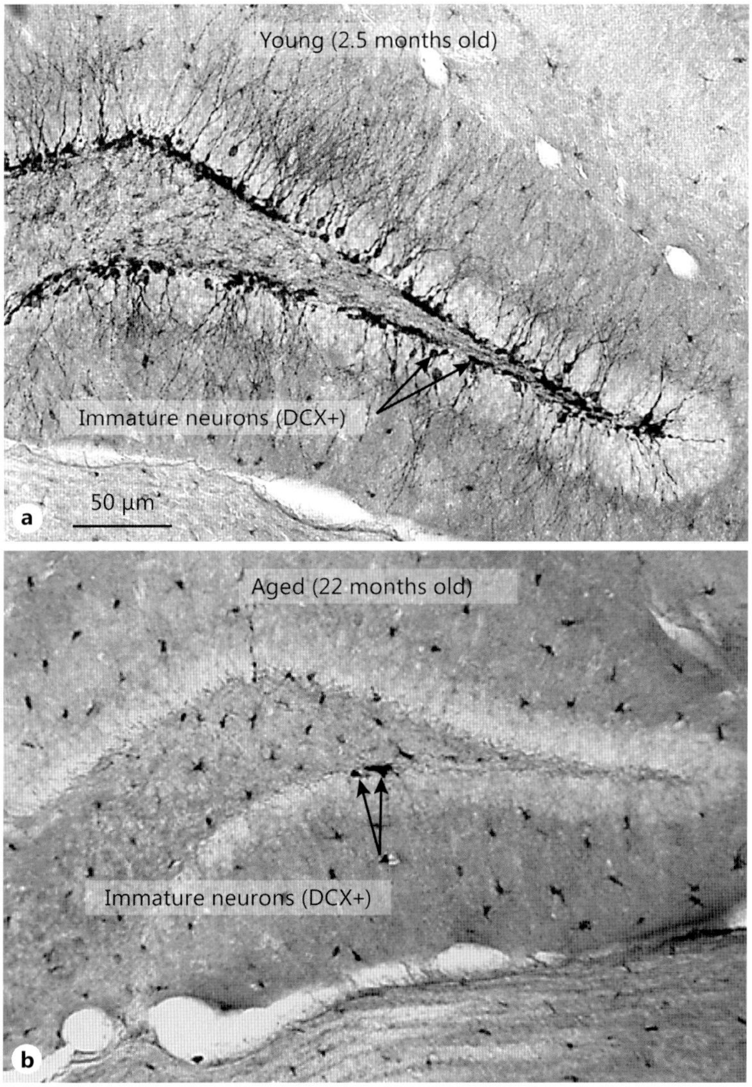

Fig. 3. Adult hippocampal neurogenesis dramatically declines with age. Photographs of the granule layer of the dentate gyrus stained for the doublecortin (DCX) marker of immature neurons of a young adult (**a**) and an aged mouse (**b**).

for the effect of age using statistical methods (Fig. 3). Levels decrease dramatically from young adult to middle aged to aged, with barely detectable levels at the end of the life span [18–20]. This has important implications for understanding the functional significance of adult hippocampal neurogenesis because neurogenesis is at such low levels in aged relative to younger subjects, yet aged subjects' brains are still broadly functional. Further, neurogenesis would be predicted to play a differential role in psychiatric illnesses depending on the time point during development that the psychiatric illness presents itself.

Functional Significance of Adult Hippocampal Neurogenesis

Because of the large literature establishing the role of the hippocampus in learning and memory, it is often assumed that adult hippocampal neurogenesis serves a specific function in learning and memory. It is well established that new neurons are more excitable than older neurons and have lower thresholds for displaying long-term potentiation [13, 21, 22]. Moreover, the idea that a new neuron might be more moldable to experience than an older neuron is intuitive and consistent with much of the evidence. However, the idea that adult neurogenesis plays a specific role in learning and memory as opposed to merely a result of prolonged development of a brain region is debatable [23–30]. One piece of evidence consistent with the latter idea is the precipitous drop-off in neurogenesis with age, much faster than any decline in learning and memory. Hence, if neurogenesis plays a specific role in a learning function, then aged subjects would be expected to be completely impaired in that domain, but no such complete impairment occurs with aging.

One method that has been used to evaluate whether neurogenesis plays a specific function in learning and memory is to abolish neurogenesis and measure the response on a learning and memory task. Older studies used crude methods to abolish neurogenesis that likely induced side effects that impaired learning and memory, such as administering chemotherapy or exposing the brain to irradiation [25, 31–33]. Modern approaches to specifically ablate neurogenesis include transgenic models. For example, in one version of these models, the nestin promotor is used to drive the expression of a viral-derived enzyme that converts exogenously administered ganciclovir into a product that incorporates into the DNA of dividing cells. The incorporation of the phosphorylated ganciclovir into the DNA drives the cell toward programmed cell death. Using these methods, neurogenesis can be specifically ablated without inducing high levels of inflammation or other side effects [34]. Recently, a meta-analysis concluded that when neurogenesis is ablated using these more specific methods, no impact on learning and memory is detected [30]. One possible explanation is that the hippocampus is able to compensate for the loss of neurons by extending synapses in older neurons, hence masking the true contribution of the new neurons to the learning and memory performance. One way to get around the problem of compensation is to use optogenetics to label cohorts of new neurons with inhibitory opsins [35] or to use inhibitory designer receptors such as inhibitory DREADDs to inactivate cohorts of new neurons instantaneously, on command [36], when the animal is acquiring a task or recalling a memory. The advantage of this method is that rather than being completely eliminated, new neurons are allowed to integrate into the circuit and are then manipulated directly during the learning or memory event. This way, the hippocampus has no time to compensate for the loss of the function of these new neurons, so it is expected to more robustly reveal their functional significance. One such study reports that inactivating a small cohort of new neurons labeled with 2 injections of retrovirus using an optogenetics strategy can completely abolish learning and memory performance on the Morris water maze spatial learning task [35]. These data are certainly incredible, if true, but will have to be replicated before any strong conclusions can be drawn at this time.

Positive Regulators of Neurogenesis

Exercise

Besides the youthfulness of the animal, the factor that arguably produces the most robust increases in levels of adult hippocampal neurogenesis is aerobic exercise [37]. This has been established in rodents [16, 38, 39] with supportive data in humans [40]. Allowing mice access to a running wheel increases adult hippocampal neurogenesis by 2- to 5-fold depending on the mouse strain and amount of running performed by the mouse [16]. Importantly, the effect of running on increasing neurogenesis is not a result of the environmental enrichment of having a wheel or sensory deprivation of not having any enrichment. Multiple studies have now confirmed that a highly enriched environment does not increase adult hippocampal neurogenesis by very much unless it includes an aerobic exercise component [41, 42]. Aerobic exercise itself strongly activates the hippocampus, which is likely related to the induction of neurogenesis detected in this region [43, 44]. These discoveries have a few important implications for understanding the relationship between adult hippocampal neurogenesis and psychiatric illnesses. First, if the illness is associated with changes in physical activity, then parallel changes in neurogenesis would be expected to follow. Second, if the illness is associated with differential activation of the dentate gyrus independent of physical activity, then corresponding changes in neurogenesis would also be expected to follow.

Antidepressant Drugs
It has been established that antidepressants increase adult hippocampal neurogenesis in rodents [45–49]. There has been some speculation that the increased neurogenesis that occurs from antidepressants is related to their therapeutic effect for treating major depression [46, 47]. This is largely because it takes approximately 3–4 weeks for a new neuron to integrate into the circuit and establish functional connections in rodent models, which is approximately how long it takes for antidepressants to ameliorate depressive symptoms in humans. However, it must be remembered that the 3- to 4-week differentiation/integration time is for mice not humans. In at least one nonhuman primate, the crab-eating macaque monkey, it was estimated that the time for a newly formed cell to differentiate into a functional neuron takes more than 6 months [50]. Hence, if the differentiation time in humans is on a similar timescale as the macaque, then increased neurogenesis is unlikely to contribute to the ameliorating effects of antidepressants.

Genetics
Mouse studies have established that adult hippocampal neurogenesis is strongly influenced by genotype [16, 51]. Certain mouse strains display much higher levels of neurogenesis than others under the same environmental conditions. The implication for psychiatric disorders is that genetic predisposition for mental illnesses would be expected to covary with levels of neurogenesis to the extent that the individuals with the illness are genetically defined.

Negative Regulators of Neurogenesis

Drugs of Abuse
Extensive literature has established that repeated administration of drugs of abuse, such as alcohol [52–55], cocaine [56–59], and opiates [60–62], decreases adult hippocampal neurogenesis, as reviewed elsewhere [63–65]. However, therapeutic doses of amphetamines increase neurogenesis in at least one mouse model [66]. The implication for a role in psychiatric illnesses is that psychiatric illnesses are often comorbid with drug abuse, and hence reduced neurogenesis would be expected to follow.

Stress
One of the first factors discovered to regulate levels of adult hippocampal neurogenesis was psychological stress [67, 68]. Unlike brief physical stress, which occurs, for example, with exercise, psychological stress decreases neurogenesis [69]. This has potentially important implications for understanding the relationship between adult neurogenesis and psychiatric disorders, as psychological stress often accompanies mental illness. The body's physiological response to stress is mediated by the hypothalamic pituitary adrenal (HPA) axis (see section on depression) and the sympathetic nervous system. Collectively, these systems coordinate the endocrine and autonomic responses through the release of cortisol,

epinephrine, and norepinephrine, which mobilize the necessary resources to handle and subsequently recover from a stressful event. Generally, activation of the stress response is adaptive, but exposure to intense and/or prolonged stressful experiences can have detrimental effects.

Psychological stressors have a well-established link to the development of psychiatric disorders, as stressful experiences often precede the onset of a depressive or psychotic episode [70]. Given that not all individuals exposed to chronic and/or extreme stressors go on to develop a psychiatric disorder, the general consensus is that a genetic predisposition, environmental factors, or other causes of intrinsic vulnerability create a subgroup of at-risk individuals [70]. Therefore, stressful experiences likely act in combination with preexisting states to impair mental health. Evidence of alterations in the stress response have been demonstrated in multiple psychiatric diseases. A sizable proportion of depressed individuals show elevated plasma levels of cortisol compared to nondepressed controls [71]. Stress is commonly used to induce symptoms of depression in animal models, as chronic stress is known to recapitulate depression-like symptoms such as anhedonia, which is often assessed by sucrose consumption [70]. In schizophrenia, measurement of salivary cortisol levels has shown that cortisol is elevated during the first schizophrenic episode [72]. At-risk subjects and those experiencing their first psychotic episode also report higher levels of perceived stress compared to controls [72]. Moreover, imaging studies have shown that the volume of the pituitary is increased during the first episode as well as in individuals that are considered at high risk of developing schizophrenia, indicating abnormalities in a key structure of the HPA axis [72]. Interestingly, the increased pituitary volume is only detected in high-risk individuals that later go on to develop psychosis. Schizophrenic individuals generally do not show increased pituitary volume compared with controls, indicating that alterations in the HPA axis may occur as the disease progresses [72]. Accompanying these basal changes are an increased responsiveness to stressful events. Schizophrenics were shown to experience an elevated emotional reaction and higher cortisol levels in response to daily life stress [73].

Stress has been convincingly shown to negatively impact the hippocampus, as chronic elevations in cortisol are associated with atrophy of the hippocampus [74]. This vulnerability is related to the hippocampal role in regulating activation of the HPA axis, as the hippocampus has a high density of the corticosterone glucocorticoid and mineralocorticoid receptors [70]. The presence of these receptors may make the hippocampus particularly reactive to stress. For instance, overwhelming evidence has shown that stress impairs hippocampal neurogenesis [70], as a number of studies, utilizing various stressors, report decreases in cell proliferation, survival, and differentiation [75–80]. In nonhuman primates, social isolation for 1 or 3 weeks has been shown to decrease new cell proliferation as well as the number of new cells that differentiate into new neurons, as shown by decreased colabeling of bromodeoxyuridine (BrdU) and doublecortin (DCX) [81]. Further, exposure to a social defeat paradigm has shown that defeated rats continue to display a reduced number of DCX-positive cells 3 months after stress exposure [82]. Ultimately, the evidence demonstrates that hippocampal neurogenesis is susceptible to stress-induced alterations.

While numerous types of stressors can reduce hippocampal neurogenesis, variations in the predictability and controllability of the stressor have been shown to modulate the effects of stress on neurogenesis. For instance, exposure to uncontrollable foot shocks has been shown to decrease cell proliferation [78]. However, if the animal can learn to avoid the foot shock, no deficits in cell proliferation or new cell survival are observed [83]. Whereas if the stressor is predictable, even if it is chronic in nature, there is evidence that neurogenesis may actually be stimulated. Parihar et al. [84] reported that exposing rats to 5 min of daily restraint stress at the same time of the day for 28 days produced a decrease in anxiety- and depression-like behaviors. Moreover, predictable chronic stress increased proliferation of new cells in the subgranular zone and increased the proportion of DCX-positive cells, indicating enhanced new neuron production. This increase in neurogenesis may relate to stress-induced memory enhancement, as glucocorticoids can enhance memory under select conditions [85]. These results demonstrate that the nature of the stress exposure alters the effects of the stressor on hippocampal neurogenesis, as the ability to predict or even control exposure to the stress helps buffer the negative effects.

Exposure to pre- or perinatal stress can have enduring effects on hippocampal neurogenesis [80]. For example, Mirescu et al. [86] found that perinatal stress in the form of daily maternal separation for 2 weeks produced a prolonged reduction in new cell proliferation. Maternal separation occurred on postnatal days 1–14, but BrdU was not administered to the offspring until they were 60–70 days old. Assessment of cell proliferation 2 h after BrdU administration showed a reduction in the number of BrdU-positive cells in the offspring that experienced maternal separation for 180 min for 2 weeks. New cell survival was also reduced 1 week after BrdU administration, but this deficit was no longer present 3 weeks after BrdU administration. No changes in cell proliferation or survival were seen if the maternal separation only lasted for 15 min, indicating that the severity of the stressor is a key factor in determining whether persistent deficits in proliferation develop [86]. Moreover, Mirescu et al. [86] stated that perinatal stress modulated the response to a stressor in adulthood (i.e., fox odor), as deficits in cell proliferation were evident in the nonstressed control offspring whereas stressed offspring failed to show a further reduction in proliferation. This lack of a decrease in proliferation may relate to the fact that proliferation in the perinatal-stress group was already attenuated in the absence of an additional stressor and may be indicative of a comprised response to adulthood stress [86]. Prenatal stress has been reported to produce similar deficits in neurogenesis. Exposing pregnant nonhuman primates to a 6-week period of stress (i.e., random acoustic startle) during either the early or late stage of pregnancy produced enduring changes in behavior, elevated cortisol levels, and a decrease in the number of BrdU-positive cells in the offspring. Assessment of the behavioral and physiological changes in the prenatally stressed and nonstressed offspring occurred when the animals were 2.5–3 years of age, indicating the prenatal effects persisted for an extended length of time confirming long-lasting changes following early-life stressors. Potentially, early-life stressors that predispose individuals to psychiatric disorders may set the stage for altered neurogenesis that persists well beyond the initial exposure.

Stress-induced reductions in neurogenesis may develop through multiple mechanisms. Direct administration of corticosterone to rats has been reported to decrease cell proliferation, as measured by the number of Ki-67-positive cells and the density of DCX-positive cells in the granular cell layer [87]. Further, corticosterone injections decreased the proportion of new cells that differentiated into neurons, as the proportion of cells that colabel with BrdU and the mature neuron marker NeuN (neuronal nuclei) is reduced in corticosterone-treated rats [88]. Interestingly, the influence of corticosterone on neuronal differentiation appears to depend on the timing of exposure, and 9 daily injections of corticosterone reduced neuronal differentiation when it was administered later in cell development (i.e., days 18–27 after division), whereas treatment during days 0–8 or 9–17 after BrdU incorporation did not affect cell differentiation. However, longer corticosterone (i.e., 27 days) treatment administered at any point after a cell was born was found to decrease the proportion of BrdU- and NeuNcolabeled cells [88]. These data potentially indicate that subchronic exposure to corticosterone has differential effects on neuronal differentiation depending on the age of the cell, whereas chronic elevations in corticosterone are able to reduce new neuron production independent of the age of the cell. Moreover, removal of the adrenal gland which prevents the stress-induced increase in corticosterone was shown to block the decrease in new cell proliferation following exposure to fox odor [89]. These data confirm a role for glucocorticoids in mediating the effects of stress on neurogenesis.

An additional route through which stress may inhibit hippocampal neurogenesis is through alterations in neurotrophic factors, which typically enhance the proliferation and survival of new cells [70]. Exposure to either acute or chronic stress has been found to decrease levels of brain-derived neurotrophic factor (BDNF) [70, 90, 91]. These effects are believed to result from glucocorticoids acting directly within the hippocampus, as direct administration of corticosterone has been shown to decrease BDNF expression in the hippocampus via mineralocorticoid and glucocorticoid receptor activation [92]. Acute immobilization stress was also found to decrease hippocampal expression of nerve growth factor (NGF) and neurotrophin-3 as well as decrease expression of the tyrosine protein kinase receptors TrKA, TrKB, and TrkC that mediate the effects of these neurotrophins [91]. Similarly, exposure to chronic unpredictable stress was found to decrease protein levels of vascular endothelial growth factor [93]. Given the central role neurotrophic factors play in sustaining hippocampal neurogenesis, stress-induced reductions in neurotrophic factors and their cognate receptors are likely a major contributor to the alterations in neurogenesis.

Lastly, stress-induced increases in inflammation have been implicated in mediating the effects of stress on neurogenesis. Prior work has shown that activation of the immune system can inhibit aspects of hippocampal neurogenesis (see section on inflammation) [94]. Research by Koo and Duman [95] demonstrated that stress-induced release of the proinflammatory cytokine interleukin (IL)-1β contributes to the inhibitory effects of stress on hippocampal neurogenesis. Inactivating IL-1β, by administering an antagonist to the IL-1 receptor, was found to block the stress-induced decrease in new cell proliferation. In agreement, IL-1 receptor-knockout mice were protected from stress-induced reductions in cell proliferation [95]. Similar protective effects were found with pharmacological inhibition of nuclear factor (NF)-κB, which is a transcription factor that regulates expression of multiple inflammatory-related genes. Inhibition of NF-κB prevented the reductions in new cell proliferation following acute or chronic stress exposure [96]. Collectively, these data highlight activation of the immune system as an additional mechanism through which stress may inhibit hippocampal neurogenesis.

Inflammation

Psychiatric disorders have been traditionally viewed as diseases of neurons, but recent advances have demonstrated that glial cells, such as microglia and astrocytes, likely contribute to psychiatric conditions. Microglia are the innate immune cells that mediate the inflammatory response within the brain (i.e., neuroinflammation). In the healthy brain, microglia express a ramified or resting phenotype that allows the cells to monitor the surrounding microenvironment for signs of injury or cellular distress [97]. If an infection or injury is detected, these cells become activated and express the classic inflammatory phenotype often referred to as M1 activation. Expression of the M1 phenotype is characterized by increased cell proliferation and physical transformations that include retraction of the long processes and an increase in the size of the cell body [97, 98]. Additionally, M1 microglia secrete a variety of inflammatory molecules, including cytokines, chemokines, and reactive oxygen species [98, 99]. These chemical mediators produced by microglia aid in coordinating the neuroinflammatory response by activating and attracting other cells to the site of injury as well as altering the activity of nonimmune cells including neurons [97]. While not considered an immune cell, astrocytes contribute to the neuroinflammatory response. Astrocytes are activated by inflammatory molecules produced by microglia and subsequently release cytokines and chemokines, upregulate expression of glial fibrillary acidic protein (GFAP), and proliferate [100, 101]. During injury or infection, temporary activation of the neuroinflammatory response is largely beneficial, but prolonged activation can disrupt normal cellular function and even cause cell death [102, 103].

In regard to neurogenesis in adulthood, neuroinflammation has been shown to disrupt the production, survival, and integration of new neurons born into the granule cell layer of the hippocampus [104]. The initial evidence that inflammation impaired hippocampal neurogenesis was provided by Monje et al. [94] and Ekdahl et al. [105]. Both reports employed the bacterial endotoxin lipopolysaccharide (LPS) to simulate a bacterial infection and induce an inflammatory response. Ekdahl et al. [105] found that chronic intracortical administration of LPS significantly decreased survival of new neurons in the hippocampus but had no effect on the proliferation of new cells. In agreement, Monje et al. [94] discovered that intraperitoneal administration of LPS had no effect on cell proliferation, but significantly reduced survival and neuronal differentiation, as shown by decreased DCX expression. Though the routes of LPS administration varied between these experiments, in both instances a neuroinflammatory response would develop as activation of immune cells within the periphery is readily transmitted to the brain and stimulates M1 activation in microglia [106]. Ekdahl et al. [105] described a negative correlation between the number of activated microglial cells and new hippocampal neurons. Moreover, administration of a microglial cell inhibitor (minocycline) was found to negate the inflammation-induced reductions in neurogenesis [105]. Collectively, these studies provided the foundational evidence that inflammation can negatively regulate hippocampal neurogenesis.

More recent work indicates that there are dose-dependent and individual factors that influence the effects of LPS on hippocampal neurogenesis. For instance, administration of a lower dose of LPS to young and aged mice selectively decreased the number of new neurons in the aged mice but had no effect on the young mice [107]. Given that aged subjects display prolonged microglial cell activation following an immune challenge [108, 109], aged subjects may be more vulnerable to inflammation-induced deficits in hippocampal neurogenesis. Exposure to an immune challenge can further suppress the already low levels of neurogenesis in the aged brain. Further, early-life immune activation was reported to increase sensitivity to adulthood exposure to LPS [110]. Mice that received an LPS injection on postnatal day 14 and a 2nd injection in adulthood had significantly fewer DCX-positive cells compared with saline-treated controls, whereas mice that received a single LPS injection at either time point showed no reductions in DCX-positive cells. Moreover, Bastos et al. [111] reported that lower doses of LPS (i.e., 0.1 and 0.5 mg/kg) had no effect on new cell survival assessed by BrdU-positive cells, whereas administration of a 1 mg/kg dose of LPS significantly decreased survival of new cells in adult mice. Taken together, these data indicate that the effects of immune activation are dose dependent, and that individual differences that alter immune activation may heighten sensitivity to inflammation-induced deficits in neurogenesis. Whether a similar sensitivity exists in psychiatric disorders has yet to be determined.

In addition to alterations in the rates of survival and cell differentiation, inflammation has been reported to alter the function and integration of new neurons into existing neural networks. Prior research has found that new neurons that undergo maturation and integration during chronic inflammation show an increase in inhibitory synaptic inputs compared with new cells developing under normal conditions and, as a result, show an increased rate and amplitude of spontaneous inhibitory postsynaptic currents [112]. Moreover, maturation during chronic inflammation caused changes in the structure of the inhibitory synapses located on the distal dendrites of new neurons that may indicate enhanced inhibitory connections [112]. These alterations in the balance of inhibitory and excitatory inputs may have enduring effects on the functional activity of new neurons that mature during chronic inflammation. However, the effects of chronic inflammation on synaptic development appear to have a short window to alter new neurons, as exposing new neurons to inflammation during the late phase of synaptic development (i.e., 4 weeks after birth) has minimal effects on excitatory and inhibitory inputs [113]. In contrast, alterations in synaptic inputs and an increase in the number of thin dendritic spines were confirmed if inflammation occurred early on in synaptic development, indicating that new neurons are more vulnerable to inflammation early in the maturation process. In accordance, Belarbi et al. [114] stated that new neurons produced during chronic inflammation show reduced ability to participate in hippocampal networks. Specifically, in the control group, exploration of a novel environment induced expression of the immediate-early gene Arc in a greater proportion of new neurons relative to the preexisting mature granule neurons. In contrast, new neurons that matured in chronic inflammation presented similar levels of Arc expression as mature neurons, indicating that inflammation may decrease the ability of new neurons to participate in hippocampal networks that respond to spatial information [114]. Exposing new neurons to inflammation during early phases of integration may have the potential to impact the normal functions mediated by the granule cell layer.

The deficits in hippocampal neurogenesis seen following immune activation are likely mediated by the chemical messengers secreted by activated microglia and astrocytes. Prior research has shown that overexpression or direct administration of individual cytokines can impair hippocampal neurogenesis. For instance, administration of the proinflammatory cytokines IL-1β, IL-6, or tumor necrosis factor (TNF)-α inhibits proliferation, survival, and neuronal differentiation of new cells in the hippocampus [94, 95, 115–118]. These chemical messengers may act directly on neural progenitor cells (NPCs), as NPCs express several cytokine receptors, including IL-1β and TNF-α [115, 119]. A recent study reported that exposing rat hippocampal NPCs to IL-1β in vitro produced a significant reduction in the number of new cells that differentiated into serotonergic neurons [120]. Research by Green and Nolan [121] indicates that IL-1β negatively regulates NPC proliferation in culture by increasing levels of glycogen synthase kinase (GSK)-3β and decreasing levels of the orphan nuclear receptor tailless homologue (TLX). Overexpression of GSK-3β has been shown to degrade β-catenin which participates in stimulating proliferation of NPCs [122]. TLX is expressed on cells undergoing division, and its overexpression increases neuronal differentiation. Inhibiting GSK-3β via a small-molecule inhibitor was found to block the IL-1β-induced reductions in TLX expression and NPC proliferation, indicating that GSK-3β participates in IL-1β-induced deficits in neurogenesis [115]. In addition, exposing differentiating cells to TNF-α in culture decreases the proportion of cells that differentiate into new neurons and increases the proportion that become astrocytes [119]. However, these effects are only seen if cells are exposed to TNF-α during differentiation and not during proliferation, indicating differential effects depending on the timing of exposure. Overall, findings indicate elevated levels of proinflammatory cytokines disrupt various stages of hippocampal neurogenesis.

Inflammation has been implicated as a risk factor for several psychiatric disorders [123, 124]. For instance, research in humans and animal models indicates that early-life infection increases the risk of developing schizophrenia [123, 125–128]. Moreover, adults that suffer from a severe infection or an autoimmune disorder are at greater risk for developing schizophrenia [129]. The specific mechanisms through which these immune events increase susceptibility for schizophrenia is still a matter of debate, but the data have repeatedly confirmed that immune activation is a clear risk factor that may contribute to the onset and/or symptoms expressed in the disease.

Beyond being a risk factor, abnormalities in immune activation appear to be present during the course of schizophrenia. Several studies have noted increased peripheral levels of inflammatory molecules as measured in serum samples from schizophrenics [130]. A meta-analysis by Miller et al. [131] delineated the role of the disease status and antipsychotic treatment in the expression of peripheral inflammation. Schizophrenic individuals experiencing an acute relapse or their first episode of psychosis show increased serum levels of several proinflammatory cytokines, including IL-6 and TNF-α, and reduced levels of anti-inflammatory cytokines. In contrast, stable medicated schizophrenics did not show higher IL-6 levels versus controls. Moreover, levels of IL-6 were found to be positively correlated with psychopathology scores and were reduced by treatment with antipsychotic medication, though only a limited number of studies have reported such results [131]. Taken together, these findings indicate that peripheral inflammation occurs in schizophrenia, but that the current status of the disease likely modulates the inflammatory profile.

Similar immune disturbances have been noted in the brains of schizophrenics. Levels of IL-6 have been reported to be increased in the cerebrospinal fluid of schizophrenics compared with controls [132, 133]. Additionally, increased microglial cell and astrocyte activation has been shown in schizophrenics, potentially indicating the presence of neuroinflammation [124, 134, 135]. However, this increase in inflammation is only detected in a subset of schizophrenic individuals. In a recent meta-analysis of the levels of neuroinflammation in postmortem samples from schizophrenics, half of the published studies found increased microglial cell activation, whereas the remaining studies found a decrease or no change. These differences in findings may relate to the clinical status of the disease as high levels of paranoia were associated with greater microglial activation, and a similar relationship was noted for peripheral inflammation [131, 135]. In accordance, heterogeneity in the presence of astrogliosis has also been reported in postmortem brains from individuals with schizophrenia [135]. Work by Catts et al. [134] demonstrated that astrogliosis was only detected in schizophrenics that showed other indices of inflammation, whereas in the absence of neuroinflammation expression of glial fibrillary acidic protein and astrocyte morphology did not differ between schizophrenics and controls. Lastly, elevated levels of IL-6 in the cerebrospinal fluid were found to be specific to schizophrenics that showed a delayed response to antipsychotics, suggesting that the presence of neuroinflammation may affect treatment responsiveness [133]. Though factors such as disease status and treatment responsiveness, for example, may help explain the differential expression of inflammation in schizophrenia, what remains unclear is how the presence or absence of inflammation contributes to the progression of the disease and symptom severity.

Inflammation has also been implicated in the pathophysiology of major depressive disorder (MDD) [136]. Similar to schizophrenia, the risk of developing MDD is greater in individuals with a chronic disease characterized by immune activation. For instance, individuals diagnosed with an autoimmune disease, such as multiple sclerosis or rheumatoid arthritis, are more likely to experience MDD [137, 138]. Moreover, 20–40% of individuals treated with the cytokine interferon-α for chronic hepatitis C subsequently develop MDD, and symptoms of depression are reduced following treatment with standard antidepressants [139]. Indeed, aspects of the behavioral changes associated with immune activation, commonly known as sickness behavior, parallel the symptoms such as altered sleep, lethargy, reduced appetite, and impaired cognitive function commonly seen in depressed individuals [139]. Animal research has demonstrated that stimulating an immune response increases expression of depression-like symptoms, as an acute immune challenge is reported to increase behavioral indices of depression in several tests of emotionality, including the forced swim test and tail suspension test [140]. Additionally, acute immune activation has been shown to produce anhedonia-like behavior, as transient inflammation has been shown to decrease consumption of an appetitive sucrose solution 24 and 48 h after the inflammatory response has dissipated [141]. In accordance, research with healthy adults has shown that transient immune activation produces an increase in anxiety and depressed mood [142]. Taken together, these data indicate that inflammation may elicit symptoms of MDD. Whether inflammation only elicits a depressive episode when it interacts with other risk factors (e.g., genetic/life stress) of MDD or whether inflammation can indiscriminately predispose individuals to MDD is presently unclear. However, a few studies indicate polymorphisms in select genes, such as IL-6 and serotonin, may alter susceptibility to cytokine-induced depression [143, 144]. Clarifying whether inflammation triggers MDD primarily in vulnerable individuals may aid in identifying populations that may benefit from anti-inflammatory therapeutics as well as facilitate preventative measures.

Assessment of peripheral markers of inflammation has shown that approximately a third of individuals with MDD have elevated levels of several inflammatory markers. For example, meta-analyses have shown that serum levels of the inflammatory molecules C-reactive protein, IL-6, and TNF-α are elevated in individuals with MDD [145, 146]. However, several studies have reported no changes in cytokine levels with MDD [145, 147], potentially indicating that inflammation may define a subgroup of individuals with MDD. There is some evidence that the presence of inflammation may negatively affect treatment responsiveness, as elevated levels of IL-6 and TNF-α were associated with treatment resistance [148, 149], although these findings have not been consistently reported [136]. More research is needed to elucidate whether the presence of inflammation represents a unique MDD phenotype and how inflammation may alter the response of the patient to antidepressant interventions.

While the precise role of inflammation in the onset and progression of psychiatric disorders is complex, the evidence points to the existence of clusters of individuals within these disease entities that show clear evidence of inflammation. This persistent increase in inflammatory molecules in these individuals likely represents a route through which deficits in neurogenesis may develop. Inflammation has clear effects on survival and neuronal differentiation, as even acute induction of an inflammatory response has been shown to inhibit hippocampal neurogenesis. Given the ability of inflammatory molecules to act directly on NPCs and negatively regulate their proliferation, the expression of a chronic inflammatory state likely creates an environment that suppresses neurogenesis. Gaining a clear understanding of whether inflammation is a causal or consequential factor of these diseases may aid in clarifying the role of inflammation in the pathophysiology of psychiatric diseases.

Neurogenesis in Schizophrenia

Schizophrenia is a chronic neuropsychiatric disease that is estimated to affect approximately 1% of the total population [150]. Schizophrenia is characterized by the presentation of negative symptoms (i.e., flat affect and social withdrawal), positive symptoms (i.e., delusions, hallucinations, and disordered thought), and cognitive deficits [150]. Abnormalities in neural development have been proposed to contribute to the disorder, as early life events, or being born with select genetic risk factors has been associated with an elevated risk of developing schizophrenia [151]. The first psychotic episode typically occurs in late adolescence or early adulthood [151]. The delayed onset of schizophrenia likely indicates the involvement of a second event (e.g., stress and/or substance abuse) that interacts with genetic and/or other etiological risk factors to trigger the expression of the disease [151].

Neurochemical abnormalities in the production or signaling mechanisms of key neurotransmitters, such as dopamine and glutamate, in select regions of the brain are believed to mediate many of the positive and negative symptoms in schizophrenia [151]. The majority of the current antipsychotic compounds are aimed at normalizing these neurochemical imbalances; while beneficial these drugs often have aversive side effects and fail to fully alleviate all of the symptoms of schizophrenia [150]. Beyond altered neurochemistry, schizophrenia is believed to be a connectivity disorder, as abnormal neural activity has been noted in the default mode network as well as a loss of white matter [152, 153]. Additionally, structural abnormalities have been noted in several regions of the brain, as enlargement of the ventricles and volumetric reductions in the prefrontal cortex, thalamus, hippocampus, and other structures have been reported [151, 154]. Collectively, these structural and neurochemical abnormalities likely contribute to dysfunctional communication within and across regions of the brain. Gaining an understanding of how these diffuse neural changes contribute to disease development and expression of individual symptoms is expected to aid in identifying alterative or additional therapeutic treatments to prevent and/or alleviate symptoms in schizophrenic individuals.

Alterations in hippocampal structure and/or function are common in many psychiatric disorders. In schizophrenia, for instance, the overall volume of the hippocampus is reduced versus healthy controls [154]. While these volume reductions are not found in all schizophrenics, reduced hippocampal volume is the most common structural abnormality reported [152, 154]. There is some evidence that the volume changes predominantly occur in the anterior division of the hippocampus and less so in the posterior division, but these regional differences are not consistently observed [152]. The changes in hippocampal volume are often present early in the disease, as volume reductions are noted in individuals experiencing their first episode of schizophrenia [155, 156]. How far in advance of disease onset the volume reductions are present is unclear, though a smaller hippocampal volume in children with schizophrenic parents has been reported to be predictive of the schizophrenia risk, potentially indicating these changes may exist long before the first episode [152]. While the volume reductions are often present at the onset of the disease, there is some evidence that hippocampal volume loss is progressive. In a study by Chakos et al. [155], hippocampal volume was significantly reduced in individuals presenting with schizophrenia for 10 or more years versus those with a disease duration of less than 5 years. A recent study reported that early volume loss in schizophrenia occurs in the CA1 and the dentate gyrus of the hippocampus [157]. Individuals diagnosed with schizophrenia 10 or more years earlier show volume reductions in all subregions of the hippocampus, indicating the volume loss may become widespread with increased disease duration [157]. Currently, the cause of these volume reductions is debated, but there is support for white matter loss as well as modest changes in gray matter that may contribute [152].

Accompanying these volume reductions are deficits in hippocampal neurogenesis. Analysis of cell proliferation in hippocampal samples from postmortem brains of schizophrenics and matched controls has demonstrated that schizophrenics show reductions in Ki-67-positive cells, indicating lower rates of proliferation [158, 159]. These deficits in new cell proliferation appear to be independent of disease duration as well as the age of onset, as no differences in Ki-67-positive cells were noted between those diagnosed at an early or later age [159]. While the decrease in hippocampal neurogenesis does not appear to be the primary mediator of the reduced volume, these changes may relate to the functional deficits seen in schizophrenic individuals. Several reports have linked the abnormalities in hippocampal function to the declarative memory deficits seen in schizophrenics [154]. For instance, imaging studies have shown that activation of the hippocampus is reduced in schizophrenics relative to controls when recalling information, indicating that the hippocampus is not being engaged during memory retrieval. Moreover, abnormalities in the interactions between the hippocampus and the prefrontal cortex have been proposed to contribute to the expression of delusions and hallucinations [154], as the hippocampus along with other regions shows increased activation prior to the onset of a visual hallucination [160]. The precise role of hippocampal neurogenesis in mediating the cognitive deficits associated with schizophrenia has yet to be fully elucidated. However, Schreiber and Newman-Tancredi [161] have suggested that the delusions seen in schizophrenics may relate to abnormalities in the ability to distinguish present from prior experiences that may contribute to the distortion of reality. As noted, hippocampal neurogenesis has been implicated in participating in pattern separation [162]. Inhibition of hippocampal neurogenesis via irradiation has been shown to impair performance on a spatial discrimination task when the two stimuli were located close together, whereas no deficits were seen if the stimuli were spaced farther apart [162]. These data indicate that new neurons in the dentate gyrus may aid in partitioning stimuli or events that have overlapping information, thus allowing the formation of separate memories for similar experiences. The potential side effects of using irradiation to inhibit hippocampal neurogenesis must be considered when evaluating these data. Though limited, recent evidence has shown deficits in pattern separation in schizophrenic individuals. However, further work is needed to dissociate whether these impairments result from a true deficit in pattern separation or alterations in basic recognition and/or response bias in schizophrenics [163, 164]. Based on the putative link between hippocampal neurogenesis and memory function, it seems plausible that the reduction in hippocampal neurogenesis could contribute to the cognitive impairments associated with schizophrenia. However, as our understanding of the role of hippocampal neurogenesis in memory function continues to evolve with future research, the selective emphasis on the role of neurogenesis in cognitive function in schizophrenia will have to be revisited.

The existing animal models of schizophrenia have also confirmed deficits in hippocampal neurogenesis. While modeling a complex disorder like schizophrenia in a rodent is challenging, current approaches are based on known neurochemical abnormalities and environmental factors associated with schizophrenia in humans. For instance, based on the higher rates of schizophrenia following maternal infection, one animal model employs the use of the viral mimetic polyriboinosinic-polyribocytidylic acid [poly(I:C)] to stimulate an inflammatory response in pregnant dams. Offspring from poly(I:C)-exposed dams show cognitive deficits and behavioral abnormalities associated with schizophrenia that can be attenuated by antipsychotic treatment [165, 166]. Assessment of hippocampal neurogenesis in the rodent maternal infection model has shown that offspring from poly(I:C)-treated dams show decreased new cell survival as measured by the number of

BrdU-positive cells 21 days after administration [167]. Further, the density of granule cells was reduced in the poly(I:C) offspring relative to controls on postnatal day 100, potentially indicating the reduced cell survival may have long-term effects on the granular cell layer. Work by Zhang and Van Praag [165] confirmed that poly(I:C) offspring show reductions in new cell survival and overall volume of the dentate gyrus. Further, the dendritic morphology of new neurons was less complex in poly(I:C) offspring relative to controls. In addition, mature and adult-born granule cells of poly(I:C) offspring showed abnormal intrinsic properties, as cells showed a decreased threshold to generate an action potential as well as a decline in the overall number of action potentials produced while input resistance was increased relative to control offspring. These data collectively indicate that animal models that mimic the behavioral and structural changes in schizophrenia also provide evidence of attenuated hippocampal neurogenesis and persistent abnormalities in granular cell activity. Further work is needed to determine whether the cognitive deficits resulting from prenatal immune activation relate to the depression of hippocampal neurogenesis.

Additional support for a potential role of neurogenesis in the pathology of schizophrenia arises from genetic studies. One of the main genes associated with an increased risk of developing schizophrenia, namely DISC1 (disrupted-in-schizophrenia 1), participates in new cell integration [168]. During embryonic development, DISC1 is widely expressed within the brain, but as the brain completes maturation, DISC1 expression is primarily localized to the hippocampus, including the dentate gyrus [169]. Moreover, assessment of cell-specific expression of DISC1 in the hippocampus demonstrated that DISC1 was present in cells expressing the proliferation marker Ki-67 or the early postmitotic neuron marker calretinin [170]. Use of virus-mediated RNA interference to selectively inhibit DISC1 in hippocampal neurons revealed that DISC1 plays a key role in new neuron integration. Selective knockdown of DISC1 in newly born neurons was found to alter cell migration, as new neurons deficient in DISC1 migrated farther than normal cells as well as showed abnormal orientation within the granule cell layer [170]. Additionally, DISC1-deficient neurons showed morphological abnormalities and accelerated development, as cells showed faster synapse formation and enlarged cell bodies compared to new neurons with normal DISC1 expression [170]. These data indicate that DISC1 normally acts as a regulator of new cell integration in terms of the rate of development and cell migration, removing the brakes from this process by downregulating DISC1 appears to adversely affect new neuron integration that may have long-lasting implications on the functional capacity of these cells.

Research has also shown that DISC1 regulates progenitor cell proliferation [171]. DISC1 has been shown to influence cell proliferation via its effects on GSK-3β, which is a serine/threonine protein kinase that is expressed at high levels in the brain [172]. GSK-3β is expressed when a cell is at rest; via phosphorylation, GSK-3β inhibits the activity of several signaling molecules and transcription factors such as β-catenin [172]. β-catenin is known to regulate the pool of progenitor cells, as deletion of β-catenin was found to reduce NPCs, whereas continuous expression increased the progenitor pool [122]. In its normal state, DISC1 directly inhibits GSK-3β, which releases β-catenin from the inhibitory control of GSK-3β and stimulates progenitor cell proliferation. In instances where DISC1 activity is impaired, GSK-3β is continually active, and the ability of β-catenin to induce proliferation is suppressed. Research has shown that the decrease in progenitor cell proliferation in DISC1-knockdown mice can be reversed by administration of a GSK-3β inhibitor or continuous β-catenin expression, indicating that a failure to inhibit GSK-3β mediates the deficits in progenitor cells in case of disrupted DISC1 [171]. Alterations in the DISC1 gene may represent one route through which hippocampal neurogenesis is impaired in schizophrenia.

In addition to DISC1, other genetic risk factors of schizophrenia are reported to disrupt hippocampal neurogenesis. Select single nucleotide polymorphisms of the G-protein-coupled receptor SREB2 (superconserved receptor expressed in brain 2) are associated with a higher risk of developing schizophrenia [173]. Transgenic mice that overexpress SREB2 show behavioral indices of schizophrenia, including deficits in cognitive function and sensorimotor gating [173]. Chen et al. [174] have reported that selective overexpression of SREB2 decreased new cell survival, as the number of BrdU-positive cells was significantly reduced 30 days after BrdU administration in the SREB2 transgenic mice. Further, the number of DCX-positive cells was also reduced with overexpression of SREB2, indicating deficits in neuronal differentiation. As complementary evidence, SREB2-knockout mice showed increased neuronal survival as indicated by higher numbers of BrdU-positive cells and colocalization of BrdU and NeuN. Alterations in SREB2 levels were also found to alter performance in behavioral tasks associated with hippocampal neurogenesis, as mice overexpressing SREB2 showed impaired pattern separation at a medium distance in a spatial task compared with control mice. In contrast, SREB2-knockout mice showed enhanced performance in the spatial pattern separation task at the closest distance compared with control mice [174]. These data confirm a role for SREB2 as a negative regulator of hippocampal neurogenesis and support the contention that alterations in neurogenesis contribute to schizophrenia.

Chromosomal deletions that dysregulate the function of several genes have also been reported to increase the risk of schizophrenia. For instance, monoallelic microdeletions on chromosome 22 at the 22q11.2 region have been associated with a higher incidence of schizophrenia as approximately one fourth of the individuals with this deletion develop schizophrenia [175]. It is estimated that the activity of the 28 genes that reside within this chromosome segment are disrupted by this deletion. Dgcr8 (DiGeorge syndrome chromosomal region 8) is one of the genes affected by this chromosomal deletion, which has been shown to regulate production of microRNA and has been implicated in the risk of developing schizophrenia [176]. Research has demonstrated that knockdown of Dgcr8 in male mice produced deficits in spatial working memory as measured by spontaneous alternation in the Y-maze test and a trend towards impaired retention of spatial memory in the water maze [176]. In addition, knockdown of Dgcr8 inhibited cell proliferation in the subgranular zone of the dentate gyrus, as measured by lower levels of Ki-67- and BrdU-positive cells. The proliferation deficits were accompanied by reductions in DCX-positive cells indicating reductions in new neuron production [176]. These data again confirm a link between genetic risk factors associated with schizophrenia and alterations in hippocampal neurogenesis and strengthen the argument that employing treatments that enhance hippocampal neurogenesis may have beneficial effects in the context of schizophrenia.

There is indication that parts of the beneficial effects of current pharmacological treatments used in schizophrenia relate to their ability to stimulate hippocampal neurogenesis. Current antipsychotics are subclassified into typical and atypical drugs. Typical antipsychotics, such as haloperidol, reduce positive symptoms of schizophrenia by antagonizing the D_2 dopamine receptor [161]. However, the development of motor abnormalities is common with these compounds given the role of dopamine in the initiation of movement. The atypical antipsychotics target both the positive and negative symptoms of schizophrenia by altering activity of multiple receptors, such as serotonin receptor subtypes, in addition to antagonizing the D_2 receptor [131].

The reduced incidence of motor abnormalities as well as the ability of these drugs to treat both positive and negative symptoms has made atypical antipsychotics the most commonly prescribed medication for schizophrenia. While research with haloperidol showed no effects on adulthood hippocampal neurogenesis, findings indicate that atypical antipsychotics may stimulate neurogenesis [177]. For example, in a mouse model of schizophrenia, maternal poly(I:C) decreased new cell survival, as measured by the number of BrdU-positive cells, in offspring up to postnatal day 57 but not at later time points. Risperidone treatment significantly increased cell survival in offspring of dams exposed to either poly(I:C) or saline [167]. Further, olanzapine was reported to increase cell proliferation in the hippocampus of adult rats [178]. The increase in proliferation was dependent on the duration of treatment, as subchronic administration lasting 7 days had no effect on cell proliferation, whereas 21 days of olanzapine administration increased cell proliferation [178]. Research by Halim et al. [179] confirmed that chronic (i.e., 28-day) treatment with the atypical drug clozapine, but not haloperidol, increased cell proliferation in the hippocampus. However, neither compound altered new cell survival when BrdU cells were assessed 3 weeks after BrdU administration. Both clozapine and olanzapine, but not haloperidol, were able to attenuate deficits in cell proliferation following phencyclidine administration [180, 181]. Though the data are limited and inconsistencies exist, taken together, the results indicate that atypical drugs may stimulate neurogenesis by increasing cell proliferation [90]. Further work is needed to clarify which features (e.g., dosage, duration of treatment, receptor subtype targeted, or model system) of atypical drugs may stimulate proliferation.

Collectively, the data support a role for the hippocampus in mediating some of the symptoms, particularly cognitive dysfunctions, seen in schizophrenic individuals. The precise link between specific reductions in hippocampal neurogenesis and cognitive dysfunction remains unclear. Regardless, the evidence indicates that schizophrenia is characterized by deficits in neurogenesis. Reductions in new cell proliferation and neurogenesis have been repeatedly demonstrated in both animal models and postmortem assessment of hippocampal neurogenesis. Some of the strongest evidence for the contribution of altered neurogenesis to the pathogenesis of schizophrenia comes from the established relationship between genetic risks of schizophrenia and the regulation of hippocampal neurogenesis. A sizable number of genes associated with developing schizophrenia have been shown to play a central role in the proliferation and survival of progenitor cells. While current antipsychotic drugs may support hippocampal neurogenesis by stimulating cell proliferation, whether the enhanced proliferation translates into lasting changes in survival or cell differentiation is unclear. Potentially, the use of alternative or additive interventions, such as aerobic exercise, that increase cell survival may have beneficial effects in alleviating cognitive deficits associated with schizophrenia. Prior work has shown that increasing aerobic fitness via regular exercise in schizophrenic individuals enhances cognitive function [182]. Given that exercise causes a potent increase in hippocampal neurogenesis, it is possible that enhancing neurogenesis is one mechanism through which exercise exerts its beneficial effects.

Neurogenesis in Major Depressive Disorder

MDD is estimated to affect 20% of individuals at some point in their lifetime and is characterized by recurrent episodes of enduring despair and behavioral abnormalities [183]. The symptoms of depression include changes in mood (e.g., excessive sadness, hopelessness, irritability, and anhedonia), physiological functions (e.g., sleep disturbances and altered appetite), energy and motivation, and cognitive function (e.g., memory impairment and rumination) [70, 183]. Diagnosis requires that a constellation of depressive symptoms persists for at least 2 weeks. While several subtypes of depressive disorders have been described, the two major forms include unipolar depression (i.e., major depression) and bipolar depression, which is distinguished by episodes of mania or hypomania in addition to depressive episodes. Development of a depressive disorder appears to arise from a complex interaction between environmental factors, such as stress and childhood trauma, and genetic factors. While depression is a heritable disease, identification of reliable risk genes or chromosome regions has been challenging, although variations in serotonin transporters have been linked to a greater disease risk under select conditions [184]. This lack of consistent genetic risk factors highlights the polygenetic nature of depression. Current pharmacological treatments target monoamine neurotransmitter systems (e.g., serotonin, dopamine, and norepinephrine), as deficits in monoamines are believed to be an underlying cause of depression. However, additional factors likely contribute to the pathophysiology of depression given that not all depressed individuals show improvements with antidepressant treatment, and treatment responsiveness is often delayed even though monoamine levels are rapidly altered [183]. Identifying the physiological changes that accompany the normalization of monoamine levels is expected to improve treatment effectiveness.

Along with other limbic system structures and regions of the prefrontal cortex, the hippocampus has been implicated in mediating depressive symptoms. Evidence for the involvement of the hippocampus in depression has come from imaging studies that often show reduced hippocampal volume in depressed individuals relative to controls [157]. Further, the decrease in hippocampal volume has been shown to be greater in individuals with longer periods of time without treatment [185]. The precise mechanisms driving these volume changes are still under investigation, but increases in neuronal and glial cell death, deficits in hippocampal neurogenesis, and reductions in dendritic complexity likely contribute [95].

Additionally, the hippocampus has been associated with depression because of its role in regulating the stress response produced by the HPA axis [80]. The end result of HPA axis activation is the release of cortisol (corticosterone in rodents) into the bloodstream by the adrenal cortex. Stressful experiences stimulate the release of corticotrophin-releasing hormone and vasopressin from the paraventricular nucleus of the hypothalamus, which subsequently induces the release of adrenocorticotrophic hormone from the anterior pituitary. This hormone travels through the bloodstream to stimulate the release of cortisol from the adrenal gland [80]. The hippocampus is part of the negative feedback mechanism that limits HPA activation, as elevated levels of cortisol circulate throughout the body and act as a self-regulating factor by inhibiting further activation of the HPA axis, in part by binding to glucocorticoid receptors in the hippocampus [186]. Excessive and/or prolonged elevations in cortisol are known to induce atrophy in the hippocampus that ultimately impairs the negative feedback function and results in reduced inhibition of cortisol release [74]. Hyperactivity of the HPA axis is often observed in depression, as higher levels of circulating cortisol are detected in a large proportion of depressed individuals [91]. These elevated cortisol levels are believed to result from impaired negative feedback and contribute to volume reductions in the hippocampus and expression of depressive symptoms.

An additional abnormality within the hippocampus that has been linked with depression is a reduction in hippocampal neurogenesis. Stressful events are known to predispose individuals to MDD. The hippocampus is particularly sensitive to stress, as stress has been shown to interfere with all aspects of hippocampal neurogenesis, including proliferation, cell differentiation, and survival (see section on stress) [80, 90]. Secondly, antidepressants increase hippocampal neurogenesis. For instance, several antidepressants have been shown to increase cell proliferation in rats, but this increase was only reported after chronic and not after acute treatment [187]. Additionally, chronic treatment with the selective serotonin reuptake inhibitor (SSRI) fluoxetine increases new cell survival, as the number of BrdU-positive cells was elevated in fluoxetine-treated rats 4 weeks after BrdU administration [187]. Fluoxetine also increased hippocampal neurogenesis in nonhuman primates, as measured by the number of DCX-positive cells in both control and stress-exposed subjects [188]. Assessment of hippocampal neurogenesis in postmortem brains from individuals with MDD confirmed that antidepressants increased neurogenesis. Individuals that had been taking an SSRI within 3 months prior to death showed a significant increase in the number of NPCs compared to unmedicated depressed individuals and nondepressed controls [189, 190]. There was no difference in the number of NPCs between unmedicated depressed and nondepressed individuals. In a similar report, cell proliferation (measured by Ki-67-positive cells) did not differ between depressed and nondepressed individuals though the medication status was not reported in this study [158]. Deficits in hippocampal neurogenesis were originally proposed to mediate the pathogenesis of depression, but clear evidence of impaired neurogenesis in depressed individuals has not been found.

Moreover, animal models have demonstrated that induction of depression-like behavior may be independent of alterations in hippocampal neurogenesis. Surget et al. [79] ablated neurogenesis in the hippocampus by targeted irradiation which inhibits progenitor cell proliferation. After 5 weeks of recovery following irradiation, half of the mice experienced 5 weeks of chronic unpredictable stress (CUS). Assessment of behavioral changes revealed that, as expected, CUS increased depression-like behavior, regardless of whether mice were irradiated or not, showing that inhibiting hippocampal neurogenesis did not enhance the response to stress. Moreover, irradiated mice that were not exposed to CUS showed no evidence of depression-like behavior, indicating that suppression of hippocampal neurogenesis is insufficient to elicit depression-like symptoms [134]. In contrast, others report that a transient reduction of neurogenesis using a genetic mouse model was associated with an increase in depression-like behavior but only when under stress [191], indicating that under some conditions suppressing neurogenesis may increase stress reactivity. Evidence that a deficit in hippocampal neurogenesis is a contributing factor to the development of depression is generally lacking, though some inconsistencies exist in the literature [183, 192, 193].

Though the role of neurogenesis as an etiological factor in depression is debated, there is stronger evidence that hippocampal neurogenesis is required for the treatment effects of antidepressants [193]. Prior work has shown that the administration of fluoxetine or the tricyclic antidepressant imipramine can block the development of behavioral indices of depression induced by CUS. However, the protective effects of fluoxetine and imipramine were absent in irradiated mice, indicating that the beneficial effects of antidepressants necessitate a significant increase in hippocampal neurogenesis [79]. Similar effects were found in nonhuman primates, as ablation of neurogenesis via irradiation prevented the effects of fluoxetine on stress-induced depression-like behavior [188]. The majority of the original studies that indicated a link between stimulation of hippocampal neurogenesis and reductions in depression-like symptoms employed irradiation to inhibit neurogenesis, which can increase inflammation as well as other confounding side effects. However, similar results were reported by Santarelli et al. [46] who ablated hippocampal neurogenesis using a genetic approach and confirmed that inhibition of neurogenesis prevented the antidepressant effects of fluoxetine in mice. Though it is important to note that inconsistent findings have been reported, the lack of an effect may relate to using unstressed animals [194]. An additional point of support is the match between the time course of the therapeutic benefits of antidepressants and hippocampal neurogenesis. Even though monoamine levels are quickly altered by antidepressants, symptom alleviation often takes several weeks to develop. This delayed effectiveness maps onto the timing of new cell maturation that at least in rodents occurs over the course of 3–4 weeks. If cell maturation follows a similar time course in humans this may strengthen the argument that hippocampal neurogenesis mediates aspects of the antidepressant effects of current medications.

The requirement of neurogenesis may, however, be limited to compounds that target monoamines, such as SSRIs, tricyclics, and monoamine oxidase, as compounds that have alternative targets can alleviate depressive symptoms even in the absence of neurogenesis. Drugs that reduce HPA axis activation, namely, corticotrophin-releasing factor 1 and vasopressin 1b receptor antagonists, are able to reduce stress-induced depression-like behavior in irradiated mice [79]. These data indicate the need to enhance neurogenesis to observe antidepressant effects may be specific to select compounds that aim to normalize levels of monoamine neurotransmitters. One possibility is that increasing neurogenesis may aid in normalizing activation of the HPA axis, given that the hippocampus is a negative regulator of the HPA axis [186]. Prior research has shown that suppression of neurogenesis in a conditional transgenic mouse model is associated with alterations in stress responses [192]. Ablation of neurogenesis delayed the recovery from acute and chronic restraint stress, as levels of corticosterone were higher at select time points in the transgenic mice than the control mice. In addition, reducing neurogenesis attenuated the ability of dexamethasone to suppress corticosterone levels [170]. The role of new neurons in regulating the HPA axis is complex, as ablation of neurogenesis via irradiation did not alter basal activity of the HPA axis as measured by dexamethasone suppression of corticosterone [79]. However, neurogenesis was found to be required for fluoxetine to normalize HPA activation following CUS, indicating that new neurons may contribute to the inhibitory influence on the HPA axis. Though more work is needed to clarify the role of new neurons in regulating the stress response, the initial findings indicate that increasing neurogenesis may represent one mechanism through which antidepressants can regulate overactivation of the HPA axis. By the same token, treatments that circumvent the hippocampus and act directly within the HPA axis may produce antidepressant effects independent of hippocampal neurogenesis.

Research has also implicated hippocampal neurogenesis as a contributing factor to the antimanic and antidepressant effects of lithium, a mood stabilizer commonly used to treat bipolar disorder [172]. Lithium has been reported to participate in the maintenance of the pool of NPCs by increasing progenitor cell proliferation [195]. In vivo work has shown that chronic lithium administration increased the number of BrdU-positive cells in the dentate gyrus [196]. Additionally, lithium has been shown to increase the total number of neurons in the dentate gyrus compared with control mice [197]. A likely mechanism through which lithium enhances hippocampal neurogenesis is through inhibition of GSK-3β [172]. Though lithium may have a number of targets within the brain, research has confirmed that lithium directly inhibits GSK-3β activity which, as noted, subsequently releases β-catenin from the inhibitory control of GSK-3β and allows β-catenin to stimulate progenitor cell proliferation [122]. Inhibition of GSK-3β likely mediates the ability of lithium to enhance neurogenesis, as overexpression of GSK-3β or deletion of β-catenin blocks lithium-induced increases in cell proliferation [172]. There is also evidence that reductions in GSK-3β may contribute to the proneurogenic effects of antidepressants (e.g., SSRIs) [172]. Overall, the data highlight a role for neurogenesis in mediating aspects of the beneficial effects of lithium.

An additional effect that likely contributes to the antidepressant-induced increase in hippocampal neurogenesis is the induction of neurotrophic and growth factors. Chronic antidepressant treatment increases expression of multiple neurotrophic and growth factors including BDNF, vascular endothelial growth factor, insulin-like growth factor, and nerve growth factor, among others [70]. Each of these molecules has been reported to have positive effects on hippocampal neurogenesis. For instance, central administration of BDNF has been found to increase new cell survival [198]. Moreover, reducing BDNF levels blocked the survival-promoting effect of the antidepressant imipramine. Chronic imipramine increased the number of BrdU-positive cells that survived 21 days after BrdU administration compared with saline-treated mice, whereas imipramine had no effects on survival in BDNF-knockdown mice [198]. Given the central role these growth and neurotrophic factors play in stimulating hippocampal neurogenesis, the induction of these factors following antidepressant treatment likely contributes to the increase in neurogenesis.

While the role of hippocampal neurogenesis in the onset of depression is in question, a wealth of evidence indicates a role for neurogenesis in the recovery process, as the antidepressant effects of the commonly employed drugs are blocked by inhibition of neurogenesis. Elucidating the functions mediated by the dentate gyrus and the new neurons born within this region in the healthy brain is expected to clarify how alterations in this process that develop in disease states may contribute to the onset and progression of disorders such as depression. Further, preclinical evidence indicates that enhancing hippocampal neurogenesis may aid in treatment-resistant cases [199]. Undoubtedly, more work is needed to elucidate the role of hippocampal neurogenesis in treatment responsiveness. Based on the existing knowledge, however, employing treatments aimed at stimulating hippocampal neurogenesis would be expected to have beneficial effects in the context of depression.

Concluding Remarks

New neurons are continuously formed in the human, nonhuman primate, and rodent adult hippocampus throughout life, though rates of neurogenesis precipitously decline with age to near zero levels at the end of the natural life span. Since its discovery in the 1960s, a large number of studies have documented numerous environmental and genetic factors which regulate adult neurogenesis. Chief among the positive regulators of neurogenesis are exercise and antidepressant drugs. Chief among the negative regulators of neurogenesis besides age are stress and inflammation. To the extent that many psychiatric disorders are comorbid with or causally related to stress and inflammation, decreased neurogenesis could be a partial contributor to the pathophysiology of the disorders. However, the functional significance of new neurons in behavior has yet to be established and is currently a hotly debated topic. Therefore, it is not clear whether changes in neurogenesis that occur alongside psychiatric illnesses are a cause or a consequence of the mediating factors such as stress, drug abuse, and inflammation which are complexly involved in the disorders. It will be important moving forward to use modern technologies capable of instantaneously inactivating cohorts of new neurons to test their functional significance in behavior and the etiology of mental illnesses.

References

1 Altman J: Are new neurons formed in the brains of adult mammals? Science 1962;135:1127–1128.
2 Rakic P: Neurogenesis in adult primate neocortex: an evaluation of the evidence. Nat Rev Neurosci 2002;3:65–71.
3 Eriksson PS, Perfilieva E, Bjork-Eriksson T, et al: Neurogenesis in the adult human hippocampus. Nat Med 1998;4: 1313–1317.
4 Kornack DR, Rakic P: Cell proliferation without neurogenesis in adult primate neocortex. Science 2001;294:2127–2130.
5 Andersen P: The Hippocampus Book. Oxford, Oxford University Press, 2007.
6 Cohen NJ, Ryan J, Hunt C, Romine L, Wszalek T, Nash C: Hippocampal system and declarative (relational) memory: summarizing the data from functional neuroimaging studies. Hippocampus 1999;9:83–98.
7 Hannula DE, Tranel D, Cohen NJ: The long and the short of it: relational memory impairments in amnesia, even at short lags. J Neurosci 2006;26:8352–8359.
8 Aimone JB, Deng W, Gage FH: Resolving new memories: a critical look at the dentate gyrus, adult neurogenesis, and pattern separation. Neuron 2011;70: 589–596.
9 Johnston ST, Shtrahman M, Parylak S, Goncalves JT, Gage FH: Paradox of pattern separation and adult neurogenesis: a dual role for new neurons balancing memory resolution and robustness. Neurobiol Learn Mem 2016;129:60–68.
10 Baker S, Vieweg P, Gao F, et al: The human dentate gyrus plays a necessary role in discriminating new memories. Curr Biol 2016;26:2629–2634.
11 Neunuebel JP, Knierim JJ: CA3 retrieves coherent representations from degraded input: direct evidence for CA3 pattern completion and dentate gyrus pattern separation. Neuron 2014;81:416–427.
12 Sahay A, Scobie KN, Hill AS, et al: Increasing adult hippocampal neurogenesis is sufficient to improve pattern separation. Nature 2011;472:466–470.
13 van Praag H, Schinder AF, Christie BR, Toni N, Palmer TD, Gage FH: Functional neurogenesis in the adult hippocampus. Nature 2002;415:1030–1034.
14 Kriegstein A, Alvarez-Buylla A: The glial nature of embryonic and adult neural stem cells. Annu Rev Neurosci 2009;32: 149–184.
15 Dayer AG, Ford AA, Cleaver KM, Yassaee M, Cameron HA: Short-term and long-term survival of new neurons in the rat dentate gyrus. J Comp Neurol 2003;460:563–572.
16 Clark PJ, Kohman RA, Miller DS, Bhattacharya TK, Brzezinska WJ, Rhodes JS: Genetic influences on exercise-induced adult hippocampal neurogenesis across 12 divergent mouse strains. Genes Brain Behav 2011;10:345–353.
17 Steiner B, Kronenberg G, Jessberger S, Brandt MD, Reuter K, Kempermann G: Differential regulation of gliogenesis in the context of adult hippocampal neurogenesis in mice. Glia 2004;46: 41–52.
18 Kohman RA, Bhattacharya TK, Wojcik E, Rhodes JS: Exercise reduces activation of microglia isolated from hippocampus and brain of aged mice. J Neuroinflammation 2013;10:114.
19 Kempermann G, Kuhn HG, Gage FH: Experience-induced neurogenesis in the senescent dentate gyrus. J Neurosci 1998;18:3206–3212.
20 Kuhn HG, Dickinson-Anson H, Gage FH: Neurogenesis in the dentate gyrus of the adult rat: age-related decrease of neuronal progenitor proliferation. J Neurosci 1996;16: 2027–2033.
21 van Praag H, Christie BR, Sejnowski TJ, Gage FH: Running enhances neurogenesis, learning, and long-term potentiation in mice. Proc Natl Acad Sci USA 1999;96:13427–13431.
22 Ge S, Yang CH, Hsu KS, Ming GL, Song H: A critical period for enhanced synaptic plasticity in newly generated neurons of the adult brain. Neuron 2007;54:559–566.
23 Kempermann G, Wiskott L, Gage FH: Functional significance of adult neurogenesis. Curr Opin Neurobiol 2004;14: 186–191.
24 Amrein I, Lipp HP: Adult hippocampal neurogenesis of mammals: evolution and life history. Biol Lett 2009;5:141–144.
25 Clark PJ, Brzezinska WJ, Thomas MW, Ryzhenko NA, Toshkov SA, Rhodes JS: Intact neurogenesis is required for benefits of exercise on spatial memory but not motor performance or contextual fear conditioning in C57BL/6J mice. Neuroscience 2008;155:1048–1058.
26 Clark PJ, Bhattacharya TK, Miller DS, Kohman RA, DeYoung EK, Rhodes JS: New neurons generated from running are broadly recruited into neuronal activation associated with three different hippocampus-involved tasks. Hippocampus 2012;22:1860–1867.
27 Lipp HP, Bonfanti L: Adult neurogenesis in mammals: variations and confusions. Brain Behav Evol 2016;87:205–221.
28 Amrein I, Isler K, Lipp HP: Comparing adult hippocampal neurogenesis in mammalian species and orders: influence of chronological age and life history stage. Eur J Neurosci 2011;34:978–987.
29 Jaholkowski P, Kiryk A, Jedynak P, et al: New hippocampal neurons are not obligatory for memory formation; cyclin D2 knockout mice with no adult brain neurogenesis show learning. Learn Mem 2009;16:439–451.
30 Groves JO, Leslie I, Huang GJ, et al: Ablating adult neurogenesis in the rat has no effect on spatial processing: evidence from a novel pharmacogenetic model. PLoS Genet 2013;9:e1003718.
31 Shors TJ, Miesegaes G, Beylin A, Zhao M, Rydel T, Gould E: Neurogenesis in the adult is involved in the formation of trace memories. Nature 2001;410:372–376.
32 Wojtowicz JM, Askew ML, Winocur G: The effects of running and of inhibiting adult neurogenesis on learning and memory in rats. Eur J Neurosci 2008;27: 1494–1502.
33 Meshi D, Drew MR, Saxe M, et al: Hippocampal neurogenesis is not required for behavioral effects of environmental enrichment. Nat Neurosci 2006;9:729–731.
34 Yu TS, Zhang G, Liebl DJ, Kernie SG: Traumatic brain injury-induced hippocampal neurogenesis requires activation of early nestin-expressing progenitors. J Neurosci 2008;28:12901–12912.
35 Gu Y, Arruda-Carvalho M, Wang J, et al: Optical controlling reveals time-dependent roles for adult-born dentate granule cells. Nat Neurosci 2012;15:1700–1706.
36 Zhu H, Roth BL: Silencing synapses with DREADDs. Neuron 2014;82:723–725.
37 van Praag H, Kempermann G, Gage FH: Running increases cell proliferation and neurogenesis in the adult mouse dentate gyrus. Nat Neurosci 1999;2:266–270.

38 Chae CH, Lee HC, Jung SL, et al: Swimming exercise increases the level of nerve growth factor and stimulates neurogenesis in adult rat hippocampus. Neuroscience 2012;212:30–37.
39 Wu CW, Chang YT, Yu L, et al: Exercise enhances the proliferation of neural stem cells and neurite growth and survival of neuronal progenitor cells in dentate gyrus of middle-aged mice. J Appl Physiol (1985) 2008;105:1585–1594.
40 Pereira AC, Huddleston DE, Brickman AM, et al: An in vivo correlate of exercise-induced neurogenesis in the adult dentate gyrus. Proc Natl Acad Sci USA 2007;104:5638–5643.
41 Mustroph ML, Chen S, Desai SC, Cay EB, DeYoung EK, Rhodes JS: Aerobic exercise is the critical variable in an enriched environment that increases hippocampal neurogenesis and water maze learning in male C57BL/6J mice. Neuroscience 2012;219:62–71.
42 Kobilo T, Liu QR, Gandhi K, Mughal M, Shaham Y, van Praag H: Running is the neurogenic and neurotrophic stimulus in environmental enrichment. Learn Mem 2011;18:605–609.
43 Clark PJ, Bhattacharya TK, Miller DS, Rhodes JS: Induction of c-Fos, Zif268, and Arc from acute bouts of voluntary wheel running in new and pre-existing adult mouse hippocampal granule neurons. Neuroscience 2011;184:16–27.
44 Clark PJ, Kohman RA, Miller DS, Bhattacharya TK, Haferkamp EH, Rhodes JS: Adult hippocampal neurogenesis and c-Fos induction during escalation of voluntary wheel running in C57BL/6J mice. Behav Brain Res 2010;213:246–252.
45 Dranovsky A, Hen R: Hippocampal neurogenesis: regulation by stress and antidepressants. Biol Psychiatry 2006;59:1136–1143.
46 Santarelli L, Saxe M, Gross C, et al: Requirement of hippocampal neurogenesis for the behavioral effects of antidepressants. Science 2003;301:805–809.
47 Jacobs BL, van Praag H, Gage FH: Adult brain neurogenesis and psychiatry: a novel theory of depression. Mol Psychiatry 2000;5:262–269.
48 Anacker C, Zunszain PA, Cattaneo A, et al: Antidepressants increase human hippocampal neurogenesis by activating the glucocorticoid receptor. Mol Psychiatry 2011;16:738–750.
49 Bambico FR, Belzung C: Novel insights into depression and antidepressants: a synergy between synaptogenesis and neurogenesis? Curr Top Behav Neurosci 2013;15:243–291.
50 Kohler SJ, Williams NI, Stanton GB, Cameron JL, Greenough WT: Maturation time of new granule cells in the dentate gyrus of adult macaque monkeys exceeds six months. Proc Natl Acad Sci USA 2011;108:10326–10331.
51 Kempermann G, Chesler EJ, Lu L, Williams RW, Gage FH: Natural variation and genetic covariance in adult hippocampal neurogenesis. Proc Natl Acad Sci USA 2006;103:780–785.
52 Golub HM, Zhou QG, Zucker H, et al: Chronic alcohol exposure is associated with decreased neurogenesis, aberrant integration of newborn neurons, and cognitive dysfunction in female mice. Alcohol Clin Exp Res 2015;39:1967–1977.
53 Taffe MA, Kotzebue RW, Crean RD, Crawford EF, Edwards S, Mandyam CD: Long-lasting reduction in hippocampal neurogenesis by alcohol consumption in adolescent nonhuman primates. Proc Natl Acad Sci USA 2010;107:11104–11109.
54 Anderson ML, Nokia MS, Govindaraju KP, Shors TJ: Moderate drinking? Alcohol consumption significantly decreases neurogenesis in the adult hippocampus. Neuroscience 2012;224:202–209.
55 Geil CR, Hayes DM, McClain JA, et al: Alcohol and adult hippocampal neurogenesis: promiscuous drug, wanton effects. Prog Neuropsychopharmacol Biol Psychiatry 2014;54:103–113.
56 Sudai E, Croitoru O, Shaldubina A, et al: High cocaine dosage decreases neurogenesis in the hippocampus and impairs working memory. Addict Biol 2011;16:251–260.
57 Noonan MA, Bulin SE, Fuller DC, Eisch AJ: Reduction of adult hippocampal neurogenesis confers vulnerability in an animal model of cocaine addiction. J Neurosci 2010;30:304–315.
58 Yamaguchi M, Suzuki T, Seki T, et al: Repetitive cocaine administration decreases neurogenesis in adult rat hippocampus. Ann NY Acad Sci 2004;1025:351–362.
59 Castilla-Ortega E, Serrano A, Blanco E, et al: A place for the hippocampus in the cocaine addiction circuit: potential roles for adult hippocampal neurogenesis. Neurosci Biobehav Rev 2016;66:15–32.
60 Eisch AJ, Barrot M, Schad CA, Self DW, Nestler EJ: Opiates inhibit neurogenesis in the adult rat hippocampus. Proc Natl Acad Sci USA 2000;97:7579–7584.
61 Bayer R, Franke H, Ficker C, et al: Alterations of neuronal precursor cells in stages of human adult neurogenesis in heroin addicts. Drug Alcohol Depend 2015;156:139–149.
62 Zhang Y, Xu C, Zheng H, Loh HH, Law PY: Morphine modulates adult neurogenesis and contextual memory by impeding the maturation of neural progenitors. PLoS One 2016;11:e0153628.
63 Kubesova A, Bubenikova-Valesova V, Mertlova M, Palenicek T, Horacek J: Impact of psychotropic drugs on adult hippocampal neurogenesis. Neurosci Res 2012;73:93–98.
64 Eisch AJ, Harburg GC: Opiates, psychostimulants, and adult hippocampal neurogenesis: insights for addiction and stem cell biology. Hippocampus 2006;16:271–286.
65 Eisch AJ, Mandyam CD: Drug dependence and addiction, II: Adult neurogenesis and drug abuse. Am J Psychiatry 2004;161:426.
66 Dabe EC, Majdak P, Bhattacharya TK, Miller DS, Rhodes JS: Chronic D-amphetamine administered from childhood to adulthood dose-dependently increases the survival of new neurons in the hippocampus of male C57BL/6J mice. Neuroscience 2013;231:125–135.
67 Gould E, McEwen BS, Tanapat P, Galea LA, Fuchs E: Neurogenesis in the dentate gyrus of the adult tree shrew is regulated by psychosocial stress and NMDA receptor activation. J Neurosci 1997;17:2492–2498.
68 Gould E, Woolley CS, McEwen BS: Adrenal steroids regulate postnatal development of the rat dentate gyrus. I. Effects of glucocorticoids on cell death. J Comp Neurol 1991;313:479–485.
69 Gould E, Tanapat P: Stress and hippocampal neurogenesis. Biol Psychiatry 1999;46:1472–1479.
70 Warner-Schmidt JL, Duman RS: Hippocampal neurogenesis: opposing effects of stress and antidepressant treatment. Hippocampus 2006;16:239–249.
71 Westrin A, Ekman R, Träskman-Bendz L: Alterations of corticotropin releasing hormone (CRH) and neuropeptide Y (NPY) plasma levels in mood disorder patients with a recent suicide attempt. Eur Neuropsychopharmacol 1999;9:205–211.

72 Shah JL, Malla AK: Much ado about much: stress, dynamic biomarkers and HPA axis dysregulation along the trajectory to psychosis. Schizophr Res 2015; 162:253–260.
73 Habets P, Collip D, Myin-Germeys I, et al: Pituitary volume, stress reactivity and genetic risk for psychotic disorder. Psychol Med 2012;42:1523–1533.
74 Sapolsky RM: Stress and plasticity in the limbic system. Neurochem Res 2003;28: 1735–1742.
75 Czeh B, Welt T, Fischer AK, et al: Chronic psychosocial stress and concomitant repetitive transcranial magnetic stimulation: effects on stress hormone levels and adult hippocampal neurogenesis. Biol Psychiatry 2002;52:1057–1065.
76 Pham K, Nacher J, Hof PR, McEwen BS: Repeated restraint stress suppresses neurogenesis and induces biphasic PSA-NCAM expression in the adult rat dentate gyrus. Eur J Neurosci 2003;17:879–886.
77 Thomas RM, Hotsenpiller G, Peterson DA: Acute psychosocial stress reduces cell survival in adult hippocampal neurogenesis without altering proliferation. J Neurosci 2007;27:2734–2743.
78 Malberg JE, Duman RS: Cell proliferation in adult hippocampus is decreased by inescapable stress: reversal by fluoxetine treatment. Neuropsychopharmacology 2003;28:1562–1571.
79 Surget A, Saxe M, Leman S, et al: Drug-dependent requirement of hippocampal neurogenesis in a model of depression and of antidepressant reversal. Biol Psychiatry 2008;64:293–301.
80 Levone BR, Cryan JF, O'Leary OF: Role of adult hippocampal neurogenesis in stress resilience. Neurobiol Stress 2015; 1:147–155.
81 Cinini SM, Barnabe GF, Galvao-Coelho N, et al: Social isolation disrupts hippocampal neurogenesis in young non-human primates. Front Neurosci 2014;8: 45.
82 Van Bokhoven P, Oomen CA, Hoogendijk WJ, Smit AB, Lucassen PJ, Spijker S: Reduction in hippocampal neurogenesis after social defeat is long-lasting and responsive to late antidepressant treatment. Eur J Neurosci 2011; 33:1833–1840.
83 Van der Borght K, Meerlo P, Luiten PG, Eggen BJ, Van der Zee EA: Effects of active shock avoidance learning on hippocampal neurogenesis and plasma levels of corticosterone. Behav Brain Res 2005;157:23–30.
84 Parihar VK, Hattiangady B, Kuruba R, Shuai B, Shetty AK: Predictable chronic mild stress improves mood, hippocampal neurogenesis and memory. Mol Psychiatry 2011;16:171–183.
85 Buchanan TW, Lovallo WR: Enhanced memory for emotional material following stress-level cortisol treatment in humans. Psychoneuroendocrinology 2001;26:307–317.
86 Mirescu C, Peters JD, Gould E: Early life experience alters response of adult neurogenesis to stress. Nat Neurosci 2004;7: 841–846.
87 Brummelte S, Galea LA: Chronic high corticosterone reduces neurogenesis in the dentate gyrus of adult male and female rats. Neuroscience 2010;168:680–690.
88 Wong EY, Herbert J: Raised circulating corticosterone inhibits neuronal differentiation of progenitor cells in the adult hippocampus. Neuroscience 2006;137: 83–92.
89 Tanapat P, Hastings NB, Rydel TA, Galea LA, Gould E: Exposure to fox odor inhibits cell proliferation in the hippocampus of adult rats via an adrenal hormone-dependent mechanism. J Comp Neurol 2001;437:496–504.
90 Balu DT, Lucki I: Adult hippocampal neurogenesis: regulation, functional implications, and contribution to disease pathology. Neurosci Biobehav Rev 2009;33:232–252.
91 Ueyama T, Kawai Y, Nemoto K, Sekimoto M, Tone S, Senba E: Immobilization stress reduced the expression of neurotrophins and their receptors in the rat brain. Neurosci Res 1997;28:103–110.
92 Hansson AC, Cintra A, Belluardo N, et al: Gluco- and mineralocorticoid receptor-mediated regulation of neurotrophic factor gene expression in the dorsal hippocampus and the neocortex of the rat. Eur J Neurosci 2000;12:2918–2934.
93 Heine VM, Zareno J, Maslam S, Joels M, Lucassen PJ: Chronic stress in the adult dentate gyrus reduces cell proliferation near the vasculature and VEGF and Flk-1 protein expression. Eur J Neurosci 2005;21:1304–1314.
94 Monje ML, Toda H, Palmer TD: Inflammatory blockade restores adult hippocampal neurogenesis. Science 2003;302: 1760–1765.
95 Koo JW, Duman RS: IL-1β is an essential mediator of the antineurogenic and anhedonic effects of stress. Proc Natl Acad Sci USA 2008;105:751–756.
96 Koo JW, Russo SJ, Ferguson D, Nestler EJ, Duman RS: Nuclear factor-κB is a critical mediator of stress-impaired neurogenesis and depressive behavior. Proc Natl Acad Sci USA 2010;107: 2669–2674.
97 Kettenmann H, Hanisch UK, Noda M, Verkhratsky A: Physiology of microglia. Physiol Rev 2011;91:461–553.
98 Colton CA: Heterogeneity of microglial activation in the innate immune response in the brain. J Neuroimmune Pharmacol 2009;4:399–418.
99 Colton CA, Gilbert DL: Microglia, an in vivo source of reactive oxygen species in the brain. Adv Neurol 1993;59:321–326.
100 Holm TH, Draeby D, Owens T: Microglia are required for astroglial Toll-like receptor 4 response and for optimal TLR2 and TLR3 response. Glia 2012;60: 630–638.
101 Facci L, Barbierato M, Marinelli C, Argentini C, Skaper SD, Giusti P: Toll-like receptors 2, -3 and -4 prime microglia but not astrocytes across central nervous system regions for ATP-dependent interleukin-1β release. Sci Rep 2014;4:6824.
102 Efremova L, Chovancova P, Adam M, Gutbier S, Schildknecht S, Leist M: Switching from astrocytic neuroprotection to neurodegeneration by cytokine stimulation. Arch Toxicol 2017; 91:231–246.
103 Kraft AD, McPherson CA, Harry GJ: Heterogeneity of microglia and TNF signaling as determinants for neuronal death or survival. Neurotoxicology 2009;30:785–793.
104 Kohman RA, Rhodes JS: Neurogenesis, inflammation and behavior. Brain Behav Immun 2013;27:22–32.
105 Ekdahl CT, Claasen JH, Bonde S, Kokaia Z, Lindvall O: Inflammation is detrimental for neurogenesis in adult brain. Proc Natl Acad Sci USA 2003; 100:13632–13637.
106 Quan N, Banks WA: Brain-immune communication pathways. Brain Behav Immun 2007;21:727–735.
107 Littlefield AM, Setti SE, Priester C, Kohman RA: Voluntary exercise attenuates LPS-induced reductions in neurogenesis and increases microglia expression of a proneurogenic phenotype in aged mice. J Neuroinflammation 2015; 12:138.

108 Godbout JP, Chen J, Abraham J, et al: Exaggerated neuroinflammation and sickness behavior in aged mice following activation of the peripheral innate immune system. FASEB J 2005;19: 1329–1331.
109 Godbout JP, Johnson RW: Age and neuroinflammation: a lifetime of psychoneuroimmune consequences. Immunol Allergy Clin North Am 2009;29: 321–337.
110 Dinel AL, Joffre C, Trifilieff P, et al: Inflammation early in life is a vulnerability factor for emotional behavior at adolescence and for lipopolysaccharide-induced spatial memory and neurogenesis alteration at adulthood. J Neuroinflammation 2014;11:155.
111 Bastos GN, Moriya T, Inui F, Katura T, Nakahata N: Involvement of cyclooxygenase-2 in lipopolysaccharide-induced impairment of the newborn cell survival in the adult mouse dentate gyrus. Neuroscience 2008;155:454–462.
112 Jakubs K, Bonde S, Iosif RE, et al: Inflammation regulates functional integration of neurons born in adult brain. J Neurosci 2008;28:12477–12488.
113 Chugh D, Nilsson P, Afjei SA, Bakochi A, Ekdahl CT: Brain inflammation induces post-synaptic changes during early synapse formation in adult-born hippocampal neurons. Exp Neurol 2013;250:176–188.
114 Belarbi K, Arellano C, Ferguson R, Jopson T, Rosi S: Chronic neuroinflammation impacts the recruitment of adult-born neurons into behaviorally relevant hippocampal networks. Brain Behav Immun 2012;26:18–23.
115 Green HF, Treacy E, Keohane AK, Sullivan AM, O'Keeffe GW, Nolan YM: A role for interleukin-1β in determining the lineage fate of embryonic rat hippocampal neural precursor cells. Mol Cell Neurosci 2012;49:311–321.
116 Vallieres L, Campbell IL, Gage FH, Sawchenko PE: Reduced hippocampal neurogenesis in adult transgenic mice with chronic astrocytic production of interleukin-6. J Neurosci 2002;22:486–492.
117 Seguin JA, Brennan J, Mangano E, Hayley S: Proinflammatory cytokines differentially influence adult hippocampal cell proliferation depending upon the route and chronicity of administration. Neuropsychiatr Dis Treat 2009;5:5–14.
118 Iosif RE, Ekdahl CT, Ahlenius H, et al: Tumor necrosis factor receptor 1 is a negative regulator of progenitor proliferation in adult hippocampal neurogenesis. J Neurosci 2006;26:9703–9712.
119 Keohane A, Ryan S, Maloney E, Sullivan AM, Nolan YM: Tumour necrosis factor-α impairs neuronal differentiation but not proliferation of hippocampal neural precursor cells: role of Hes1. Mol Cell Neurosci 2010;43:127–135.
120 Zhang K, Xu H, Cao L, Li K, Huang Q: Interleukin-1β inhibits the differentiation of hippocampal neural precursor cells into serotonergic neurons. Brain Res 2013;1490:193–201.
121 Green HF, Nolan YM: Unlocking mechanisms in interleukin-1β-induced changes in hippocampal neurogenesis – a role for GSK-3β and TLX. Transl Psychiatry 2012;2:e194.
122 Zechner D, Fujita Y, Hulsken J, et al: β-Catenin signals regulate cell growth and the balance between progenitor cell expansion and differentiation in the nervous system. Dev Biol 2003;258: 406–418.
123 Estes ML, McAllister AK: Maternal immune activation: implications for neuropsychiatric disorders. Science 2016;353:772–777.
124 Mueller N, Weidinger E, Leitner B, Schwarz MJ: The role of inflammation in schizophrenia. Front Neurosci 2015; 9:372.
125 Babulas V, Factor-Litvak P, Goetz R, Schaefer CA, Brown AS: Prenatal exposure to maternal genital and reproductive infections and adult schizophrenia. Am J Psychiatry 2006;163: 927–929.
126 Brown AS, Begg MD, Gravenstein S, et al: Serologic evidence of prenatal influenza in the etiology of schizophrenia. Arch Gen Psychiatry 2004;61:774–780.
127 Vuillermot S, Weber L, Feldon J, Meyer U: A longitudinal examination of the neurodevelopmental impact of prenatal immune activation in mice reveals primary defects in dopaminergic development relevant to schizophrenia. J Neurosci 2010;30:1270–1287.
128 Fortier ME, Luheshi GN, Boksa P: Effects of prenatal infection on prepulse inhibition in the rat depend on the nature of the infectious agent and the stage of pregnancy. Behav Brain Res 2007;181:270–277.
129 Benros ME, Nielsen PR, Nordentoft M, Eaton WW, Dalton SO, Mortensen PB: Autoimmune diseases and severe infections as risk factors for schizophrenia: a 30-year population-based register study. Am J Psychiatry 2011;168: 1303–1310.
130 Potvin S, Stip E, Sepehry AA, Gendron A, Bah R, Kouassi E: Inflammatory cytokine alterations in schizophrenia: a systematic quantitative review. Biol Psychiatry 2008;63:801–808.
131 Miller BJ, Buckley P, Seabolt W, Mellor A, Kirkpatrick B: Meta-analysis of cytokine alterations in schizophrenia: clinical status and antipsychotic effects. Biol Psychiatry 2011;70:663–671.
132 Sasayama D, Hattori K, Wakabayashi C, et al: Increased cerebrospinal fluid interleukin-6 levels in patients with schizophrenia and those with major depressive disorder. J Psychiatr Res 2013;47:401–406.
133 Garver DL, Tamas RL, Holcomb JA: Elevated interleukin-6 in the cerebrospinal fluid of a previously delineated schizophrenia subtype. Neuropsychopharmacology 2003;28:1515–1520.
134 Catts VS, Wong J, Fillman SG, Fung SJ, Shannon Weickert C: Increased expression of astrocyte markers in schizophrenia: association with neuroinflammation. Aust NZ J Psychiatry 2014;48: 722–734.
135 Trepanier MO, Hopperton KE, Mizrahi R, Mechawar N, Bazinet RP: Postmortem evidence of cerebral inflammation in schizophrenia: a systematic review. Mol Psychiatry 2016;21:1009–1026.
136 Krishnadas R, Cavanagh J: Depression: an inflammatory illness? J Neurol Neurosurg Psychiatry 2012;83:495–502.
137 Lo Fermo S, Barone R, Patti F, et al: Outcome of psychiatric symptoms presenting at onset of multiple sclerosis: a retrospective study. Mult Scler 2010; 16:742–748.
138 Dickens C, McGowan L, Clark-Carter D, Creed F: Depression in rheumatoid arthritis: a systematic review of the literature with meta-analysis. Psychosom Med 2002;64:52–60.

139 Raison CL, Capuron L, Miller AH: Cytokines sing the blues: inflammation and the pathogenesis of depression. Trends Immunol 2006;27:24–31.
140 O'Connor JC, Lawson MA, Andre C, et al: Lipopolysaccharide-induced depressive-like behavior is mediated by indoleamine 2,3-dioxygenase activation in mice. Mol Psychiatry 2009;14:511–522.
141 Frenois F, Moreau M, O'Connor J, et al: Lipopolysaccharide induces delayed FosB/DeltaFosB immunostaining within the mouse extended amygdala, hippocampus and hypothalamus, that parallel the expression of depressive-like behavior. Psychoneuroendocrinology 2007;32:516–531.
142 Reichenberg A, Yirmiya R, Schuld A, et al: Cytokine-associated emotional and cognitive disturbances in humans. Arch Gen Psychiatry 2001;58:445–452.
143 Bull SJ, Huezo-Diaz P, Binder EB, et al: Functional polymorphisms in the interleukin-6 and serotonin transporter genes, and depression and fatigue induced by interferon-α and ribavirin treatment. Mol Psychiatry 2009;14: 1095–1104.
144 Pierucci-Lagha A, Covault J, Bonkovsky HL, et al: A functional serotonin transporter gene polymorphism and depressive effects associated with interferon-α treatment. Psychosomatics 2010;51:137–148.
145 Dowlati Y, Herrmann N, Swardfager W, et al: A meta-analysis of cytokines in major depression. Biol Psychiatry 2010;67:446–457.
146 Liu Y, Ho RC, Mak A: Interleukin (IL)-6, tumour necrosis factor alpha (TNF-α) and soluble interleukin-2 receptors (sIL-2R) are elevated in patients with major depressive disorder: a meta-analysis and meta-regression. J Affect Disord 2012;139:230–239.
147 Howren MB, Lamkin DM, Suls J: Associations of depression with C-reactive protein, IL-1, and IL-6: a meta-analysis. Psychosom Med 2009;71:171–186.
148 Lanquillon S, Krieg JC, Bening-Abu-Shach U, Vedder H: Cytokine production and treatment response in major depressive disorder. Neuropsychopharmacology 2000;22:370–379.
149 Eller T, Vasar V, Shlik J, Maron E: Pro-inflammatory cytokines and treatment response to escitalopram in major depressive disorder. Prog Neuropsychopharmacol Biol Psychiatry 2008;32: 445–450.
150 Holder SD, Wayhs A: Schizophrenia. Am Fam Physician 2014;90:775–782.
151 Pino O, Guilera G, Gomez-Benito J, Najas-Garcia A, Rufian S, Rojo E: Neurodevelopment or neurodegeneration: review of theories of schizophrenia. Actas Esp Psiquiatr 2014;42:185–195.
152 Heckers S, Konradi C: Hippocampal neurons in schizophrenia. J Neural Transm (Vienna) 2002;109:891–905.
153 Haatveit B, Jensen J, Alnaes D, et al: Reduced load-dependent default mode network deactivation across executive tasks in schizophrenia spectrum disorders. Neuroimage Clin 2016;12:389–396.
154 Heckers S: Neuroimaging studies of the hippocampus in schizophrenia. Hippocampus 2001;11:520–528.
155 Chakos MH, Schobel SA, Gu H, et al: Duration of illness and treatment effects on hippocampal volume in male patients with schizophrenia. Br J Psychiatry 2005;186:26–31.
156 Steen RG, Mull C, McClure R, Hamer RM, Lieberman JA: Brain volume in first-episode schizophrenia: systematic review and meta-analysis of magnetic resonance imaging studies. Br J Psychiatry 2006;188:510–518.
157 Ho NF, Iglesias JE, Sum MY, et al: Progression from selective to general involvement of hippocampal subfields in schizophrenia. Mol Psychiatry 2017; 22:142–152.
158 Reif A, Fritzen S, Finger M, et al: Neural stem cell proliferation is decreased in schizophrenia, but not in depression. Mol Psychiatry 2006;11: 514–522.
159 Allen KM, Fung SJ, Weickert CS: Cell proliferation is reduced in the hippocampus in schizophrenia. Aust NZ J Psychiatry 2016;50:473–480.
160 Oertel V, Rotarska-Jagiela A, van de Ven VG, Haenschel C, Maurer K, Linden DE: Visual hallucinations in schizophrenia investigated with functional magnetic resonance imaging. Psychiatry Res 2007;156:269–273.
161 Schreiber R, Newman-Tancredi A: Improving cognition in schizophrenia with antipsychotics that elicit neurogenesis through 5-HT_{1A} receptor activation. Neurobiol Learn Mem 2014; 110:72–80.
162 Clelland CD, Choi M, Romberg C, et al: A functional role for adult hippocampal neurogenesis in spatial pattern separation. Science 2009;325:210–213.
163 Martinelli C, Shergill SS: Clarifying the role of pattern separation in schizophrenia: the role of recognition and visual discrimination deficits. Schizophr Res 2015;166:328–333.
164 Das T, Ivleva EI, Wagner AD, Stark CE, Tamminga CA: Loss of pattern separation performance in schizophrenia suggests dentate gyrus dysfunction. Schizophr Res 2014;159:193–197.
165 Zhang Z, van Praag H: Maternal immune activation differentially impacts mature and adult-born hippocampal neurons in male mice. Brain Behav Immun 2015;45:60–70.
166 Meyer U, Feldon J, Schedlowski M, Yee BK: Immunological stress at the maternal-foetal interface: a link between neurodevelopment and adult psychopathology. Brain Behav Immun 2006; 20:378–388.
167 Piontkewitz Y, Bernstein HG, Dobrowolny H, Bogerts B, Weiner I, Keilhoff G: Effects of risperidone treatment in adolescence on hippocampal neurogenesis, parvalbumin expression, and vascularization following prenatal immune activation in rats. Brain Behav Immun 2012;26:353–363.
168 Duan X, Chang JH, Ge S, et al: Disrupted-In-Schizophrenia 1 regulates integration of newly generated neurons in the adult brain. Cell 2007; 130:1146–1158.
169 Austin CP, Ky B, Ma L, Morris JA, Shughrue PJ: Expression of Disrupted-In-Schizophrenia-1, a schizophrenia-associated gene, is prominent in the mouse hippocampus throughout brain development. Neuroscience 2004;124:3–10.
170 Meyer KD, Morris JA: Immunohistochemical analysis of Disc1 expression in the developing and adult hippocampus. Gene Expr Patterns 2008;8:494–501.
171 Mao Y, Ge X, Frank CL, et al: Disrupted in schizophrenia 1 regulates neuronal progenitor proliferation via modulation of GSK3β/β-catenin signaling. Cell 2009;136:1017–1031.
172 Wada A: Lithium and neuropsychiatric therapeutics: neuroplasticity via glycogen synthase kinase-3β, β-catenin, and neurotrophin cascades. J Pharmacol Sci 2009;110:14–28.

173 Matsumoto M, Straub RE, Marenco S, et al: The evolutionarily conserved G protein-coupled receptor SREB2/GPR85 influences brain size, behavior, and vulnerability to schizophrenia. Proc Natl Acad Sci USA 2008;105:6133–6138.
174 Chen Q, Kogan JH, Gross AK, et al: SREB2/GPR85, a schizophrenia risk factor, negatively regulates hippocampal adult neurogenesis and neurogenesis-dependent learning and memory. Eur J Neurosci 2012;36:2597–2608.
175 Biswas AB, Furniss F: Cognitive phenotype and psychiatric disorder in 22q11.2 deletion syndrome: a review. Res Dev Disabil 2016;53–54:242–257.
176 Ouchi Y, Banno Y, Shimizu Y, et al: Reduced adult hippocampal neurogenesis and working memory deficits in the Dgcr8-deficient mouse model of 22q11.2 deletion-associated schizophrenia can be rescued by IGF2. J Neurosci 2013;33:9408–9419.
177 Wakade CG, Mahadik SP, Waller JL, Chiu FC: Atypical neuroleptics stimulate neurogenesis in adult rat brain. J Neurosci Res 2002;69:72–79.
178 Kodama M, Fujioka T, Duman RS: Chronic olanzapine or fluoxetine administration increases cell proliferation in hippocampus and prefrontal cortex of adult rat. Biol Psychiatry 2004;56:570–580.
179 Halim ND, Weickert CS, McClintock BW, Weinberger DR, Lipska BK: Effects of chronic haloperidol and clozapine treatment on neurogenesis in the adult rat hippocampus. Neuropsychopharmacology 2004;29:1063–1069.
180 Liu J, Suzuki T, Seki T, Namba T, Tanimura A, Arai H: Effects of repeated phencyclidine administration on adult hippocampal neurogenesis in the rat. Synapse 2006;60:56–68.
181 Maeda K, Sugino H, Hirose T, et al: Clozapine prevents a decrease in neurogenesis in mice repeatedly treated with phencyclidine. J Pharmacol Sci 2007;103:299–308.
182 Kimhy D, Vakhrusheva J, Bartels MN, et al: The impact of aerobic exercise on brain-derived neurotrophic factor and neurocognition in individuals with schizophrenia: a single-blind, randomized clinical trial. Schizophr Bull 2015;41:859–868.
183 Mahar I, Bambico FR, Mechawar N, Nobrega JN: Stress, serotonin, and hippocampal neurogenesis in relation to depression and antidepressant effects. Neurosci Biobehav Rev 2014;38:173–192.
184 Belmaker RH, Agam G: Major depressive disorder. N Engl J Med 2008;358:55–68.
185 Sheline YI, Gado MH, Kraemer HC: Untreated depression and hippocampal volume loss. Am J Psychiatry 2003;160:1516–1518.
186 Herman JP, Ostrander MM, Mueller NK, Figueiredo H: Limbic system mechanisms of stress regulation: hypothalamo-pituitary-adrenocortical axis. Prog Neuropsychopharmacol Biol Psychiatry 2005;29:1201–1213.
187 Malberg JE, Eisch AJ, Nestler EJ, Duman RS: Chronic antidepressant treatment increases neurogenesis in adult rat hippocampus. J Neurosci 2000;20:9104–9110.
188 Perera TD, Dwork AJ, Keegan KA, et al: Necessity of hippocampal neurogenesis for the therapeutic action of antidepressants in adult nonhuman primates. PLoS One 2011;6:e17600.
189 Boldrini M, Underwood MD, Hen R, et al: Antidepressants increase neural progenitor cells in the human hippocampus. Neuropsychopharmacology 2009;34:2376–2389.
190 Boldrini M, Hen R, Underwood MD, et al: Hippocampal angiogenesis and progenitor cell proliferation are increased with antidepressant use in major depression. Biol Psychiatry 2012;72:562–571.
191 Yun S, Donovan MH, Ross MN, et al: Stress-induced anxiety- and depressive-like phenotype associated with transient reduction in neurogenesis in adult nestin-CreERT2/diphtheria toxin fragment A transgenic mice. PLoS One 2016;11:e0147256.
192 Snyder JS, Soumier A, Brewer M, Pickel J, Cameron HA: Adult hippocampal neurogenesis buffers stress responses and depressive behaviour. Nature 2011;476:458–461.
193 Ruan L, Lau BW, Wang J, et al: Neurogenesis in neurological and psychiatric diseases and brain injury: from bench to bedside. Prog Neurobiol 2014;115:116–137.
194 Holick KA, Lee DC, Hen R, Dulawa SC: Behavioral effects of chronic fluoxetine in BALB/cJ mice do not require adult hippocampal neurogenesis or the serotonin 1A receptor. Neuropsychopharmacology 2008;33:406–417.
195 Wexler EM, Geschwind DH, Palmer TD: Lithium regulates adult hippocampal progenitor development through canonical Wnt pathway activation. Mol Psychiatry 2008;13:285–292.
196 Chen G, Rajkowska G, Du F, Seraji-Bozorgzad N, Manji HK: Enhancement of hippocampal neurogenesis by lithium. J Neurochem 2000;75:1729–1734.
197 Rajkowska G, Clarke G, Mahajan G, et al: Differential effect of lithium on cell number in the hippocampus and prefrontal cortex in adult mice: a stereological study. Bipolar Disord 2016;18:41–51.
198 Sairanen M, Lucas G, Ernfors P, Castren M, Castren E: Brain-derived neurotrophic factor and antidepressant drugs have different but coordinated effects on neuronal turnover, proliferation, and survival in the adult dentate gyrus. J Neurosci 2005;25:1089–1094.
199 Wang Y, Chang T, Chen YC, et al: Quetiapine add-on therapy improves the depressive behaviors and hippocampal neurogenesis in fluoxetine treatment resistant depressive rats. Behav Brain Res 2013;253:206–211.

Justin S. Rhodes
Department of Psychology
The Beckman Institute
University of Illinois
405 N. Mathews Avenue
Urbana, IL 61801 (USA)
E-Mail jrhodes@illinois.edu

Halaris A, Leonard BE (eds): Neuroprogression in Psychiatric Disorders.
Mod Trends Pharmacopsychiatry. Basel, Karger, 2017, vol 31, pp 152–161 (DOI: 10.1159/000470813)

The Brain-Gut Axis Contributes to Neuroprogression in Stress-Related Disorders

Kieran Rea[a] • Timothy G. Dinan[a, b] • John F. Cryan[a, c]

[a] APC Microbiome Institute and Departments of [b] Psychiatry and Neurobehavioural Science and [c] Anatomy and Neuroscience, University College Cork, Cork, Ireland

Abstract

There is a growing emphasis on the relationship between the complexity and diversity of the microorganisms that inhabit our gut (human gastrointestinal microbiota) and brain health. The microbiota-gut-brain axis is a dynamic matrix of tissues and organs including the brain, glands, gut, immune cells, and gastrointestinal microbiota that communicate in a complex multidirectional manner to maintain homeostasis. Changes in this environment may contribute to the neuroprogression of stress-related disorders by altering physiological processes including hypothalamic-pituitary-adrenal axis activation, neurotransmitter systems, immune function, and inflammatory responses. While appropriate, coordinated physiological responses, such as immune or stress responses, are necessary for survival, the contribution of repeated or chronic exposure to stress may predispose individuals to a more vulnerable state leaving them more susceptible to stress-related disorders. In this chapter, the involvement of the gastrointestinal microbiota in stress- and immune-mediated modulation of neuroendocrine, immune, and neurotransmitter systems and the consequential behavior is considered. We also focus on the mechanisms by which commensal gut microbiota can regulate neuroinflammation and further aim to exploit our understanding of their role in the effects of the microbiota-gut-brain axis on the neuroprogression of stress-related disorders as a consequence of neuroinflammatory processes.

Introduction

The Microbiome

We harbor a multitude of microorganisms in our gut (bacteria, archaea, yeasts, single-celled eukaryotes, as well as helminth parasites, and viruses, including bacteriophages) collectively referred to as the microbiota [1–4]. Along with the enteric nervous system, sympathetic and parasympathetic divisions of the autonomic nervous system, and neuroendocrine and neuroimmune components of the central nervous system, the gut microbiota forms part of a complex network termed the microbiota-gut-brain axis [5, 6]. The composition of the core bacteria that we harbor throughout our adult life is established early in our first few years of life and shaped by a number of factors including mode of delivery (vaginal or cesarean section), whether we are breastfed or bottle-fed, diet, medication (in particular antibiotic medication), and exposure to viral or bacterial infections and stress [7]. Interestingly, the critical development period including the seeding of our core microbiota and the development of the bacterial community in our gut occurs in parallel with the growth, maturation, and sprouting of neurons in the young brain, including microglia [7, 8], and a similar profile is evident in old age where a decline in microbiota complexity and diversity occurs in parallel with a decrease in neuronal and microglial complexity [9].

Host-Microbial Interaction

Gastrointestinal microorganisms are by their very definition constrained to the gut and interact with the host tissue along the alimentary canal as well as with their cohabitating microbial consortia. The epithelial layer of the gut has many functions, e.g., to regulate the absorption of nutrients and fluid from the lumen of the intestines and to serve as a physical barrier to invading pathogens or harmful substances [10].

Under homeostatic conditions, there exists a healthy, resting inflammatory tone where the microbiota stimulates cytokine and chemokine release, which in turn regulates local levels of bacteria in the gut. As well as influencing local immune responses at the epithelium, microbiota can synthesize and release neurotransmitters and short-chain fatty acids (SCFAs), as well as influence the release of neuropeptides and hormones from enteroendocrine cells of the intestines. Gut peptides such as ghrelin, gastrin, orexin, galanin, cholecystokinin, leptin, and neuropeptide Y are thought to influence peripheral neural communication and can also act centrally to influence behavior. Current hypotheses suggest that these circulating cytokines, chemokines, endocrine messengers, and microbial by-products can infiltrate the blood and lymphatic systems, or influence neural messages carried by the vagal and spinal afferent neurons to impact on centrally mediated events, including the regulation of hypothalamic-pituitary-adrenal (HPA) axis activity and neuroinflammation.

Penetrating the Blood-Brain Barrier

The blood-brain barrier is a highly capillarized structure composed of astrocytes and endothelial cells, amongst others. The permeability of this barrier can be compromised in certain physiological conditions, particularly in response to stress, infection, or injury, allowing the influx and egress of messenger molecules to influence the homeostatic environment in the brain. As with the epithelium of the gut, pattern recognition receptors on brain immune cells recognize microbial by-products, while changes in the environment as a consequence of cytokines, chemokines, and endocrine messengers ultimately result in neuroinflammation and trafficking of monocytes to the brain. Whilst the multimodal influence of microbiota on physiological events at the level of the CNS has gained credence, more recently there has been a shift to a stronger role for microbiota in regulating neuroinflammation via intervention in the recruitment of local immune regulators from the periphery to the brain.

In this chapter, we introduce the intestinal microbiota as a key regulator of stress and immune responses, facilitating adaption or habituation to these conditions, and protecting against the development and neuroprogression of stress-related and/or neuroinflammation-related diseases and dysfunction. We will first focus on the interaction of stress and neuroinflammation, as well as their reciprocal role in the neuroprogression of stress-related disorders, and then outline the role of gastrointestinal microbiota in these responses.

The Stress Response and Stress-Related Disorders

Stress may be defined as a complex dynamic condition in which the homeostasis, or the steady-state internal milieu of an organism, is disturbed or threatened [11–13]. At some stage in their life span, all organisms are exposed to factors that surpass a homeostatic threshold, resulting in a stress response, be they physical, psychological, or immunological in nature. Evolution has equipped most organisms with the biological machinery necessary to mount a defensive response to an acute stressful stimulus and to restore homeostatic balance once the stress or perceived threat has subsided. The coordination of this defensive response is governed by a pathway linking the hypothalamus, the pituitary gland, and the adrenal glands (the HPA axis), ultimately resulting in the release of behavior-altering chemicals including glucocorticoids, mineralocorticoids, and catecholamines. While a deficient or blunted HPA axis is commonly observed in the clinic in a wide range of autoimmune and inflammatory diseases, chronic or prolonged exposure to stress can contribute to the precipitation, maintenance, and/or neuroprogression of a number of neuroinflammatory- and stress-related disorders including anxiety [14, 15], depression [15, 16], bipolar disorder [17], posttraumatic stress disorder [18], Alzheimer disease [19, 20], and schizophrenia [21] amongst others.

An emerging perspective is that the neuroprogression of these disorders involves a combination of increased neurodegeneration, neuronal apoptosis, or neurotoxic susceptibility, and lowered neuroplasticity [22, 23]. Importantly, all of these physiological events are facilitated by neuroinflammatory processes and microglial activation, which in turn are regulated to some extent by stress responses.

Stress and Neuroimmune Function

Microglia – Key Regulators of Stress and Neuroinflammation

Microglia are the predominant immune cells in the CNS (5–12% of total brain cells) that can detect subtle changes in molecules in their locality [24, 25] and ultimately are responsible for neuroinflammatory processes [26]. Historically microglia were thought to be quiescent under normal physiological conditions and only activated in response to tissue damage or infection. More recently, these lifelong surviving cells have been shown to have a number of important roles beyond immune surveillance and response, including synaptic pruning, synaptic activity regulation, neurogenesis, and neurodegeneration [25, 27]. The role of the microglia in synaptic processes, including pruning and maturation, is most evident during embryonic and postnatal brain development [25, 27], while in adolescence and adulthood these roles are dampened except in learning and memory, with a decrease in microglial density noted in white, but not gray, matter as a consequence of a decrease in arbor length [28]. Neurodegenerative diseases are associated with marked alterations in the microglial status; interestingly, a decrease in microglial functionality with a concomitant increase in microglial numbers was apparent in aging [29]. Changes in microglial morphology (branching, arbor length, and spine densities) are causally involved in the neuroprogression of these neurodegenerative disorders, as the microglia are no longer as efficient in their neuroprotective, clearance, neurogenerative, and/or neuroplastic roles.

Under normal physiological conditions, microglia are highly ramified and extremely plastic; once activated, they can release a number of cytokines and chemokines, express numerous antigenic markers, regulate neurotransmitters, and undergo extreme morphological changes [29–31]. Cytokines and chemokines are involved in the repair of damaged tissue and the restoration of homeostasis [32]. However, they also play a key role in signaling between immune cells. Cytokines are generally classified as pro- or anti-inflammatory cytokines, which facilitate or inhibit inflammatory processes, respectively. The mobilization of these pro- and anti-inflammatory cytokines and chemical messengers is beneficial to the host by allowing the body to fight against infection or tissue damage. However, a delicate balance between pro- or anti-inflammatory alterations needs to be tightly regulated, as an uncontrolled and sustained response can induce significant inflammatory collateral damage and detrimental effects on mood and behavior.

Reciprocal Involvement of Stress and the Immune System

Acknowledging the intricate nature of stress and neuroinflammatory responses, it is not surprising to note that glucocorticoid receptors are highly expressed on microglia throughout the brain [33], while integration of stress-induced signaling is facilitated in the brain by microglial-mediated neuroinflammatory signaling [34, 35]. Under stressful situations, neuroendocrine pathways alter peripheral and central immune responses, leading to monocyte priming and trafficking, subsequent alterations in the microglial phenotype, and, ultimately, in neuroinflammation [35]. Neuroinflammation and the release of cytokines following microglial activation have been reported to have a causal role in the behavioral phenotype during stress [36, 37] and HPA axis activity [38], and play a key role in stress-related disorders including depression, dementia, and Alzheimer disease [39]. Intriguingly, stress-induced microglial activation occurs in discrete brain regions and in specific neuronal circuits, heavily involved in stress-mediated neuroendocrine activation [40].

Thus, in addition to their fundamental physiological significance independently, a greater appreciation of the reciprocal interaction of stress and immune responses may prove useful in understanding the etiology and neuroprogression of stress-related disorders and in developing new approaches for their treatment.

The Role of Gastrointestinal Microbiota in the Neuroprogression of Stress-Related Disorders as a Consequence of Neuroinflammatory Processes

Regulation of Stress Responses by Gastrointestinal Microbiota

It has been shown in animal models that different types of psychological stress, including maternal separation, chronic social defeat, restraint conditions, crowding, heat stress, and acoustic stress, can alter the composition of gastrointestinal microbiota [41–48]. Conversely, a number of experimental models and conditions have been employed demonstrating a key role of the microbiota in regulating stress responses, including pre- and probiotic [49–55] interventions, antibiotic administration [55–58], fecal transplantation, and the use of germ-free and specific pathogen-free animals [45, 59–65]. Limited studies have also seen beneficial health reports in humans [66, 67].

Regulation of the Neuroinflammatory Responses by Gastrointestinal Microbiota

Until very recently, the interaction between gastrointestinal microbiota and CNS-restricted immune cells had not been investigated. We now know that the gut microbiota influences the relative populations, migration, and function of various subsets of immune cells [68], and can modulate innate and adaptive immune responses at intestinal mucosal surfaces during infection, inflammation, and autoimmunity [69, 70].

Along with their established role in host defense, the constitutively active microglia are critically involved in neuronal events at various stages in development and adulthood, including synaptic remodeling to improve neuronal network signaling [71, 72]. Despite the seclusion of the microglia as a consequence of blood-brain barrier protection, a recent comprehensive study [73] has highlighted a critical role for gastrointestinal microbiota in microglial maturation, morphology, and immunological function. Under normal homeostatic conditions, a diverse gastrointestinal microbiota is necessary to maintain microglia in a healthy functional state. In contrast, the absence of a complex host microbiota (as evidenced by germ-free models, antibiotic suppression of microbiota, or use of specific pathogen-free mice) led to increased microglial populations; defects in microglial maturation, activation, and differentiation; alterations in microglial morphology (with longer processes and increased branching, terminal points, and clubbing at synaptic boutons); and a compromised immune response to bacterial or viral infection [73]. The observed alterations in the microglial phenotype were reversed with recolonization of gut microbiota. The findings from this study have redefined the ideology that the relationship between our microbiota, the immune system, and neurodevelopment is moot in adulthood; a healthy, diverse gastrointestinal microbiota is essential for the continuous preservation of healthy microglia and proper brain function throughout our life span [5].

As outlined earlier (Fig. 1), our microbiota diversity and stability decline [74], and there is a concomitant decrease in functionality and increase in number, size, and activation of microglia [29] that can lead to an increased inflammatory tone termed 'inflammaging' [75] and share common characteristics with aging. In normal aging, close cross talk with astrocytes, neurons, and other brain cells serves crucial functions as the scavenger system of the CNS, providing beneficial functions such as tissue repair in the CNS. However, chronic, dysregulated activation of microglia can lead to an increased inflammatory tone ultimately resulting in malfunction and damage of brain cells. What drives this dysregulation is not fully understood, but it seems likely that the microbiota may play a key role in this phenomenon.

Peripheral Influence of Microbiota Regulating Neuroinflammation and the Hypothalamic-Pituitary-Adrenal Axis

Emerging evidence implicates a novel neuroimmune circuit involving microglial activation and an altered sympathetic neural tone to the peripheral immune system to recruit inflammatory monocytes to the brain [34] and further reinforce stress-related behaviors. The trafficking of these monocytes to the brain appears to be governed by tumor necrosis factor-α-mediated microglial activation [76] and was reversible by probiotic intervention with VSL#3 [77]. Indeed, the mediation of certain behavioral phenotypes as a consequence of prolonged HPA activation involves the trafficking of monocytes from the spleen [34], a tissue with a high density of FFAR2 receptors [73]. These FFAR2 receptors are G_i/G_o G-protein-coupled receptors and have been localized to lymphocytes [78]. SCFAs, the natural ligands for FFARs, were shown to reverse the detrimental effects on microglial density, morphology, and maturity in the absence of a complex microbiota [73]. Taken together, these findings suggest that gastrointestinal microbiota may govern centrally mediated events indirectly through regulation of monocyte trafficking to the brain and subsequent microglial activation and HPA axis activity, possibly via SCFA-mediated mechanisms (Fig. 2).

Concluding Remarks

The gastrointestinal microbiota has recently emerged as a key regulator of stress adaption and immune responses in the body, and contributes to brain health and disease. The mechanisms by which the gut microbiota communicates with distant organs such as the brain is only beginning to emerge, but likely involves microbiota-derived signaling molecules, immune mediators, gut hormones, as well as vagal and spinal afferent neurons. The evidence presented herein suggests that our commensal gut microbiota plays a key role in

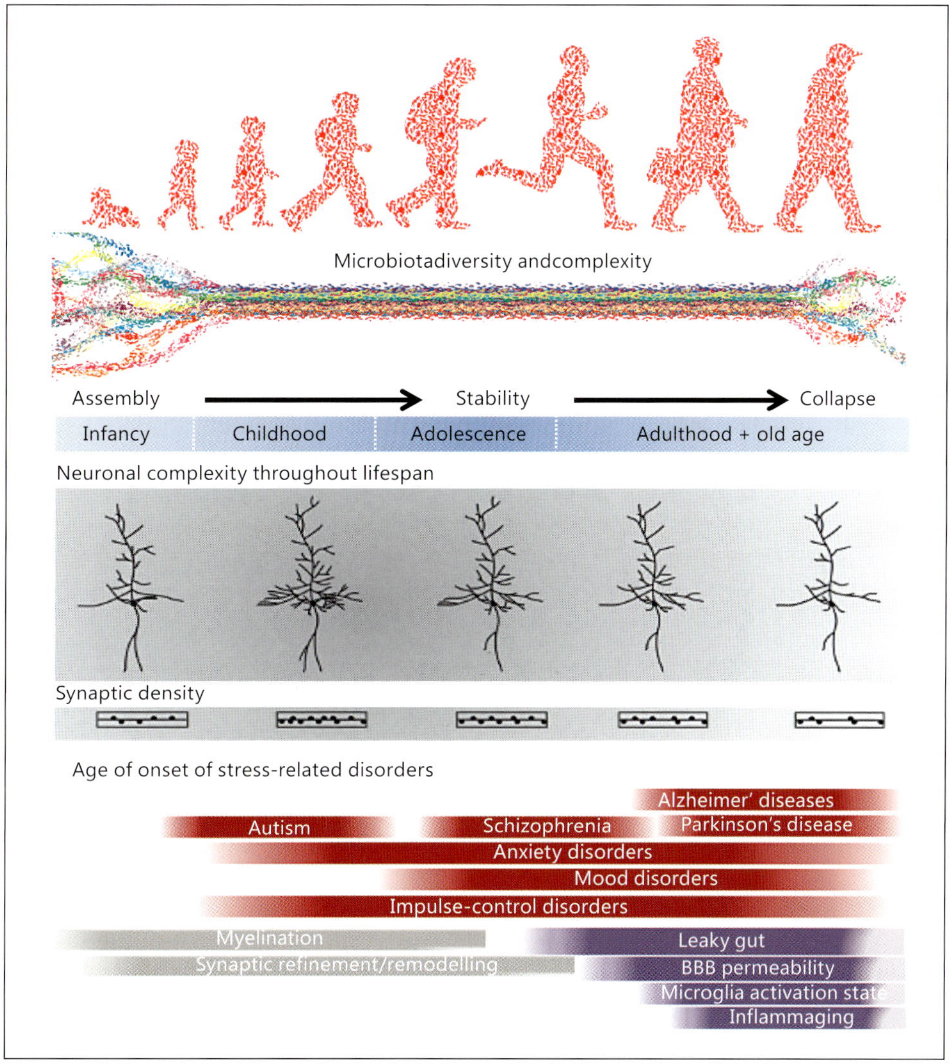

Fig. 1. Associative progression and decline of gastrointestinal microbiota with neuronal and synaptic complexity and the onset of stress-related disorders relative to age. BBB, blood-brain barrier.

facilitating the neuroprogression of stress-related disorders as a consequence of neuroinflammatory processes. The spate of recent publications in the microbiota field has transformed our understanding of how these microorganisms can contribute to our day-to-day functioning of the brain. The involvement of our gut microbiota in brain health and disease has now a credible scientific foundation, and although the vast majority of studies examining the role of microbiota in brain development and function have yielded promising results, further clinical studies are warranted. With the explosive increase in microbiota-based research and recent advances in sequencing and scanning, it is hoped that microbiota-altering therapeutics may provide a realistic target for influencing the progression of psychiatric disorders.

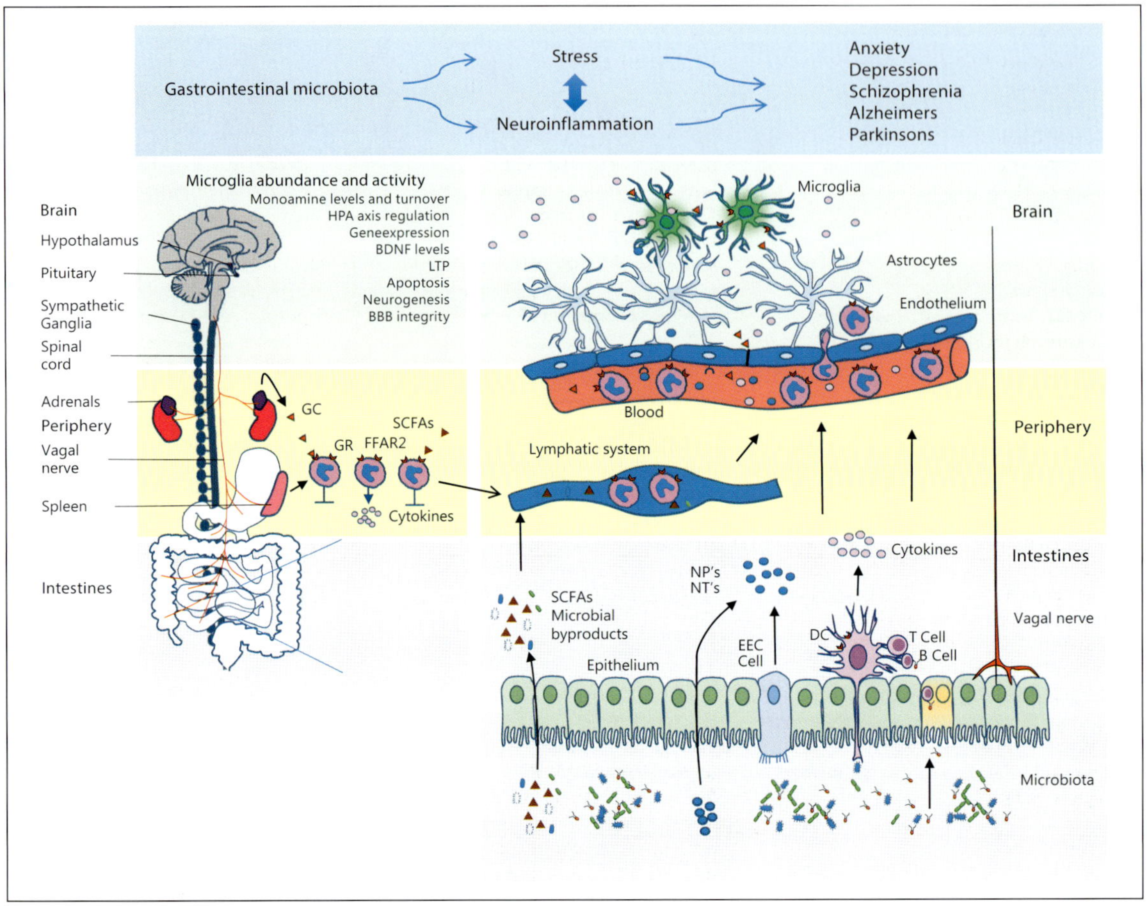

Fig. 2. Schematic for the role of the microbiota-gut-brain axis in the neuroprogression of stress-related disorders as a consequence of neuroinflammatory processes. BBB, blood-brain barrier; BDNF, brain-derived neurotrophic factor; DC, dendritic cell; EEC, enteroendocrine cell; FFAR2, free fatty acid receptor 2; GC, glucocorticoids; GR, glucocorticoid receptor; HPA, hypothalamic-pituitary-adrenal; LTP, long-term potentiation; NP, neuropeptide; NT, neurotransmitter; SCFAs, short chain fatty acids.

References

1 Eckburg PB, Bik EM, Bernstein CN, Purdom E, Dethlefsen L, Sargent M, et al: Diversity of the human intestinal microbial flora. Science 2005;308:1635–1638.

2 Gaci N, Borrel G, Tottey W, O'Toole PW, Brugere JF: Archaea and the human gut: new beginning of an old story. World J Gastroenterol 2014;20:16062–16078.

3 Scarpellini E, Ianiro G, Attili F, Bassanelli C, De Santis A, Gasbarrini A: The human gut microbiota and virome: potential therapeutic implications. Dig Liver Dis 2015;47:1007–1012.

4 Williamson LL, McKenney EA, Holzknecht ZE, Belliveau C, Rawls JF, Poulton S, et al: Got worms? Perinatal exposure to helminths prevents persistent immune sensitization and cognitive dysfunction induced by early-life infection. Brain Behav Immun 2016;51:14–28.

5 Cryan JF, Dinan TG: More than a gut feeling: the microbiota regulates neurodevelopment and behavior. Neuropsychopharmacology 2015;40:241–242.

6 Mayer EA, Knight R, Mazmanian SK, Cryan JF, Tillisch K: Gut microbes and the brain: paradigm shift in neuroscience. J Neurosci 2014;34:15490–15496.

7 Borre YE, Moloney RD, Clarke G, Dinan TG, Cryan JF: The impact of microbiota on brain and behavior: mechanisms & therapeutic potential. Adv Exp Med Biol 2014;817:373–403.

8 Borre YE, O'Keeffe GW, Clarke G, Stanton C, Dinan TG, Cryan JF: Microbiota and neurodevelopmental windows: implications for brain disorders. Trends Mol Med 2014;20:509–518.
9 Biagi E, Candela M, Fairweather-Tait S, Franceschi C, Brigidi P: Aging of the human metaorganism: the microbial counterpart. Age (Dordr) 2012;34:247–267.
10 Farhadi A, Banan A, Fields J, Keshavarzian A: Intestinal barrier: an interface between health and disease. J Gastroenterol Hepatol 2003;18:479–497.
11 De Kloet ER, Oitzl MS, Schobitz B: Cytokines and the brain corticosteroid receptor balance: relevance to pathophysiology of neuroendocrine-immune communication. Psychoneuroendocrinology 1994;19:121–134.
12 McEwen BS, Gray JD, Nasca C: 60 years of neuroendocrinology: redefining neuroendocrinology: stress, sex and cognitive and emotional regulation. J Endocrinol 2015;226:T67–T83.
13 Wilder RL: Neuroendocrine-immune system interactions and autoimmunity. Annu Rev Immunol 1995;13:307–338.
14 Furtado M, Katzman MA: Neuroinflammatory pathways in anxiety, posttraumatic stress, and obsessive compulsive disorders. Psychiatry Res 2015;229:37–48.
15 Eley TC, Stevenson J: Specific life events and chronic experiences differentially associated with depression and anxiety in young twins. J Abnorm Child Psychol 2000;28:383–394.
16 Furtado M, Katzman MA: Examining the role of neuroinflammation in major depression. Psychiatry Res 2015;229:27–36.
17 Berk M, Kapczinski F, Andreazza AC, Dean OM, Giorlando F, Maes M, et al: Pathways underlying neuroprogression in bipolar disorder: focus on inflammation, oxidative stress and neurotrophic factors. Neurosci Biobehav Rev 2011;35:804–817.
18 Bauer ME, Wieck A, Lopes RP, Teixeira AL, Grassi-Oliveira R: Interplay between neuroimmunoendocrine systems during post-traumatic stress disorder: a minireview. Neuroimmunomodulation 2010;17:192–195.
19 Rogers J, Shen Y: A perspective on inflammation in Alzheimer's disease. Ann NY Acad Sci 2000;924:132–135.
20 Ricci S, Fuso A, Ippoliti F, Businaro R: Stress-induced cytokines and neuronal dysfunction in Alzheimer's disease. J Alzheimers Dis 2012;28:11–24.
21 Schultze-Lutter F, Schimmelmann BG, Klosterkotter J, Ruhrmann S: Comparing the prodrome of schizophrenia-spectrum psychoses and affective disorders with and without psychotic features. Schizophr Res 2012;138:218–222.
22 Walker AJ, Kim Y, Price JB, Kale RP, McGillivray JA, Berk M, et al: Stress, inflammation, and cellular vulnerability during early stages of affective disorders: biomarker strategies and opportunities for prevention and intervention. Front Psychiatry 2014;5:34.
23 Streit WJ, Xue QS: Life and death of microglia. J Neuroimmune Pharmacol 2009;4:371–379.
24 Kim SU, de Vellis J: Microglia in health and disease. J Neurosci Res 2005;81:302–313.
25 Tremblay ME, Stevens B, Sierra A, Wake H, Bessis A, Nimmerjahn A: The role of microglia in the healthy brain. J Neurosci 2011;31:16064–16069.
26 Ransohoff RM, Schafer D, Vincent A, Blachere NE, Bar-Or A: Neuroinflammation: ways in which the immune system affects the brain. Neurotherapeutics 2015;12:896–909.
27 Kettenmann H, Hanisch UK, Noda M, Verkhratsky A: Physiology of microglia. Physiol Rev 2011;91:461–553.
28 Robillard KN, Lee KM, Chiu KB, MacLean AG: Glial cell morphological and density changes through the lifespan of rhesus macaques. Brain Behav Immun 2016;55:60–69.
29 Wong WT: Microglial aging in the healthy CNS: phenotypes, drivers, and rejuvenation. Front Cell Neurosci 2013;7:22.
30 Pocock JM, Kettenmann H: Neurotransmitter receptors on microglia. Trends Neurosci 2007;30:527–535.
31 Siskova Z, Tremblay ME: Microglia and synapse: interactions in health and neurodegeneration. Neural Plast 2013;2013:425845.
32 Charo IF, Ransohoff RM: The many roles of chemokines and chemokine receptors in inflammation. N Engl J Med 2006;354:610–621.
33 Sierra A, Gottfried-Blackmore A, Milner TA, McEwen BS, Bulloch K: Steroid hormone receptor expression and function in microglia. Glia 2008;56:659–674.
34 Wohleb ES, McKim DB, Sheridan JF, Godbout JP: Monocyte trafficking to the brain with stress and inflammation: a novel axis of immune-to-brain communication that influences mood and behavior. Front Neurosci 2014;8:447.
35 McKim DB, Niraula A, Tarr AJ, Wohleb ES, Sheridan JF, Godbout JP: Neuroinflammatory dynamics underlie memory impairments after repeated social defeat. J Neurosci 2016;36:2590–2604.
36 Kreisel T, Frank MG, Licht T, Reshef R, Ben-Menachem-Zidon O, Baratta MV, et al: Dynamic microglial alterations underlie stress-induced depressive-like behavior and suppressed neurogenesis. Mol Psychiatry 2014;19:699–709.
37 Wohleb ES, Powell ND, Godbout JP, Sheridan JF: Stress-induced recruitment of bone marrow-derived monocytes to the brain promotes anxiety-like behavior. J Neurosci 2013;33:13820–13833.
38 Bellavance MA, Rivest S: The HPA – immune axis and the immunomodulatory actions of glucocorticoids in the brain. Front Immunol 2014;5:136.
39 Kim YK, Na KS, Myint AM, Leonard BE: The role of pro-inflammatory cytokines in neuroinflammation, neurogenesis and the neuroendocrine system in major depression. Prog Neuropsychopharmacol Biol Psychiatry 2016;64:277–284.
40 Tynan RJ, Naicker S, Hinwood M, Nalivaiko E, Buller KM, Pow DV, et al: Chronic stress alters the density and morphology of microglia in a subset of stress-responsive brain regions. Brain Behav Immun 2010;24:1058–1068.
41 Bailey MT: Influence of stressor-induced nervous system activation on the intestinal microbiota and the importance for immunomodulation. Adv Exp Med Biol 2014;817:255–276.
42 Bailey MT, Dowd SE, Galley JD, Hufnagle AR, Allen RG, Lyte M: Exposure to a social stressor alters the structure of the intestinal microbiota: implications for stressor-induced immunomodulation. Brain Behav Immun 2011;25:397–407.
43 Bharwani A, Mian MF, Foster JA, Surette MG, Bienenstock J, Forsythe P: Structural & functional consequences of chronic psychosocial stress on the microbiome & host. Psychoneuroendocrinology 2016;63:217–227.

44 De Palma G, Collins SM, Bercik P, Verdu EF: The microbiota-gut-brain axis in gastrointestinal disorders: stressed bugs, stressed brain or both? J Physiol 2014; 592:2989–2997.
45 De Palma G, Blennerhassett P, Lu J, Deng Y, Park AJ, Green W, et al: Microbiota and host determinants of behavioural phenotype in maternally separated mice. Nat Commun 2015;6: 7735.
46 Moloney RD, Dinan TG, Cryan JF: Stress & the microbiota-gut-brain axis in visceral pain. Psychoneuroendocrinology 2015;61:8.
47 Golubeva AV, Crampton S, Desbonnet L, Edge D, O'Sullivan O, Lomasney KW, et al: Prenatal stress-induced alterations in major physiological systems correlate with gut microbiota composition in adulthood. Psychoneuroendocrinology 2015;60:58–74.
48 Galley JD, Nelson MC, Yu Z, Dowd SE, Walter J, Kumar PS, et al: Exposure to a social stressor disrupts the community structure of the colonic mucosa-associated microbiota. BMC Microbiol 2014; 14:189.
49 Desbonnet L, Garrett L, Clarke G, Kiely B, Cryan JF, Dinan TG: Effects of the probiotic *Bifidobacterium infantis* in the maternal separation model of depression. Neuroscience 2010;170:1179–1188.
50 Gareau MG, Jury J, MacQueen G, Sherman PM, Perdue MH: Probiotic treatment of rat pups normalises corticosterone release and ameliorates colonic dysfunction induced by maternal separation. Gut 2007;56:1522–1528.
51 McKernan DP, Fitzgerald P, Dinan TG, Cryan JF: The probiotic *Bifidobacterium infantis* 35624 displays visceral antinociceptive effects in the rat. Neurogastroenterol Motil 2010;22:1029–1035. e268.
52 Savignac HM, Kiely B, Dinan TG, Cryan JF: Bifidobacteria exert strain-specific effects on stress-related behavior and physiology in BALB/c mice. Neurogastroenterol Motil 2014;26: 1615–1627.
53 Palomar MM, Maldonado Galdeano C, Perdigon G: Influence of a probiotic lactobacillus strain on the intestinal ecosystem in a stress model mouse. Brain Behav Immun 2014;35:77–85.
54 Bravo JA, Forsythe P, Chew MV, Escaravage E, Savignac HM, Dinan TG, et al: Ingestion of *Lactobacillus* strain regulates emotional behavior and central GABA receptor expression in a mouse via the vagus nerve. Proc Natl Acad Sci USA 2011;108:16050–16055.
55 Ait-Belgnaoui A, Durand H, Cartier C, Chaumaz G, Eutamene H, Ferrier L, et al: Prevention of gut leakiness by a probiotic treatment leads to attenuated HPA response to an acute psychological stress in rats. Psychoneuroendocrinology 2012;37: 1885–1895.
56 Desbonnet L, Clarke G, Traplin A, O'Sullivan O, Crispie F, Moloney RD, et al: Gut microbiota depletion from early adolescence in mice: implications for brain and behaviour. Brain Behav Immun 2015;48:165–173.
57 Matsuo K, Zhang X, Ono Y, Nagatomi R: Acute stress-induced colonic tissue HSP70 expression requires commensal bacterial components and intrinsic glucocorticoid. Brain Behav Immun 2009; 23:108–115.
58 O'Mahony SM, Felice VD, Nally K, Savignac HM, Claesson MJ, Scully P, et al: Disturbance of the gut microbiota in early-life selectively affects visceral pain in adulthood without impacting cognitive or anxiety-related behaviors in male rats. Neuroscience 2014;277: 885–901.
59 Clarke G, Grenham S, Scully P, Fitzgerald P, Moloney RD, Shanahan F, et al: The microbiome-gut-brain axis during early life regulates the hippocampal serotonergic system in a sex-dependent manner. Mol Psychiatry 2013;18:666–673.
60 Sudo N, Aiba Y, Oyama N, Yu XN, Matsunaga M, Koga Y, et al: Dietary nucleic acid and intestinal microbiota synergistically promote a shift in the Th1/Th2 balance toward Th1-skewed immunity. Int Arch Allergy Immunol 2004;135: 132–135.
61 Sudo N, Chida Y, Aiba Y, Sonoda J, Oyama N, Yu XN, et al: Postnatal microbial colonization programs the hypothalamic-pituitary-adrenal system for stress response in mice. J Physiol 2004;558: 263–275.
62 Neufeld KA, Kang N, Bienenstock J, Foster JA: Effects of intestinal microbiota on anxiety-like behavior. Commun Integr Biol 2011;4:492–494.
63 Neufeld KM, Kang N, Bienenstock J, Foster JA: Reduced anxiety-like behavior and central neurochemical change in germ-free mice. Neurogastroenterol Motil 2011;23:255–264.e119.
64 Asano Y, Hiramoto T, Nishino R, Aiba Y, Kimura T, Yoshihara K, et al: Critical role of gut microbiota in the production of biologically active, free catecholamines in the gut lumen of mice. Am J Physiol Gastrointest Liver Physiol 2012; 303:G1288–G1295.
65 Nishino R, Mikami K, Takahashi H, Tomonaga S, Furuse M, Hiramoto T, et al: Commensal microbiota modulate murine behaviors in a strictly contamination-free environment confirmed by culture-based methods. Neurogastroenterol Motil 2013;25: 521–528.
66 Messaoudi S, Manai M, Kergourlay G, Prevost H, Connil N, Chobert JM, et al: *Lactobacillus salivarius*: bacteriocin and probiotic activity. Food Microbiol 2013;36:296–304.
67 Clarke G, Cryan JF, Dinan TG, Quigley EM: Review article: probiotics for the treatment of irritable bowel syndrome – focus on lactic acid bacteria. Aliment Pharmacol Ther 2012;35:403–413.
68 Dorrestein PC, Mazmanian SK, Knight R: Finding the missing links among metabolites, microbes, and the host. Immunity 2014;40:824–832.
69 Kayama H, Takeda K: Functions of innate immune cells and commensal bacteria in gut homeostasis. J Biochem 2016;159:141–149.
70 McDermott AJ, Huffnagle GB: The microbiome and regulation of mucosal immunity. Immunology 2014;142:24–31.
71 Schafer DP, Lehrman EK, Kautzman AG, Koyama R, Mardinly AR, Yamasaki R, et al: Microglia sculpt postnatal neural circuits in an activity and complement-dependent manner. Neuron 2012; 74:691–705.
72 Schafer DP, Stevens B: Microglia function in central nervous system development and plasticity. Cold Spring Harb Perspect Biol 2015;7:a020545.
73 Erny D, Hrabe de Angelis AL, Jaitin D, Wieghofer P, Staszewski O, David E, et al: Host microbiota constantly control maturation and function of microglia in the CNS. Nat Neurosci 2015;18:965–977.

74 Biagi E, Nylund L, Candela M, Ostan R, Bucci L, Pini E, et al: Through ageing, and beyond: gut microbiota and inflammatory status in seniors and centenarians. PLoS One 2010; 5:e10667.
75 Franceschi C, Capri M, Monti D, Giunta S, Olivieri F, Sevini F, et al: Inflammaging and anti-inflammaging: a systemic perspective on aging and longevity emerged from studies in humans. Mech Ageing Dev 2007;128:92–105.
76 D'Mello C, Le T, Swain MG: Cerebral microglia recruit monocytes into the brain in response to tumor necrosis factorα signaling during peripheral organ inflammation. J Neurosci 2009;29: 2089–2102.
77 D'Mello C, Ronaghan N, Zaheer R, Dicay M, Le T, MacNaughton WK, et al: Probiotics improve inflammation-associated sickness behavior by altering communication between the peripheral immune system and the brain. J Neurosci 2015;35:10821–10830.
78 Nilsson NE, Kotarsky K, Owman C, Olde B: Identification of a free fatty acid receptor, FFA2R, expressed on leukocytes and activated by short-chain fatty acids. Biochem Biophys Res Commun 2003; 303:1047–1052.

John F. Cryan, MRIA
APC Microbiome Institute, Departments Anatomy and Neuroscience
University College Cork
386 Western Gateway Building
Cork T12 XF62 (Ireland)
E-Mail j.cryan@ucc.ie

Halaris A, Leonard BE (eds): Neuroprogression in Psychiatric Disorders.
Mod Trends Pharmacopsychiatry. Basel, Karger, 2017, vol 31, pp 162–176 (DOI: 10.1159/000470814)

Pharmacological and Nonpharmacological Interventions to Arrest Neuroprogression in Psychiatric Disorders

Fotini Boufidou[a] • Angelos Halaris[b]

[a]Laboratory of Biopathology, First Department of Psychiatry, National and Kapodistrian University of Athens Medical School, Athens, Greece; [b]Department of Psychiatry and Behavioral Neuroscience, Stritch School of Medicine, Loyola University Chicago, Chicago, IL, and Loyola University Medical Center, Maywod, IL, USA

Abstract

The concept of neuroprogression describes the progressive course of the disorder and stresses the progressive, recurrent, and chronic course of the disease entity under consideration. It subsumes clinical manifestations of the disease process and may also entail morphological, biochemical, neurochemical, immunological, physiological, and genetic aspects that contribute to the progressive course of the disease in question. In an attempt to identify the appropriate agent or method that could arrest neuroprogression in psychiatric patients, we conducted an evaluation of the use of anti-inflammatory drugs under the perspective of current pharmacological and neurophysiological data. This evaluation included the use of nonsteroidal anti-inflammatory drugs (NSAIDs) as adjunctive treatment to conventional pharmacotherapy as well as the use of natural products exerting anti-inflammatory properties (i.e., ω-3 fatty acids) given as adjunctive or monotherapeutic treatments in less severe cases. The therapeutic significance of nonpharmacological methods, such as psychotherapy, physical exercise, and body-mind therapies, was also considered and will be discussed in this chapter. In conclusion, the role of psychotropic and select anti-inflammatory drugs in arresting neuroprogression is a very promising new frontier in psychiatric research and clinical practice. Modulators of a specific prostanoid synthase or receptor across the cyclooxygenase (COX)-2 downstream pathway along with new multitarget NSAIDs are expected to be tested by the pharmaceutical industry as potential agents to antagonize neuroprogression. Meanwhile, salicylates and selective COX-2 inhibitors could still be used in carefully selected subgroups of patients. Psychotherapy and nonpharmacological, stress-relieving methods should be considered as adjunctive tools to aid in arresting neuroprogression.

Introduction

The significant clinical implications of neuroprogression in psychiatric disorders as well as the biological mechanisms underlying neuroprogression have been extensively described in previous chapters of this volume. Likely pathophysiological substrates include a sustained proinflammatory state, increased oxidative stress, and deficits in neuroprotection and neuroplasticity. There is also a complex interplay with the monoaminergic, cholinergic, glutamatergic, and GABAergic neurotransmitter systems that require further elucidation. The possibility of ameliorating or even reversing such pathophysiological mechanisms through pharmacological, nonpharmacological, or combined interventions should be explored, and the need to arrest neuroprogression can be seen as an emerging new priority in treating and even preventing the chronicity of psychiatric and neuropsychiatric disorders.

Up to now, besides conventional therapies, the only "antineuroprogressive" pharmacological interventions for treating psychiatric disorders could be considered to be the administration of anti-inflammatory/antioxidant agents as adjunctive therapy, as for ethical reasons such agents could not be given alone as monotherapy, especially in severe cases. This past experience will be briefly presented in the current chapter, and information about drugs that have been used will be provided. Additionally, side effects and considerations about possible implications for human homeostasis will be discussed. Attempts for nonpharmacological interventions and the significance of conventional therapies in arresting neuroprogression will also be discussed.

Pharmacological Interventions to Arrest Neuroprogression

Anti-Inflammatory Drugs and Their Potential Role in Arresting Neuroprogression

Nonsteroidal anti-inflammatory drugs (NSAIDs) are a large family of anti-inflammatory drugs that have been widely used for the treatment of different inflammatory conditions, acting mainly by inhibiting the two isoforms of the enzyme cyclooxygenase (COX) on the arachidonic acid cascade and thus reducing downstream prostaglandin synthesis. NSAIDs have been classified according to their selectivity with regard to COX-1/COX-2 inhibition [1]: salicylates (i.e., acetylsalicylic acid, ASA, diflunisal, and salicylamide) due to their irreversible acetylation of COX-1 and COX-2 enzymes, nonselective NSAIDs (i.e., ibuprofen, naproxen, ketoprofen, dexibuprofen, tolfenamic acid, and piroxicam) because of their nonselective inhibition of either COX-1 or COX-2, nonselective COX inhibitors (i.e., diclofenac, etodolac, meloxicam, and nabumetone) with a more pronounced COX-2 inhibition as compared to nonselective NSAIDs, and selective COX-2 inhibitors (coxibs; i.e., celecoxib, rofecoxib, valdecoxib, parecoxib, and etoricoxib).

Signs of chronic, low-grade inflammation have been repeatedly found in major psychiatric disorders since the early 1990s [2]. In the beginning of the new millennium, this knowledge led some researchers to perform, quite successfully, clinical trials with NSAIDs used as adjunctive therapy to conventional antipsychotic and antidepressant medications. Besides these clinical attempts of using NSAIDs for the benefit of psychiatric patients, presented in the following section, there are several other reasons supporting the use of NSAIDs and especially ASA and selective coxibs as antineuroprogressive agents:

In major psychiatric disorders, there have been numerous findings of elevated proinflammatory mediators directly or indirectly associated with the arachidonic acid cascade:

In depression, prostaglandin E_2 (PGE_2) levels have been found to be elevated both in serum and cerebrospinal fluid [3] as well as in the saliva of patients [4]. Moreover, in vitro studies have shown an increased PG production from patients' lymphocytes [5], while antidepressant medications inhibit it [6]. PGs play a central role in the inflammatory response partly because of their vasodilating properties while they enhance proinflammatory cytokine production, such as interleukin (IL)-6 and tumor necrosis factor (TNF)-α. These cytokines have been repeatedly found to be elevated in depressed patients [7–9] as has been the expression of COX-2. Galecki et al. [10] have shown that mRNA expression of COX-2 and phospholipase A_2-IIA, a rate-limiting enzyme that generates arachidonic acid, is indeed increased in patients with recurrent depression. In schizophrenia, an immune imbalance with an overactivation of type 2 responses and a lack of type 1 responses as a result of chronic inflammation has been repeatedly described [11]. This pathophysiologically important polarization of the immune system may be ameliorated by a selective COX-2 inhibitor [12].

Inflammatory mediators can impair neurotransmission and interfere with the function of the hypothalamic-pituitary-adrenal axis:

Elevated levels of PGE_2 and proinflammatory cytokines may have significant impact on tryptophan and kynurenine metabolism resulting in impaired serotonergic and glutamatergic neurotransmission and consequences on neuronal plasticity. It has been shown that in the PG-mediated control of kynurenic acid synthesis opposite actions by COX-1 and COX-2 are involved [13]. PG can also cause an overactivation of the hypothalamic-pituitary-adrenal axis [5] resulting in increased secretion of cortisol and consequently enhanced tryptophan 2,3-dioxygenase activity with further diversion of tryptophan metabolism [6].

Safety of ASA and selective coxibs:

There is a century-long experience in the use of aspirin as an anti-inflammatory agent while its antithrombotic effects have been proven most valuable in the prevention of cardiovascular and cerebrovascular diseases. On the other hand, selective coxibs represent a new generation of NSAIDs with a targeted mechanism of action set to become the first coxibs of minimal gastrotoxicity [15].

Non-anti-inflammatory beneficial effects of selective coxibs:

Studies have shown that coxibs, besides acting as anti-inflammatory agents, exhibit antidepressant effects, and they may also improve cognition in schizophrenia [16]. In the depression model of the olfactory bulbectomized rat, a decrease in hypothalamic cytokine levels and improved behavior were observed after celecoxib treatment [17]. Interestingly, in a large clinical trial with the participation of 2,228 patients suffering from osteoarthritis, rofecoxib, in addition to exerting analgesic and anti-inflammatory effects, also showed significant antidepressant effectiveness [18]. In another study, mice with a genetic overexpression of COX-2 showed more prominent deficits in cognition, which were attenuated by celecoxib [19]. These findings are in accordance with the results of a clinical trial with celecoxib add-on to amisulpride in first-episode schizophrenia where the efficacy of therapy in the celecoxib group was especially pronounced in improving cognition [20].

Clinical Trials with Adjunctive Anti-Inflammatory Treatment in Psychiatric Disorders

Selective and Nonselective COX Inhibition in the Treatment of Schizophrenia

Müller et al. [21] carried out the first prospective, randomized, double-blind, placebo-controlled study with the COX-2 inhibitor, celecoxib, as add-on treatment to risperidone in acute exacerbation of schizophrenia. A significant therapeutic effect of add-on celecoxib was observed. This finding of a clinical advantage of COX-2 inhibition could not be replicated in a second study by the same group of investigators [22]. Further analysis of the data revealed that the outcome depends on the duration of the disease [23]. In subsequent clinical studies with a similar design, positive results of adjunctive celecoxib treatment were obtained in a Chinese population of first-episode schizophrenic patients [24] and in an Iranian population of chronic schizophrenic patients [25]. However, no advantage of celecoxib was found in continuously ill schizophrenic patients. [26]. Finally, the last study of Müller's team with first-episode schizophrenic patients showed a beneficial effect of celecoxib add-on to amisulpride, and this finding underlines the effectiveness of short-term treatment with a COX-2 inhibitor in the early stages of schizophrenia [20]. Laan et al. reported the results of a clinical trial

Table 1. Double-blind, randomized, placebo-controlled, clinical trials of selective cyclooxygenase (COX)-inhibitors and acetylsalicylic acid (ASA) add-on in schizophrenia

Authors	Diagnosis (patients, *n*)	Disease course and duration	Trial duration	Concomitant drug	COX-2 inhibitor	Outcome
Müller et al. [21] (2002)	Schizophrenia (50)	Not specified mean 5.9 years	5 weeks	Risperidone (flexible dose)	Celecoxib 400 mg/day	Significant advantage of celecoxib
Rappard and Müller [22] (2004)	Schizophrenia (270)	≤10 years	11 weeks	Risperidone (flexible dose)	Celecoxib 400 mg/day	No advantage of celecoxib
Rapaport et al. [26] (2005)	Schizophrenia (38)	Continuously ill mean 20 years	8 weeks	Risperidone or olanzapine (constant dose)	Celecoxib 400 mg/day	No advantage of celecoxib
Zhang et al. [24] (2006)	Schizophrenia (40)	First manifestation	12 weeks	Risperidone (flexible dose)	Celecoxib 400 mg/day	Significant advantage of celecoxib
Akhondzadeh et al. [25] (2007)	Schizophrenia (60)	Chronic type (active phase)	8 weeks	Risperidone (fixed dose)	Celecoxib 400 mg/day	Significant advantage of celecoxib
Müller et al. [20] (2010)	Schizophrenia (49)	First manifestation	6 weeks	Amisulpride (flexible dose)	Celecoxib 400 mg/day	Significant advantage of celecoxib
Laan et al. [27] (2010)	Schizophrenic spectrum disorders (70)	Mean 4.1 years	3 months	Olanzapine, clozapine, risperidone	ASA 1,000 mg/day	Significant advantage of ASA

with ASA as an adjunct to antipsychotic medication in which a significant therapeutic advantage of ASA was shown [27]. Interestingly, treatment efficacy was more pronounced in patients with higher anti-inflammatory cytokine production from stimulated peripheral blood cells. Data from these placebo-controlled clinical trials are summarized in Table 1 (adopted from Müller [28]). A meta-analysis of the clinical effects of NSAIDs in schizophrenia revealed significant effects on total schizophrenic and positive and negative symptoms [29], while another meta-analysis that included several unpublished data found that aspirin was superior to placebo and beneficial in schizophrenic patients with a short duration of the disease or schizophrenics in their first episode [30]. However, these authors outline the limited database and the need for further studies especially in first-episode patients. In this last, extended meta-analysis, the safety of the NSAIDs used in schizophrenia trials was summarized: it appears that no clinically significant side effects by NSAID were observed, and there was no significant difference between groups with respect to side effects noted during the trials.

Selective and Nonselective COX Inhibition in Treating Major Depression and Bipolar Disorder

As in schizophrenia, Müller et al. [31] were the first to perform a prospective, randomized, double-blind, placebo-controlled study of celecoxib as an add-on to reboxetine in patients suffering from an acute depressive episode. In this study, celecoxib significantly reduced depressive symptomatology. No cardiovascular events or side effects were observed in the celecoxib group although the authors report some cases of increased blood pressure that led to subject exclusion from the study. In a second similar study, the combination of fluoxetine and celecoxib showed significant superiority over fluoxetine alone in an Iranian population [32]. Side effects did not differ between groups, and there was no pharmacokinetic interaction between celecoxib and fluoxetine. Three years later, Abbasi et al. [33] tried a lower dose of celecoxib as an add-on to sertraline in a third study of similar design. The celecoxib group had significantly better response and remission rates, while IL-6 serum levels were positively correlated to Hamilton Rating Scale for Depression (Ham-D) scores. Side effects did not differ between the two groups.

In a meta-analysis, Na et al. [34] evaluated the above-mentioned studies and determined that the patients receiving adjunctive celecoxib had significantly higher mean changes in their Ham-D scores between baseline and end point measurements compared with those receiving placebo, and also showed better remission and response rates. The authors concluded that adjunctive treatment with celecoxib can be a promising strategy for patients with depressive disorders.

The last of the Iranian studies of celecoxib augmentation of sertraline in the treatment of 30 drug-naïve, first-episode, depressed women was published in 2014 [35]. The results suggested that celecoxib, even at the low dose of 200 mg/day also used by Abbasi et al. [33], hastens the onset of the therapeutic action of sertraline and increases response and remission rates. Side effects did not differ between groups and did not include any cardiovascular or gastrointestinal problems.

In 2013, Raison et al. [36] published data from a different study. These authors sought to determine whether repeated intravenous administration of infliximab, a monoclonal antibody directed against TNF, would improve the mood of 60 treatment-resistant depressed patients. No overall difference in Ham-D score changes between the infliximab and the placebo group was found. However, change in Ham-D scores favored infliximab-treated patients with a baseline high-sensitivity C-reactive protein concentration >5 mg/L and placebo-treated patients with a baseline concentration ≤5 mg/L. Dropouts and adverse events were limited and did not differ between groups. It should be noted that infliximab is too large as a molecule to cross the blood-brain barrier; therefore, its positive effects mainly reflect the anti-inflammatory action of the drug in the periphery.

As far as bipolar disorder is concerned, one study of adjunctive celecoxib in the treatment of depressive or mixed episodes has been published showing that celecoxib produces a rapid-onset antidepressant effect and that it was well tolerated [37]. Data from these clinical trials are summarized in Table 2. In a recently completed study of treatment-resistant bipolar patients with celecoxib add-on to escitalopram similar findings were obtained with significant differences in outcomes of the celecoxib add-on randomized group of patients compared to the placebo add-on group [Halaris et al., in preparation].

ASA as adjunctive treatment to conventional antidepressants has been evaluated in two open studies. In the first study of Mendlewicz et al. [38], a shortened onset of action of selective serotonin reuptake inhibitors (SSRIs) was achieved in 24 nonresponder depressed patients using ASA augmentation. Three years later, Galecki et al. [39] performed a larger study of ASA as adjunctive to fluoxetine and concluded that combined therapy has the same efficacy as fluoxetine monotherapy, although the administration of ASA in depressed patients decreased oxidative stress parameters over a 3-month period. It should be mentioned that patients in this study were not equally divided in study groups while data on their gender are missing.

Gallagher et al. [40] performed a pharmacovigilance study on antidepressant response in 1,528 patients with major depression exposed to NSAIDs and found that exposure to NSAIDs, with the exception of COX-2 inhibitors and

Table 2. Double-blind, randomized, placebo-controlled clinical trials of selective cyclooxygenase (COX) and tumor necrosis factor (TNF)-α inhibitor add-on in affective disorders

Authors	Diagnosis (patients, *n*)	Trial duration	Concomitant drugs	Anti-inflammatory drug	Outcome
Müller et al. [31] (2006)	Major depression (40)	6 weeks	Reboxetine and lorazepam	Celecoxib 400 mg/day	Significant advantage of the COX-2 inhibitor
Nery et al. [37] (2008)	Bipolar disorder I/II, major depression, or mixed episode (28)	6 weeks	Mood stabilizers, antipsychotics	Celecoxib 400 mg/day	Shorter onset of action of concomitant drugs
Akhondzadeh et al. [32] (2009)	Major depression (40)	6 weeks	Fluoxetine	Celecoxib 400 mg/day	Significant advantage of the COX-2 inhibitor
Abbasi et al. [33] (2012)	Major depression (40)	6 weeks	Sertraline	Celecoxib 200 mg/day	Significant advantage of the COX-2 inhibitor
Raison et al. [36] (2013)	Major depression, treatment resistant (60)	12 weeks	Not specified antidepressants (*n* = 37), free of medication (*n* = 23)	Infliximab 5 mg/kg, weeks 1, 2, 6	Significant advantage of TNF-α inhibitor only when initial CRP ≥5 mg/L
Majd et al. [35] (2014)	Major depression, first episode (30)	8 weeks	Sertraline	Celecoxib 200 mg/day	Significant advantage of the COX-2 inhibitor

CRP, C-reactive protein.

salicylates, was associated with a greater likelihood of depression classified as treatment resistant compared with depression classified as responsive to SSRIs. This association was not found when NSAIDs were limited to COX-2 inhibitors and salicylates. Finally, an extended pharmacoepidemiological study on the combined use of SSRIs and NSAIDs and psychiatric outcomes among 123,351 Danish SSRI users showed that low-dose ASA reduced the risk of psychiatric contact in general and of psychiatric contact due to depression [41].

Natural Products Exerting Anti-Inflammatory and Anti-Oxidative Properties and Their Potential Role in Arresting Neuroprogression

A growing number of observational and epidemiological studies have suggested that mental illness is associated with reduced dietary intake and/or cellular abundance of ω-3 polyunsaturated fatty acids (PUFA) [42, 43]. PUFAs are classified into n-3 (or ω-3) and n-6 (or ω-6) groups. Eicosapentaenoic acid (EPA) and docosahexaenoic acid (DHA), the major bioactive components of n-3 PUFAs, are not efficiently synthesized in humans and should, therefore, be obtained directly from the diet, particularly by consuming fish [44]. In the brain, n-3 and n-6 fatty acids regulate membrane fluidity, neurotransmission, membrane-bound enzymes, and cellular signal transduction [45, 46], and, as they cooperate and compete with each other, changes in their concentrations are significantly related to neuronal, glial, and immune cell functions [47]. Cellular and molecular studies suggest that EPA protects neurons against neurotoxicity through anti-inflammation and up-regulates anti-oxidant and anti-apoptosis gene expression and neurotrophins and tyrosine protein kinase B receptors [48–50]. EPA also maintains the balance in immune functions by reducing membrane arachidonic acid (an n-6 PUFA) and downstream PGE_2 synthesis [44]. Lu et al. [51] showed how DHA can modulate oxidative reactions and inflammatory cytokine production in microglial and neuronal cells.

Clinical Trials with Natural Products Exerting Anti-Inflammatory Properties in Psychiatric Disorders

ω-3 Fatty Acids in Treating Schizophrenia

The phospholipid hypothesis of schizophrenia [52] postulates that a variety of mental health problems could result from abnormalities of the phospholipid structure of neuronal membranes. However, clinical trials of ω-3 fatty acids in schizophrenia, despite being mostly of high quality, have produced inconsistent results and small effect sizes of doubtful clinical significance [43]. A recent meta-analysis of randomized controlled trials on symptomatic outcome revealed no beneficial effect of EPA augmentation in subjects with established schizophrenia. However, the authors underscore that no conclusion can be drawn for medium- to long-term effects of EPA in schizophrenia, in particular on relapse prevention in the early course of psychotic disorders [53]. Interestingly, in a randomized, double-blind, placebo-controlled study, a 12-week intervention with ω-3 PUFA given as monotherapy reduced the risk of progression to psychosis in 81 young individuals at ultrahigh risk of psychotic disorder [54].

ω-3 Fatty Acids in Treating Major Depression and Bipolar Disorder

Compared to schizophrenia, there is more convincing evidence in support of beneficial effects of ω-3 in mood disorders [43]. A recent meta-analysis [55] of 11 and 8 trials conducted on patients with a DSM-defined diagnosis of major depressive disorder and patients with depressive symptomatology but no diagnosis of major depressive disorder, respectively, demonstrated a significant clinical benefit of ω-3 PUFA treatment compared to placebo. Use of EPA rather than DHA, and the use of ω-3 PUFA adjunctively rather than monotherapeutically, influenced the final clinical efficacy. Similarly, a meta-analysis of 3 randomized controlled trials on adult patients with bipolar disorder demonstrated that ω-3 PUFA was effective in ameliorating depressive symptoms [55]. In the latest meta-analysis conducted by Mocking et al. [56], randomized placebo-controlled trials were pooled in order to assess the effect of ω-3 PUFA supplementation on symptoms of major depressive disorder. The results suggested a beneficial overall effect of ω-3 PUFA supplementation, especially with higher doses of EPA and in participants taking antidepressants adjunctively. The authors proposed that potential long-term biochemical side effects of high-dose add-on EPA supplementation should be carefully monitored. However, in their review of all clinical trials up to 2007 in a large range of patients with psychiatric disorders, Ross et al. [43] mentioned that ω-3 PUFA were well tolerated by both children and adults with mild gastrointestinal effects being the only consistently reported adverse event.

Nonpharmacological Interventions to Arrest Neuroprogression

Given the influence of stress and stress-induced activation of the sympathetic nervous system, behavioral interventions that address psychological and autonomic reactivity to stress, including psychotherapy and exercise, may be efficacious both in terms of treatment and prevention. As far as psychotherapy is concerned, there are obvious difficulties in performing clinical trials on its effects on the immune system, and the first study of this kind has only recently been published [57]. Interestingly, the study showed that recovery from depression in the absence of pharmacotherapy occurred with concomitant normalization of various cytokine levels that had been found to be initially increased. On the other hand, physical exercise, especially aerobic, is thought to enhance production and release of neurotrophins, mainly brain-derived neurotrophic factor, and also induce a cascade of molecular and cellular processes that favor angiogenesis, neurogenesis, and synaptogenesis [58]. Additionally, exercise is believed to enhance and preserve cognitive functions, such as memory, processing speed, attention, and executive functions [59, 60]. Exercise also benefits subjects suffering from mild cognitive impairment or early-stage dementia [60]. Women who reported being physically active at any point over the life span, and especially during adolescence, have a lower likelihood of cognitive impairment in late life [61]. The prefrontal and medial temporal cortices, particularly the hippocampus, are sensitive to exercise with trophic responses leading to increased volume and increases in hippocampal cerebral blood flow [62]. Exercise is considered to be protective for Alzheimer disease [63] and also an adjunctive treatment and a preventive measure in Parkinson disease [64].

Patients with serious mental illnesses may also benefit from exercise: it has been shown to effectively treat depression and anxiety disorders along with enhancing cognition [65, 66] while a Cochrane systematic review found exercise to significantly improve negative symptoms along with physical health in schizophrenic patients [67]. However, one recent review found that exercise programs had no significant effect on symptoms in adults with serious mental illness [68].

During the last two decades, mind-body therapies (MBTs), including yoga, meditation, tai chi, and qigong, have received increasing awareness and attention from the scientific community seeking to understand the safety and efficacy of these widely used practices. Tai chi, qi gong, and yoga are multidimensional behavioral therapies that integrate moderate physical activity, deep breathing, and meditation to promote stress reduction and relaxation, while meditation includes more integrative, mindfulness-based, stress reduction programs in the absence of physical activity. Considering the multiple benefits of yoga, the World Psychiatric Organization has declared the 21st of June as the "International Yoga Day" [69]. Morgan et al. [70] conducted the first comprehensive review of available controlled trials on the effects of MBTs on the immune system. This study revealed that MBTs do have a moderate effect in reducing C-reactive protein levels while there is evidence of increasing immune response to vaccination.

Meditation and yoga, along with physical exercise, sleep, and music, belong to a group of brain stimulation methods that are supposed to improve brain function and cognition, called "nonpharmacological cognitive enhancers." Their effects on cognition have been reviewed by Sachdeva et al. [71] who showed that there is converging evidence on several levels that nonpharmacological cognitive enhancers benefit and preserve cognition throughout the life span and across various neuropsychiatric disorders.

Discussion

The anti-inflammatory and potentially antineuroprogressive agents that have been used as adjuncts to conventional pharmacotherapy act on the metabolism of arachidonic acid. Arachidonic acid is a basic constituent of cell membranes. After its release from membrane phospholipids by phospholipases, it is metabolized via three pathways: the COX branch, the lipoxygenase branch, and the cytochrome P_{450} branch. Modulators of the arachidonic acid cascade have been the focus of research for the treatment of inflammation and pain for several decades [72]. COX or PGH synthases are membrane-bound enzymes converting

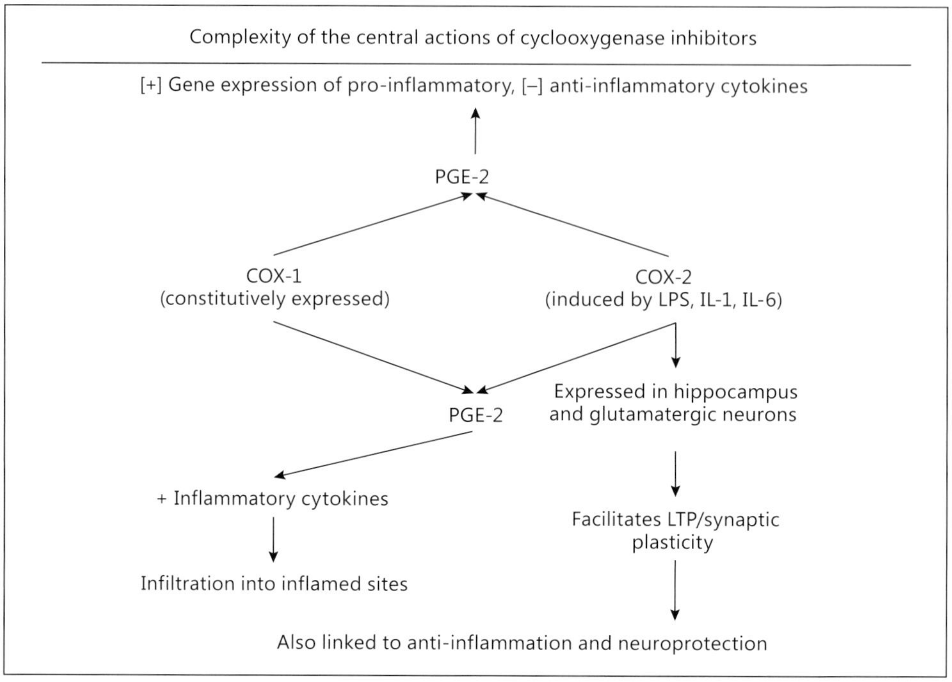

Fig. 1. Differential expression and central actions of cyclooxygenase (COX)-1 vs. COX-2 [97, 98]. IL, interleukin; LPS, lipopolysaccharide; PGE_2, prostaglandin E_2; LTP, long-term potentiation; [+], stimulation; [–], inhibition.

arachidonic acid to PGH_2. They exist in two isoforms: COX-1, a constitutively expressed enzyme responsible for basal homeostatic PG synthesis, and COX-2, the inducible form stimulated by cytokines, growth factors, and proinflammatory molecules [73] (Fig. 1). A third isoform (COX-3 or COX-1b) was first described in canines, but its physiological role remains unknown [74]. COX converts arachidonic acid into bioactive prostanoids. Prostanoids, as well as the whole range of eicosanoids synthesized by COX, are powerful lipid mediators that exert a variety of biological effects, including gastroprotection, vasodilation, and vasoconstriction.

Because COX-2 is typically induced by inflammatory stimuli in the majority of tissues, it has been considered as the best target for anti-inflammatory drugs. Coxibs were developed in the 1990s, and early trials comparing coxibs versus traditional NSAIDs could confirm that coxibs at doses with similar analgesic efficacy had less gastrointestinal toxicity [75, 76]. However, a few years later, selective COX-2 inhibition was associated with an increased risk of severe cardiovascular adverse events due to rather unexpected consequences on the vasoprotective prostacyclin (PGL_2) biosynthesis [77], which led to the voluntary withdrawal of some selective COX-2 inhibitors from the market [78]. Nowadays, the US Food and Drug Administration requires that the summaries of product characteristics of all NSAIDs carry a boxed warning about the risk of cardiovascular disease [79], whereas the Committee for Medicinal Products for Human Use of the European Medicines Agency has decided that coxibs (but not traditional NSAIDs) should be contraindicated in patients with coronary heart disease or stroke and used with caution in patients

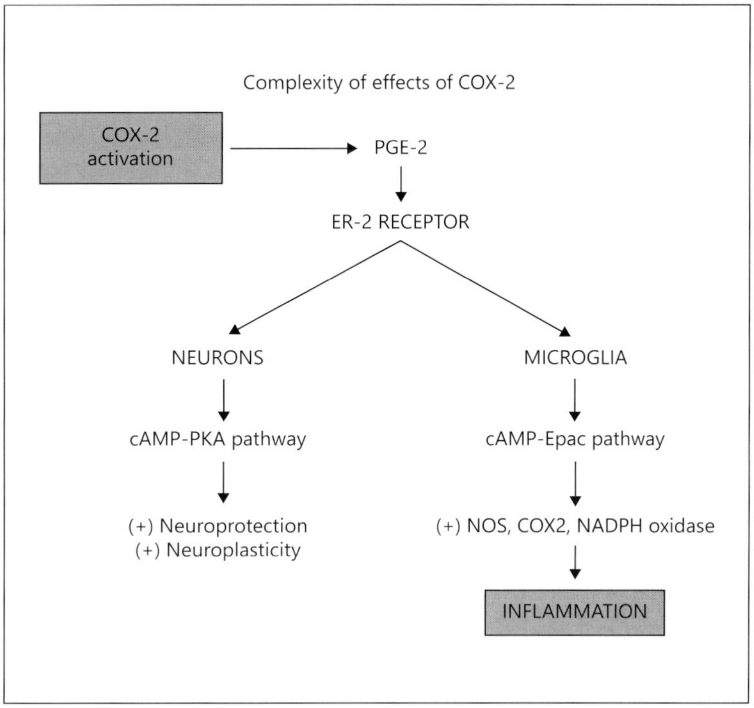

Fig. 2. Differential effects of cyclooxygenase (COX)-2 in neurons vs. microglia [92]. Epac, exchange protein directly activated by cAMP; PGE_2, prostaglandin E_2; PKA, protein kinase A; NOS, nitric oxide synthase.

with risk factors for coronary heart disease [80, 81]. An extended meta-analysis that was performed in 2013 suggested that although NSAIDs increase vascular and gastrointestinal risk, the size of these risks can be predicted, which could help guide clinical decision making [82].

In the brain, opposite to the majority of the other tissues, both COX-1 and COX-2 are constitutively expressed. However, they differ in their regulatory mechanisms, cell localization, and function. COX-2 is detected in synaptic dendrites and excitatory terminals, particularly in the cortex, hippocampus, and amygdala, with both neuronal and vascular associations, whereas COX-1 is expressed by microglia and perivascular cells [83, 84]. The contribution of COX-2 to neuropathology seems to be associated with its involvement in synaptic modification (Fig. 2) [85]. COX-2 has also been implicated in important physiological functions such as synaptic activity, long-term potentiation, long-term synaptic depression, and memory consolidation [86, 87]. Furthermore, COX-2, but not COX-1, can oxygenate endocannabinoids, which represent an important metabolic pathway in neurons to regulate excitatory synaptic transmission [88]. At the end of the past decade, accumulated information suggested that COX-1 is more than previously believed actively involved in brain injury induced by proinflammatory stimuli while, on the other hand, novel COX-2-derived lipid mediators with anti-inflammatory and proresolution properties had been discovered in the lipid field underlining the neuroprotective role of COX-2 [89]. As a result, the classical view that considers COX-1 as the isoform merely responsible for physiological production of prostanoids and COX-2 as the major proinflammatory isoform had been questioned. These new data, along with the reports for cardiovascular complications in patients receiving selective COX-2 inhibitors, led researchers in the field to consider targeting COX-2 in depression as a not viable therapeutic approach that can even aggravate the pathophysiology underpinning depression [90] and proposed aspirin as a therapeutic potential for mental illness [91]. However, the potential risks associated with prolonged use of salicylates, especially at doses sufficiently high to produce adequate COX inhibition in the brain, are a major limiting factor. At that time, identifying the specific roles of COX-1 and COX-2 in CNS inflammation and their coupling to downstream receptors emerged as a new priority that could lead to revisit therapeutic strategies and open new approaches for the treatment of neuroinflammatory conditions. In the following years, new research data have shown that modulation of a specific prostanoid synthase or receptor could provide therapeutic alternatives to NSAIDs for treating COX-2-governed pathological conditions. Among the COX-2 downstream signaling pathways, $PTGER_2$ (PGE_2 receptor EP_2 subtype) appears to be a crucial mediator of many physiological and pathological events [92]. It appears that neuronal EP_2 activation promotes acute neuroprotection, neuronal survival, and neuronal plasticity through cAMP/protein kinase A signaling. Conversely, glial – especially microglial – EP_2 activation often leads to secondary neurotoxicity and neuronal injury via upregulation of inflammatory mediators such as COX-2, inducible nitric oxide synthase, and NOX in chronic brain inflammation [93]. It is possible that transient and early delivery of EP_2 potentiators might provide neuroprotection in acute neuronal injuries from excitotoxic conditions, such as ischemic and hemorrhagic strokes [94], whereas delayed inhibition of EP_2 via selective antagonists would be expected to reduce brain inflammation and injury in neurological disorders associated with chronic inflammation. However, further research is needed in order to elucidate the conditions under which multiple EP_2 signaling pathways are engaged. In order to overcome the gap that coxibs had left, besides modulating COX-2 downstream selective targets, pharmaceutical research has also been focused on drugs interfering with several targets and the design of multitarget ligands. Designed multiple ligands targeting various steps of the arachidonic acid cascade with only low-dose treatment are expected to show increased therapeutic effectiveness and lower side effects. Particularly, it is expected that partial inhibition of the individual targets will be sufficient for the pharmacological effects and will still allow enough residual activity for maintenance of physiological functions of the individual pathway [72].

In attempting to identify the appropriate agent or method to arrest neuroprogression, nonpharmacological interventions, although not yet fully confirmed in controlled clinical trials, should not be ignored. Similarly, psychotherapy, physical exercise, yoga, and the rest of MBTs positively influence brain function by reducing the level of stress thus counteracting stress-related neuroprogressive processes. At the same time, there is converging evidence that successful pharmacotherapy, including the appropriate use of augmentation strategies with coxibs, may act in a similar way [95].

Conclusions

In an attempt to identify the appropriate agent for arresting neuroprogression, the evaluation of anti-inflammatory drugs that have already been used for the benefit of psychiatric patients, besides conventional pharmacotherapy, is a reasonable and urgent approach. These drugs, as almost all NSAIDs, act by inhibiting the two isoforms of COX both in the brain and in the periphery. However, the experience with coxibs has uncovered the need for therapeutic alternatives to NSAIDs for treating COX-2-governed pathological conditions. Therefore, current pharmacological research has focused on targeting COX-2 downstream prostanoid signaling instead of a generic block of the entire COX-2 cascade. It appears that neuronal EP_2 activation promotes acute neuroprotection and neuroplasticity through cAMP/protein kinase A signaling, while microglial EP_2 activation leads to neuroinflammation. While waiting for new pharmaceutical products, the positive clinical effects of selective COX-2 inhibitors should not be ignored, and carefully selected subgroups of patients could still benefit from them. This is particularly important once critical issues pertaining to the required length of coadministration with an antidepressant agent and any potential long-term effects on neurons and the cardiovascular system have been solved. As is true for inflammatory and degenerative brain diseases [96], selecting the appropriate antineuroprogressive agent is a complex decision reflecting the pathoanatomical complexity of neuroprogression itself. Numerous issues contribute to this problem. Several cell types, including resident cells and infiltrating blood cells, can respond to a number of specific inflammatory, oxidative, apoptotic, or excitotoxicity signals. Such signals can be specifically related to a disease or to a stage of a disease. Furthermore, the consequences of the peripheral actions of candidate drugs on central inflammation are not yet clear.

Natural products such as ω-3 fatty acids acting mainly in the initial steps of the inflammatory cascade seem to have minimal effect on human homeostasis. Given that ω-3 PUFAs are generally beneficial to health and without clinically relevant adverse effects, their preventive use in neuroprogression also merits consideration.

Besides drugs, there is converging evidence on several levels that stress-relieving methods, including physical exercise, psychotherapy, and MBTs, benefit brain function throughout the life span and across various neuropsychiatric disorders. Most of these methods are cost effective and relatively free of adverse effects, and, therefore, they could also be used as adjunctive therapeutic tools to arrest neuroprogression.

References

1 Knights KM, Mangoni AA, Miners JO: Defining the COX inhibitor selectivity of NSAIDs: implications for understanding toxicity. Expert Rev Clin Pharmacol 2010;3:769–776.
2 Myint A-M: Inflammation, neurotoxins and psychiatric disorders; in Halaris A, Leonard BE (eds): Inflammation in Psychiatry. Mod Trends Pharmacopsychiatry. Basel, Karger, 2013, vol 28, pp 61–74.
3 Calabrese JR, Skwerer RG, Barna B, Gulledge AD, Valenzuela R, Butkus A, Subichin S, Krup NE: Depression, immunocompetence, and prostaglandins of the E series. Psychiatry Res 1986;17:41–47.
4 Ohishi K, Ueno R, Nishino S, Sakai T, Hayaishi O: Increased level of salivary prostaglandins in patients with major depression. Biol Psychiatry 1988;23:326–334.
5 Song C, Leonard BE: Fundamentals of Psychoneuroimmunology. Chichester, Wiley, 2006.
6 Yaron I, Shirazi I, Jundovich R, Levartovsky D, Caspi D, Yaron M: Fluoxetine and amitriptyline inhibit nitric oxide, prostaglandin E_2, and hyaluronic acid production in human synovial cells and synovial tissue cultures. Arthritis Rheum 1999;42:2561–2568.
7 Lanquillon S, Krieg JC, Bening-Abu-Shach U, Vedder H: Cytokine production and treatment response in major depressive disorder. Neuropsychopharmacology 2000;22:370–379.
8 Hestad KA, Tonseth S, Stoen CD, Ueland T, Aukrust P: Raised plasma levels of tumor necrosis factor alpha in patients with depression: normalization after electroconvulsive therapy. J ECT 2003;19:183–188.
9 Muller N, Schwarz MJ: Role of the cytokine network in major psychoses. Adv Mol Cell Biol 2004;31:999–1031.
10 Galecki P, Galecka E, Maes M, Chamielec M, Orzechowska A, Bobilska K, Lewilski A, Sjemraj J: The expression of genes encoding for COX-2, MPO, iNOS and $sPLA_2$-IIA in patients with recurrent disorder. J Affect Disord 2012;138:360–366.
11 Müller N, Schwarz MJ: Immune system and schizophrenia. Curr Immunol Rev 2010;6:213–220.
12 Müller N, Ulmschneider M, Scheppach C, Schwarz MJ, Ackenheil M, Möller H-J, Grubber R, Riedel M: COX-2 inhibition as a treatment approach in schizophrenia: immunological considerations and clinical effects of celecoxib add-on therapy. Eur Arch Psychiatry Clin Neurosci 2004;254:14–22.
13 Schwieler L, Erhardt S, Erhardt C, Engberg G: Prostaglandin-mediated control of rat brain kynurenic acid synthesis – opposite actions by COX-1 and COX-2 isoforms. J Neural Transm (Vienna) 2005;112:863–872.

14 Salter M, Pogson CI: The role of tryptophan 2,3-dioxygenase in the hormonal control of tryptophan metabolism in isolated rat liver cells. Effects of glucocorticoids and experimental diabetes. Biochem J 1985;229:499–504.
15 Hawkey CJ: COX-2 inhibitors. Lancet 1999;353:307–314.
16 Müller N, Riedel M, Schwarz MJ, Engel RR: Clinical effects of COX-2 inhibitors on cognition in schizophrenia. Arch Psychiatry Clin Neurosci 2005;255:149–151.
17 Myint AM, Steinbusch HW, Coeghegan L, Luchtman D, Kim YK, Leonard BE: Effect of the COX-2 inhibitor celecoxib on behavioural and immune changes in an olfactory bulbectomised rat model of depression. Neuroimmunomodulation 2007;14:65–71.
18 Collantes-Esteves E, Fernandez-Perrez C: Improved self-control of osteoarthritis pain and self-reported health status in non-responders to celecoxib switched to rofecoxib: results of PAVIA, an open-label post-marketing survey in Spain. Curr Med Res Opin 2003;19:402–410.
19 Melnikova T, Savonenko A, Wang Q, Liang X, Hand T, Wu L, Kaufmann WE, Vehmas A, Andreasson KI: Cyclooxygenase-2 activity promotes cognitive deficits but not increased amyloid burden in a model of Alzheimer's disease in a sex-dimorphic pattern. Neuroscience 2006;141:1149–1162.
20 Müller N, Krause D, Dehning S, Musil R, Shennach-Wolff R, Obermeier M, Möller H-J, Klauss V, Schwarz MJ, Riedel M: Celecoxib treatment in an early stage of schizophrenia: results of a randomized, double-blind, placebo-controlled trial of celecoxib augmentation of amisulpride treatment. Schizophr Res 2010;118–124.
21 Müller N, Riedel M, Scheppach C, Brandstätter B, Sokullu S, Krampe K, Ulmschneideir M, Engel RR, Möller H-J, Schwarz MJ: Beneficial antipsychotic effects of celecoxib add-on therapy compared to risperidone alone in schizophrenia. Am J Psychiatry 2002;159:1029–1034.
22 Rappard F, Müller N: Celecoxib add-on therapy does not have beneficial antipsychotic effects over risperidone alone in schizophrenia. Neuropsychopharmacology 2004;29(suppl 1):222.
23 Müller N, Ulmschneider M, Scheppach C, Schwarz MJ, Ackenheil M, Möller HJ, Gruber R, Riedel M: COX-2 inhibition as a treatment approach in schizophrenia: immunological considerations and clinical effects of celecoxib add-on therapy. Eur Arch Psychiatry Clin Neurosci 2004;254:14–22.
24 Zhang Y, Chun Chen D, Long Tan Y, Zhou DF: A double-blind, placebo-controlled trial of celecoxib added to risperidone in first-episode and drug-naïve patients with schizophrenia. Eur Arch Psychiatry Clin Neurosci 2006;256(suppl 2):50.
25 Akhondzadeh S, Tabatabaee M, Amini H, Ahmadi Abhari SA, Abbasi SH, Behnam B: Celecoxib as adjunctive therapy in schizophrenia: a double-blind, randomized and placebo-controlled trial. Schizophr Res 2007;90:179–185.
26 Rapaport MH, Delrahim KK, Bresee CJ, Maddux RE, Ahmadpour O, Dolnak D: Celecoxib augmentation of continuously ill patients with schizophrenia. Biol Psychiatry 2005;57:1594–1596.
27 Laan W, Grobbee DE, Selten JP, Heijnen CJ, Kahn RS, Burger H: Adjuvant aspirin therapy reduces symptoms of schizophrenia spectrum disorders: results from a randomized, double-blind, placebo-controlled trial. J Clin Psychiatry 2010;71:520–527.
28 Müller N: The role of anti-inflammatory treatment in psychiatric disorders. Psychiatr Danub 2013;25:292–298.
29 Sommer IE, de Witte L, Begemann M, Kahn RS: Nonsteroidal anti-inflammatory drugs in schizophrenia: ready for practice or a good start? A meta-analysis. J Clin Psychiatry 2012;73:414–419.
30 Nitta M, Kishimoto T, Müller N, Weiser M, Davidson M, Kane JM, Correll CU: Adjunctive use of nonsteroidal anti-inflammatory drugs for schizophrenia: a meta-analytic investigation of randomized controlled trials. Schizophr Bull 2013;39:1230–1241.
31 Müller N, Schwarz MJ, Dehning S, Douhe A, Cerovecki A, Goldstein-Müller B, Spellmann I, Hetzel G, Maino K, Kleindienst N, Moller HJ, Arolt V, Riedel M: The cycooxygenase-2 inhibitor celecoxib has therapeutic effects in major depression: results of a double-blind randomized, placebo controlled, add-on pilot study to reboxetine. Mol Psychiatry 2006;11:680–684.
32 Akhondzadeh S, Jafari S, Raisi F, Nasehi AA, Ghoreishi A, Salehi B, Mohhebi-Rasa S, Raznahan M, Kamalippour A: Clinical trial of adjunctive celecoxib treatment in patients with major depression: a double blind and placebo controlled trial. Depress Anxiety 2009;26:607–611.
33 Abbasi SH, Hosseini F, Modabbernia A, Ashrafi M, Akhondzadeh S: Effect of celecoxib add-on treatment on symptoms and serum IL-6 concentrations in patients with major depressive disorder: randomized double-blind placebo-controlled study. J Affect Disord 2012;141:308–314.
34 Na KS, Lee KJ, Lee JS, Cho YS, Jung HY: Efficacy of adjunctive celecoxib treatment for patients with major depressive disorder: a meta-analysis. Prog Neuropsychopharmacol Biol Psychiatry 2014;48:79–85.
35 Majd M, Hashemian F, Hosseini SM, Shariatpanahi MV, Sharifi A: A randomized, double-blind placebo-controlled trial of celecoxib augmentation of sertraline in treatment of drug-naïve depressed women: a pilot study. Iran J Pharm Res 2014;14:891–899.
36 Raison CL, Rutherford RE, Woolwine BJ, Shuo C, Schettler P, Drake DF, Haroon E, Miller AH: A randomized controlled trial of the tumor necrosis factor antagonist infliximab for treatment-resistant depression: the role of baseline inflammatory biomarkers. JAMA Psychiatry 2013;70:31–41.
37 Nery FG, Monkul ES, Hatch JP, Fonseca M, Zunta-Soares GB, Frey BN, Bowden CL, Soares JC: Celecoxib as an adjunct in the treatment of depressive or mixed episodes of bipolar disorder: a double-blind, randomized, placebo-controlled study. Hum Psychopharmacol 2008;23:87–94.
38 Mendlewicz J, Kriwin P, Oswald P, Souery D, Alboni S, Brunello N: Shortened onset of action of antidepressants in major depression using acetylsalicylic acid augmentation: a pilot open-label study. Int Clin Psychopharmacol 2006;21:227–231.
39 Galecki P, Szemraj J, Bienkiewiez M, Zboralski K, Galeeka E: Oxidative stress parameters after combined fluoxetine and acetylsalicylic acid therapy in depressive patients. Hum Psychopharmacol 2009;24:277–286.

40 Gallagher PJ, Castro V, Fava M, et al: Antidepressant response in patients with major depression exposed to NSAIDs: a pharmacovigilance study. Am J Psychiatry 2012;169:1065–1072.
41 Köhler O, Petersen L, Mors O, Gasse C: Inflammation and depression: combined use of selective serotonin reuptake inhibitors and NSAIDs or paracetamol and psychiatric outcomes. Brain Behav 2015;5:e00338.
42 Peet M: Omega-3 polyunsaturated fatty acids in the treatment of schizophrenia. Isr J Psychiatry Relat Sci 2008;45: 19–25.
43 Ross BM, Seguin J, Sieswerda LE: Omega-3 fatty acids as treatments for mental illness: which disorder and which fatty acid? Lipids Health Dis 2007;6:21.
44 Su K-P: Biological mechanism of antidepressant effect of omega-3 fatty acids: how does fish oil act as a "mind-body interface"? Neurosignals 2009;17:144–152.
45 Gupta S, Knight AG: Saturated long chain fatty acids activating inflammatory signaling in astrocytes. J Neurochem 2012;120:1060–1071.
46 Weylandt KH, Chiu CY, Gomolka B, Waechter SF, Wiedenmann B: Omega-3 fatty acids and their lipid mediators: towards an understanding of resolvin and protectin formation. Prostaglandins Other Lipid Mediat 2012;97:73–82.
47 Song C, Zhao S: Omega-3 fatty acid eicosapentanoic acid. A new treatment for psychiatric and neurodegenerative diseases: a review of clinical investigations. Expert Opin Investig Drugs 2007; 16:1627–1638.
48 Kou W, Luchtman D, Song C: Eicosapentanoic acid increases cell viability and expression of neurotrophin receptors in retinoic acid and brain-derived neurotrophic factor differentiated SH-SY5Y cells. Eur J Nutr 2008;47:104–113.
49 Taepavarapruk P, Song C: Reductions of acetylcholine release and nerve growth factor expression are correlated with memory impairment induced by interleukin-1β administrations: effects of omega-3 fatty acid EPA treatment. J Neurochem 2010;112:1054–1064.
50 Luchtman DW, Meng Q, Song C: Ethyl-eicosapentaenoate (E-EPA) attenuates motor impairments and inflammation in the MPTP-probenecid mouse model of Parkinson's disease. Behav Brain Res 2012;226:386–396.
51 Lu DY, Tsao YY, Leung YM, Su KP: Docosahexaenoic acid suppresses neuroinflammatory responses and induces heme oxygenase-1 expression in BV-2 microglia: implications of antidepressant effects for ω-3 fatty acids. Neuropsychopharmacology 2010;35:2238–2248.
52 Horrobin DF: The membrane phospholipid hypothesis as a biochemical basis for the neurodevelopmental concept of schizophrenia. Schizophr Res 1998;30: 193–208.
53 Fusar-Poli P, Berger G: Eicosapentaenoic acid interventions in schizophrenia: meta-analysis of randomized, placebo-controlled studies. J Clin Psychopharmacol 2012;32:179–185.
54 Amminger GP, Schäfer MR, Papageorgiou K, Klier CM, Cotton SM, Harrigan SM, Mackinnon A, McGorry PD, Berger GE: Long-chain omega-3 fatty acids for indicated prevention of psychotic disorders: a randomized, placebo-controlled trial. Arch Gen Psychiatry 2010;67:146–154.
55 Grosso G, Pajak A, Marventano A, Castellano S, Galvano F, Bucolo C, Drago F, Caraci F: Role of omega-3 fatty acids in the treatment of depressive disorders: a comprehensive meta-analysis of randomized clinical trials. PLoS One 2014; 9:e96905.
56 Mocking RJ, Harmsen I, Assies J, Koeter MW, Ruhé HG, Schene AH: Meta-analysis and meta-regression of omega-3 polyunsaturated fatty acid supplementation for major depressive disorder. Transl Psychiatry 2016;6:e756.
57 Dahl J, Ormstad H, Aass HC, Sandvik L, Malt UF, Andreassen OA: Recovery from major depressive disorder episode after non-pharmacological treatment is associated with normalized cytokine levels. Acta Psychiatr Scand 2016;134:40–47.
58 Coelho FG, Gobbi S, Andreatto CA, Corazza DI, Pedroso RV, Santos-Galduroz RF: Physical exercise modulates peripheral levels of brain-derived neurotrophic factor (BDNF): a systematic review of experimental studies in the elderly. Arch Gerontol Geriatr 2013;56: 10–15.
59 Hillman CH, Erickson KI, Kramer AF: Be smart, exercise your heart: exercise effects on brain and cognition. Nat Rev Neurosci 2008;9:58–65.
60 Smith PJ, Blumenthal JA, Hoffman BM, Cooper H, Strauman TA, Welsh-Bohmer K, et al: Aerobic exercise and neurocognitive performance: a meta-analytic review of randomized controlled trials. Psychosom Med 2010;72:239–252.
61 Middleton LE, Barnes DE, Lui LY, Yaffe K: Physical activity over the life course and its association with cognitive performance and impairment in old age. J Am Geriatr Soc 2010;58:1322–1326.
62 Berchicci M, Lucci G, DiRusso F: Benefits of physical exercise on the aging brain: the role of the prefrontal cortex. J Gerontol A Biol Sci Med Sci 2013;68: 1337–1341.
63 Kaur J, Garnawat D, Bhatia MS, Sachdev M: Rehabilitation in Alzheimer's disease. Delhi Psychiatry J 2013;16:166–170.
64 Kaur J, Sharma S, Sachdev M, Mittal J: Rehabilitation of patients with Parkinson disease. Delhi Psychiatry J 2012;15: 398–401.
65 Rimer J, Dwan K, Lawlor DA, Greig CA, McMurdo M, Morley W, et al: Exercise for depression. Cochrane Database Syst Rev 2012;7:CD004366.
66 Jayakody K, Gunadasa S, Hosker C: Exercise for anxiety disorders: systematic review. Br J Sports Med 2014;48:187–196.
67 Gorczynski P, Faulkner G: Exercise therapy for schizophrenia. Cochrane Database Syst Rev 2010;5:CD004412.
68 Pearsall R, Smith DJ, Pelosi A, Geddes J: Exercise therapy in adults with serious mental illness: a systematic review and meta-analysis. BMC Psychiatry 2014;14: 117.
69 From the UN General Assembly: United Nations declares 21 June as International Day of Yoga. New Delhi, United Nations Information Centre for India and Bhutan, 2014, www.unic.org.in/display.php?E=13712 &K=Yoga.
70 Morgan N, Irwin MR, Chung M, Wang C: The effects of mind-body therapies on the immune system: meta-analysis. PLoS One 2014;9:e100903.
71 Sachdeva A, Kumar K, Anand KS: Non pharmacological cognitive enhancers – current perspectives. J Clin Diagn Res 2015;9:VE01–VE06.
72 Meirer K, Steinhilber D, Proschak E: Inhibitors of the arachidonic acid cascade: interfering with multiple pathways. Basic Clin Pharmacol Toxicol 2014;114:83–91.

73 O'Banion MK: Cyclooxygenase-2: molecular biology, pharmacology, and neurobiology. Crit Rev Neurobiol 1999;13:45–82.
74 Kis B, Snipes JA, Busija DW: Acetaminophen and the cyclooxygenase-3 puzzle: sorting out facts, fictions, and uncertainties J Pharmacol Exp Ther 2005;315:1–7.
75 Bombardier C, Laine L, Reicin A: Comparison of upper gastrointestinal toxicity of rofecoxib and naproxen in patients with rheumatoid arthritis. VIGOR Study Group. N Engl J Med 2000;343:1520–1528.
76 Schnitzer TJ, Burmester GR, Mysler E: Comparison of lumiracoxib with naproxen and ibuprofen in the Therapeutic Arthritis Research and Gastrointestinal Event Trial (TARGET), reduction in ulcer complications: randomized controlled trial. Lancet 2004;364:665–674.
77 Patrignani P, Patrono C: Cyclooxygenase inhibitors: from pharmacology to clinical read-outs. Biochim Biophys Acta 2015;1851:422–432.
78 Funk CD, FitzGerald GA: COX-2 inhibitors and cardiovascular risk: J Cardiovasc Pharmacol 2007;50:470–479.
79 US Food and Drug Administration: Analysis and recommendations for agency action regarding non-steroidal anti-inflammatory drugs and cardiovascular risk. http://www.fda.gov/Drugs/DrugSafety/PostmarketDrugSafetyInformationforPatientsandProviders/.
80 European Medicines Agency: Press release: European Medicines Agency review concludes positive benefit-risk balance for non-selective NSAIDs. http://www.ema.europa.eu/docs/en_GB/document_library/Press_release/2009/12/WC500017362.pdf, 2006.
81 European Medicines Agency: Press release: European Medicines Agency concludes action on COX-2 inhibitors. http://www.ema.europa.eu/docs/en_GB/document_library/Press_release/2010/01/WC500059088.pdf, 2005.
82 Coxib and traditional NSAID Trialists' (CNT) Collaboration: Vascular and upper gastrointestinal effects of non-steroidal anti-inflammatory drugs: meta-analyses of individual participant data from randomized trials. Lancet 2013;382:769–779.
83 Kaufmann WE, Worley PF, Pegg J, Bremer M, Isakson P: COX-2, a synaptically induced enzyme, is expressed by excitatory neurons at postsynaptic sites in rat cerebral cortex. Proc Natl Acad Sci USA 1996;93:2317–2321.
84 Kaufmann WE, Andreasson KI, Isakson P, Worley PF: Cyclooxygenases and the central nervous system. Prostaglandin 1997;54:601–624.
85 Yang H, Chen C: Cyclooxygenase-2 in synaptic signaling. Curr Pharm Des 2008;14:1443–1451.
86 Stefanovic B, Bosetti F, Silva AC: Modulatory role of cyclooxygenase-2 in cerebrovascular coupling. Neuroimage 2006;32:23–32.
87 Hewett SJ, Bell SC, Hewett JA: Contributions of cyclooxygenase-2 to neuroplasticity and neuropathology of the central nervous system. Pharmacol Ther 2006;112:335–357.
88 Slanina KA, Schweitzer P: Inhibition of cyclooxygenase-2 elicits a CB1-mediated decrease of excitatory transmission in rat CA1 hippocampus. Neuropharmacology 2005;49:653–659.
89 Aïd S, Bosetti F: Targeting cyclooxygenases-1 and -2 in neuroinflammation: therapeutic implications. Biochimie 2011;93:46–51.
90 Maes M: Targeting cyclooxygenase-2 in depression is not a viable therapeutic approach and may even aggravate the pathophysiology underpinning depression. Metab Brain Dis 2012;27:405–413.
91 Berk M, Dean O, Drexhage H, McNeil JJ, Moylan S, O'Neil A, Davey CG, Sanna L, Maes M: Aspirin: a review of its neurobiological properties and therapeutic potential for mental illness. BMC Med 2013;11:74.
92 Jiang J, Dingledine R: Prostaglandin receptor EP2 in the crosshairs of anti-inflammation, anti-cancer, and neuroprotection. Trends Pharmacol Sci 2013;34:413–423.
93 Andreasson K: Emerging roles of PGE2 receptors in models of neurological disease. Prostaglandins Other Lipid Mediat 2010;91:104–112.
94 Mohan S, Ahmad AS, Glushakov AV, Chambers C, Doré S: Putative role of prostaglandin receptor in intracerebral hemorrhage. Front Neurol 2012;3:145.
95 Halaris A, Myint AM, Savant V, Meresh E, Lim E, Guillemin G, Hoppensteadt D, Fareed J, Sinacore J: Does escitalopram reduce neurotoxicity in major depression? J Psychiatr Res 2015;66–67:118–126.
96 Minghetti L: Cyclooxygenase-2 (COX-2) in inflammatory and degenerative brain diseases. J Neuropathol Exp Neurol 2004;63:901–910.
97 Bertolini A, Ottani A, Sandrini M: Selective COX-2 inhibitors and dual acting anti-inflammatory drugs: critical remarks. Curr Med Chem 2002;9:1033–1043.
98 Aoki T, Narumiya S: Prostaglandins and chronic inflammation. Trends Pharmacol Sci 2012;33:304–311.

Fotini Boufidou, MD
Laboratory of Biopathology, Eginition Hospital, UoA
Vas. Sophias 72-74 GR, Athens 11528 (Greece)
E-Mail fboufidou@med.uoa.gr

Author Index

Subject Index